# SHIFTING SENSE
## LOOKING BACK TO THE FUTURE
# IN SPATIAL PLANNING

Faculty of Architecture
Delft University of Technology

# SHIFTING SENSE
## LOOKING BACK TO THE FUTURE
# IN SPATIAL PLANNING

E.D. Hulsbergen, I.T. Klaasen, I. Kriens (eds)

Techne Press / 2005

Design/**Science**/Planning

**Series Editor:**
Dr. I.T. Klaasen, Faculty of Architecture (Spatial Planning), Delft University of Technology, The Netherlands, I.T.Klaasen@bk.tudelft.nl



**Published and distributed by:**
Techne Press, Amsterdam, The Netherlands
http://www.technepress.nl

Printed in The Netherlands

**Shifting Sense**
**Looking back to the future**
**in Spatial Planning**

Keywords: spatial planning, regeneration, network, space-time, strategy, infrastructure, ICT, environment, urban research, urban policy, new charter of athens

**ISBN-13: 978908590043**
**ISBN-10: 9085940044**

| | |
|---|---|
| Editors: | Edward D. Hulsbergen, Ina T. Klaasen, Iwan Kriens, Faculty of Architecture (Spatial Planning), Delft University of Technology, The Netherlands |
| Translation and language editing: | Susan M. van der Werff-Woolhouse, English Text & Dialogue Services, Hilversum |
| Lay-out: | Matthias C. Huijgen, Ana Maria Fernández-Maldonado |
| Editing assistance: | Matthias C. Huijgen, Pieneke Leys, Tonny van Velzen |
| Cover: | Iwan Kriens |

This publication is produced within the frame of the research programme of the Delft Centre for Sustainable Urban Areas and financed by the Chair of Spatial Planning and the Dean of the Faculty of Architecture Delft University of Technology, The Netherlands

# Contents

## Part IV Strategies

# Preface

The contributions to *Shifting Sense in Spatial Planning* have been submitted by a diverse group of authors: researchers from our faculty and from other universities, both in the Netherlands and abroad, and practitioners in urban planning and design. The editors of the book are staff members of the Chair of Spatial Planning, one of the nucleus Chairs of Urbanism in the Faculty of Architecture of Delft University of Technology.

The publication of this book has been organised to coincide with the retirement of Professor Paul Drewe, the Chair-holder of Spatial Planning since 1973. In submitting their chapters, the contributors had two motivations in common: to write about their research or practical experiences in the field of spatial planning and to do so as a gesture of respect and affection for Paul Drewe's work and personality. References to his work run through the book in an inspiring thread.

The aim of this book is threefold. To begin with, to focus on past, present and future perspectives of spatial planning, and its development as a discipline in the context of urban planning and design (urbanism). Secondly, to produce a book for the new generation of students and colleagues, also those in related disciplines, that would also serve as a source of inspiration for urban planning and design practitioners who take an interest in the faculty's work on spatial planning. Thirdly, to honour Paul Drewe who, for more than three decades, explored existing and developed new fields of urban planning and design research. His concern was to further national and international co-operation, thereby encouraging, sustaining and setting the scenes for the educational and research programme of his Chair.

Spatial planning as a discipline of urbanism has changed drastically in recent decades. The constant is the study of the relation between societal and spatial aspects, between societal needs and spatial conditions, and the attention given to new technological developments. The changes are reflected in the history of the Chair. Moreover, within the context of urbanism, Spatial Planning at Delft is different from Chairs of the same name elsewhere. The many definitions of space lead to widely differing points of departure and fields of application. The authors of this book view 'space' as a workable reality, rather than a mere social construction, geographical entity or tool. Thus they are not restrained by conventional lines of development and exploitation of space, but can work with an open mind, to invent new possibilities, meanings and uses of space. This latter is the core identity of the Faculty of Architecture (and Urban Design and Planning) of the Delft University of Technology. The year 2005 is also marked by the Faculty's 100th anniversary of this domain and mission. The publication of this book fits in very well with this celebration too.

Prof. Ir. Hans Beunderman MBA, Dean of the Faculty of Architecture

# Introducing *Shifting Sense*

## Objective

With ***Shifting Sense***, *Looking Back to the Future* **in Spatial Planning**, the main aim of the Chair of Spatial Planning in the Faculty of Architecture of Delft University of Technology is to contribute to the discussion about content and procedures in this field of study. Although the starting point is the work of the Chair itself, the majority of the contributions comes from respected experts from other Chairs, from other universities and from practitioners in the field.

Secondly, this book will establish a starting point for the future of the Chair of Spatial Planning, one of the central Chairs of Urbanism. Paul Drewe has held this Chair for over thirty years and has therefore directed, over a long period, the extensive changes, visible in retrospect, in its content. His retirement is used here as an occasion for drawing up a balance – and making a broad profile – of what the essence of spatial planning is within Urbanism, and how it differs from the other central Chairs, the Chair of Urban Design and the Chair of Landscape Architecture.
The third objective is to offer to the new generation of planners and designers, and to our students, more and more of whom come from distant countries, a book worth reading.

## Title

***Shifting Sense***, *Looking Back to the Future* **in Spatial Planning** is a daring title. It will come as no surprise to learn that it has been discussed for a long time. Literally dozens of titles were proposed and rejected, not least because of the varied and broad content of the book. *Shifting* and *Sense* are both concepts with many meanings. In combination, they form a hybrid. 'Sense' has to do with intelligence, perception, feelings, purpose, and discernment (with 'nonsense' as an antonym). 'Shifting' is related to changing, moving, and adapting (with 'static' and 'rigid' as antonyms). The title *Shifting Sense* indicates that it is the intention here for the chapters in this book to be seen as a bundling of initiatives; on the one hand, to describe and explain the rationality, meaning, emotion, direction of the change, moves, and adaptations that have been made so far, and, on the other, to indicate in which direction the discipline of Spatial Planning is expected to go.

The history of Spatial Planning research reveals a number of shifts in focus. What will remain in research and education is the attention that is given to the relation between societal and spatial aspects, the actual and potential role of urban planning and design, the interest in policies, actors and instruments, and, of course, methodological issues. The shifts relate to the interpretation and focus of the content. In the 1970s, the main attention was on urban population developments, housing conditions and urban economy. During the 80s, European cities and cross-national co-operation came more into focus, but also urban phenomena such as vulnerability and deprivation, unemployment and mobility. In the 90s, the new information technologies, the network city and the need for new urban concepts in relation to changing urban complexity became the fields of study. At the beginning of the 21st century, time–space relations, urban modelling and the links between top–down planning and bottom–up initiatives were added to the list of research topics. In our view, 'shifting sense' is a hallmark of the spatial planning undertaken within the context of universities; the shifts themselves are societally induced and anticipatory.

## Themes

The book is comprised of four parts. These can be considered as four main themes, each of which has its own demonstrable history, a present-day meaning and an anticipated future relevance. The division into four parts does not imply that the themes are clearly demarked. A concept such as 'network', for instance, is dealt with in various chapters, albeit from different points of view. The overlaps cause the chapters to interlink with one another in all sorts of ways.

The theme of Part One is: 'Societal changes and their effects on the use of space'. The societal questions in spatial planning and the aspects of design that are dealt with here are: designing and planning for the changing and reduced mobility of the elderly and for vulnerable groups in deprived urban areas; accommodating economic changes brought about by (top) location development; changing technology and its use; cross-national co-operation in logistics and in knowledge about integral forms of planning.

Part Two focuses on the opposite starting point, with the theme: 'Spatial developments and their societal effects'. The discussion in this section is about urban and regional developments and how they affect societal life. A selection of experiences and theoretical and practical points of departure from various countries and cultures makes it clear that more precision is needed in the *spatial scale* of the subjects and questions that are dealt with, and with that a thorough knowledge of the spatial changes that are taking place, so that their effects can be fully understood and used as lessons to help find solutions for situations elsewhere.

Part Three is: 'Networks'. For these, space is usually the point of departure. The historical perspective of networks has already been dealt with in Part Two. Here, in Part Three, attention is given to the various ways in which theories about spatial problems are formulated and applied. It will become clear that networks – ranging from technical, functional and users' networks to natural networks and patterns – are understood as actual concepts and tools for urban planning and design. Linking public-space networks with the time-space problems of various groups of citizens, illustrates their importance for future spatial-planning problems.

Part Four is headed: 'Strategies'. In this final section, the spotlight is on strategies and the role of the spatial planner. Here, particular attention is given to the process side of future assignments. The towns and regions, the areas where spatial planning is applied, call for effective strategies to meet the changing needs of both citizens and businesses. Apart from strategies for transforming urbanised areas sustainably, attention is also given to the adaptation and positioning of problems within an economic environment that is oriented more and more towards a European and global scale; directed more towards competition than complementarity.

## Contributors

The contributors are together a rather diverse group: they are affiliated with different universities (Barcelona, Delft, Gent, Liverpool, Paris, San Antonio, Utrecht), active as (public and private) practitioners in their various fields, and are from different countries and cultures. The initiators of this project are all members of the Chair of Spatial Planning. The remaining twenty or more contributors were invited to contribute to the book because of their relation with the Chair of Spatial Planning and their highly valued expertise, capacities, interests and challenging visions on various areas that have come under discussion by the Chair over the years.

This diversity of contributors and backgrounds has also had its influence on the language. Although English is the language of communication, the usage of words and the meanings given to them and the sentence construction may still vary. In editing the chapters, much attention has been given to preventing confusion of concepts and to increasing readability, while respecting the contributors' writing styles. It goes without saying that, with such a variety of contributors, this in itself has been an extensive task. In order to increase the accessibility of this book a References Index has also been added.

Finally, making this book, with all its efforts, obstacles, strokes of luck and, not least, pleasure in realising the design, has been a complex, collective job. The Chair that initiated this project has felt greatly honoured by the contributors' unconditional response in sending us their submissions.

# Spatial Planning at Delft University of Technology - A Historic Overview

Jan den Draak and Edward Hulsbergen

## Introduction

The Chair of Spatial Planning was held by Paul Drewe for more than three decades. During this period, not only did the name change – up to 1985, it was known as *stedebouwkundig onderzoek* [urban planning and design research] – but the contents of the discipline also changed. In this chapter we sketch the history of the Chair and the changing context which it served. Our aim is to highlight developments, rather than give a full and chronological review. To gain an understanding of how the Chair was positioned in 1973, at the time when Paul Drewe became the Chair-holder, it is relevant to look at the early history of the discipline of urban planning and design research and at how it became part of the curriculum of the Faculty of Architecture at Delft University of Technology [*Technische Universiteit Delft*] (see sections 2 and 3). The history, as it relates to Drewe, is described decade by decade (see section 4) up to the present day. We end this introductory overview with some notes on future perspectives (see section 5).

## Urban planning and design research up to the 1960s

The roots of the discipline of urban planning and design research lie in the distant past. As long ago as 1889, Sitte, in his book *Der Städtebau nach seinen künstlerischen Grundsätzen* criticised the one-sided economic-technical character of the unimaginative town planning of the time. He viewed town planning as an artistic task, coupled with perception of the urban environment and the needs of city-dwellers. For him, these were the points of departure for spatial design. Sitte put forward proposals for basing urban planning practices on scientific procedures. In his opinion, when planning an urban area, the first requirement was a preparatory study comprised of: a forecast of the anticipated increase in population fifty years on from the date of the study, developments in traffic, an estimate of the public amenities and housing needed, and their positioning and connections, bearing in mind the characteristics of the areas to be developed. Only after completing this preparatory study should a design be made (Sitte, 1889:137-142).

The Scottish biologist, sociologist and urbanist Geddes followed on from Sitte by trying to break through the one-sided planning of the time that was based only on technical and design principles. He argued for "a full and thorough survey of country and town, village and city, as preparatory to all town planning and city design" (Geddes, 1915:356-357). The main point of this 'survey before plan' is that it facilitates the study of a variety of population aspects. Quantitative data became a characteristic part of this preliminary study. As Van der Cammen (1979:58) typifies: "Reporting had the character of a one-off affair, the handing over of the report (to the designers). In this way, the researcher is a link in the classical production scheme of town and country planning: researchers supply building stones on demand, in the form of figures about (expected) socio-spatial developments; designers process these into the 'semi-manufactured article', the urban design, whereupon, finally, the policy officer drapes it all with a juridically-administratively justified coat and a covering letter. It was not until in the seventies that some of the planning research reports began to show the structure of Geddes' general outline."

In the Netherlands, De Casseres joined Geddes in largely agreeing with his views. He advocated carrying out a 'civic survey' primarily to study population developments, social relations and economic conditions. He considered statistics as an indispensable tool, also as a means for predicting future trends (De Casseres, 1926:12-17). In the second half of the 1920s, De Casseres' ideas gained ground, also in the practice of urban design. The most famous example is the municipality of Amsterdam, where, in 1928, a Town Planning Department (and research office) was established as part of the 'Works and Public Buildings Department'. Other large municipalities followed suit, and urban planning and design research became institutionalised. The civil engineers Van Lohuizen and Angenot were pioneers in promoting 'preliminary scientific research' as a foundation for urban design, thereby following Geddes' principles.

The civil engineers were soon succeeded by social-geographers, educated in Amsterdam and Utrecht, who dominated urban research between the 1940s and the 1960s. To satisfy the demands from practice, the research mainly concentrated on surveys and forecasts of population needs. The Utrecht geographer De Smidt mentioned that the strength of these studies was that they produced a respectable structural analysis of population, housing and sources of subsistence; much less attention was given to other functions. Research and spatial design did not interrelate easily (De Smidt, 1986:9).

In the second half of the 1960s, drastic changes occurred in both spatial planning and in the position of urban planning and design research. The main changes were the politicising of plans and the shift of blueprint planning to process planning. The results of scientific analyses of societal developments and spatial trends were no longer accepted as established facts and trends, but as information that could be manipulated according to political views. The politicising of the plan gave way to the politicising of plan-making: there were no 'final plans' any more, but a subdivision of the process of preparing plans into a number of distinct phases, all of which became subject to political decision-making (Van der Cammen & De Klerk, 2003:241). These developments induced changes in the function and character of urban research. 'Analysis and forecasts of activities' use of space was replaced by making *the exploration of problems* subservient to participation and consensus processes. In this way, planning research contributed to shifting the 'aims-means' rationality of planning in the direction of the political rationality of the negotiating table and, by so doing, undermined its own position in the planning process" (*ibid.*:293).

## 1948-1972: 25 years of urban research within architectural education at Delft University of Technology

Urban planning and design research in the Faculty of Architecture at Delft University of Technology started in 1948 with the appointment of Van Lohuizen as Professor by Special Appointment. In his inaugural lecture, entitled *De eenheid van het stedebouwkundig werk* [The unity of town planning work], he stated that an urban design comes, or should come, into being: "by co-operation between workers of different ability, education and attitude, prominent among which are the urban designers, the researchers and engineers in various fields (...). The prerequisite is that all of them must possess the consciousness and ability to identify with the way of thinking and needs of the others, and that they all have great mutual respect for each others' work" (Van Lohuizen, 1948). In his opinion, research should primarily serve the tangible plan, not just the development of theory. It is clear that Van Lohuizen distinguished two disciplines: urban design, on the one hand, and urban research, on the other. His views no doubt inspired him to strive for what he called a 'laboratory of urban planning and design research'.

Van Lohuizen, who died in 1956, was an important pace-maker of urban research in the Faculty, which, compared to other faculties, did not have a long tradition of research.

In 1957, Van Lohuizen was succeeded by Kruijt, who had joined the Faculty first as a university lecturer, later following his dissertation in 1960, becoming Professor of Urban Research and Sociology. Meanwhile, in 1959, the *Instituut voor Stedebouwkundig Onderzoek* (ISO) [Institute of Town Planning Research] was founded, and Kruijt became its director. The ISO's task was one of fundamental research in the field of spatial planning and urban design. In Kruijt's inaugural lecture (Kruijt, 1961), he discussed the urbanisation of the Netherlands. It was strongly based on statistical data about changes in the urbanisation process and the spatial dispersion of the population since 1870. Even at this time, he drew attention to the downsides of continuing suburbanisation.

Then, in the first half of the 60s, a significant change took place: the urban research part of Kruijt's Chair was transformed into an independent Chair. Angenot became Professor by Special Appointment of this Chair, as well as director of ISO. Angenot was a civil engineer and, just like Van Lohuizen, he had also been a pioneering advocate of carrying out 'preparatory scientific research' to underpin the urban plan. In the second half of the 60s, this view lost its importance. However, Angenot had a wide outlook on urban research: he initiated urban-physics research, for example. The ISO researchers had a varied background too, not only technical and spatial, but also socio-scientific. In his inaugural lecture (Angenot, 1963), Angenot focused on the expanding city, and especially on the growing need for, and consumption of, space, not only for housing, but also for recreation, industry and traffic. As he stated, this growth was not only a consequence of increasing urban populations (which was already at an end) and the number of employed people, but was mainly triggered by the undeniably increasing need, per inhabitant, for space and accommodation. Referring to developments in some parts of the USA, Angenot drew attention to the dangers of 'urban explosion' and to the governmental actions on all levels, national, provincial and municipal, that would be necessary to prevent the harmful consequences of this development.

In the second half of the 1960s, Heimans (also a civil engineer) was appointed as reader in town planning research. His inaugural lecture included themes such as paying more attention to humanising the spatial environment, democratisation in town and country planning, the need to apply rational planning methods and to retain flexibility in the planning process. "A conscientious interplay of research, planning and administrative measures is a first-order requirement within this framework, an interplay, the actions of which form the environment into an indivisible whole " (Heimans, 1970:14). Angenot's valedictory lecture (1971) referred not so much to the changes in the discipline of research, as to the conditions for good and effective environmental policy and development, and he warned against the consequences of neglecting them.

To round off the reflection on the early history and position of the discipline of town and country research, we will now focus on the preparations made for Angenot's succession. In choosing a new full professor, the Faculty decided on a number of criteria. The emphasis was placed on societal aspects, the position of research in the town and country planning process, on methodology and aspects of methodological techniques, and on the more 'traditional' aspects of town and country planning. On the whole, this profile aligned well with developments in the discipline. Nevertheless, it took a long time to round off the appointment process. In the course of 1973, Drewe was appointed. With a background in sociology and economics, Drewe had worked until then at the *Nederlands Economisch Instituut* (NEI) [Dutch Economics Institute] of the Erasmus University Rotterdam, as a researcher into socio-economic topics, such as European migration policy, the labour market, urban problems, and segregation. With Paul Drewe's appointment, the Chair-holder was once more a researcher from the social sciences.

## The Chair of Spatial Planning and its environment, 1973–2005

### The 1970s

Inaugural Lectures offer new professors the opportunity of expressing their views on their discipline within the context of the Faculty they represent. Drewe's point of departure was the importance of making aims explicit, both in spatial planning (as an expression of a general tendency towards democratisation) and in research training. With respect to spatial planning, Drewe states that it is essential to discern its inherent political nature (Drewe, 1974:10). As examples, he mentions the choices that have to be made between economic and ecological interests, between present and future generations, and between individual and collective interests (1974:13). He criticises the one-sided attention given to the spatial dimension. Efficient spatial planning, he says, needs multi-disciplinary co-operation and the integration of spatial and social measures in implementing plans. He also gives attention to advanced methods and research techniques and puts out a plea for research on how urban planners and designers go about their work. The planning process is important for town and country research, but with an emphasis on substantial planning theory.

In 1975, Faludi was appointed for the Urban Design Chair (procedural) Planning Theory. Theoretically, the Chairs of Drewe and Faludi, together with Kruijt's Urban Sociology Chair, could complement each other quite well, as educational disciplines. However, Faludi was rather opposed to the existing planning syllabuses: "The question (…) to be asked is: on what does planning education base its identity? Apart from the deceptive, but obvious, principle that planning should spring from a thorough knowledge of the object of planning, there is a tendency in planning education to rely too much on substantial knowledge" (Faludi, 1975:18). The way Faludi wanted to develop his planning theory met with almost no response from the other Chairs. After a few years, he left to join the University of Amsterdam. In a biographical note to his inaugural lecture there, he mentions that his relation with the Faculty of Architecture did not go smoothly. In his view, developing planning as a self-supportive discipline had not been given enough priority (Faludi, 1978:26). Although his Chair in Delft was discontinued, the course on planning processes continued as *Planvorming* (De Boer, 1990) under De Boer's Design Chair, until his retirement in 1989. After that, the subject was transferred to Drewe's Chair, but no longer as compulsory part of the degree course.

Drewe's teaching staff were always drawn from varied backgrounds – technical, spatial and social – and there were links with ISO. After Angenot's retirement, the sociologist Den Draak became director-secretary of ISO, while Drewe became a member of the ISO board, later taking on the Chairmanship. The research topics undertaken by both the Chair, and the Institute as a whole, were very varied, although almost all of them included spatial aspects such as concentration/dispersion and mixing/separation. Examples of ISO research themes included housing and living in post-war districts, the infrastructural aspects of shopping centres, spatial abstraction in traffic and transport planning, and models to combat noise and wind. In the 1970s and early 80s, Drewe concentrated his research on demographic aspects, the segregation of minorities in large cities and the ethnic component of migration, while his staff continued to conduct research on economic, infrastructural and methodological questions. Research financed by third parties was hardly ever undertaken in the Faculty, as many feared that this would curtail their independent scientific research. However, by the second half of the 1970s, contract research had become a substantial part of the research programme, with Drewe playing a stimulating role in this process. It should be mentioned here, that despite the range of research subjects, the Design Chairs in those years were already questioning the usefulness for their design tasks.

## The 1980s

It was relatively quiet at the beginning of the 80s with respect to the courses offered by the Faculty, the research undertaken there and the organisation of the Faculty. In the Chair's educational programme, cyclical planning processes and the position of research in the different phases was given a more explicit and prominent place. In Dutch practice, 'theory *in* planning' gained attention at the expense of 'theory *of* planning', as propagated by Faludi and others.

Half way through the 80s, the Faculty adopted a new plan for Chair-holders and associate professors, and with it, the name of the Chair changed to Spatial Planning. That the word 'research' was dropped from the name of the Chair was no coincidence; it was in line with changes in thinking about research. In a university, it might be expected that all staff members, including the 'designers' would carry out research. Therefore, monopolising research to a specific Chair no longer fitted in with the image of a modernising university: and this was not the only change. During Drewe's Chair holding, the contents of the educational programme, and the research that was carried out, had also evolved.

We do not want to go too far here into the changes that took place at the end of the 80s in how the educational programme was organised and the changing and limiting effects this had on the organisation and contents of spatial planning. Suffice is it to say that, within the framework of what was known as 'problem-based learning', the subjects offered within the syllabuses were reorganised and adapted. What was remarkable – and this continued into the following reorganisation in the 90s, and later – was the gradual decrease of guidance on how to conduct research. Within the context of Urban Planning and Design, the Chair of Spatial Planning became responsible for a (quarter) module, called 'Society', which focused predominantly on research and reporting. The staff members who gave this course, also participated in other educational modules. The Chair's research programme included subjects such as innovative environments, urban economy, information and communication technology applications, unemployment, vulnerability and urban deprivation. The research included not only methodological issues, but the spatial implications and effects of each topic were also investigated. In ISO, the main focus was on route-choice behaviour, the effects of increasing leisure, living in high densities, and the planning of urban amenities.

A new system for financing the research programmes of both the Chair and the Institute was introduced in 1983. This had the positive effect of encouraging more exchange of information, tuning and incidental co-operation with the Urban Design Chairs. Design professors, such as Van Tol and Veltmeyer, had a positive attitude towards research and its results, and De Boer stressed the importance of including spatial planning in the educational programme. In 1986, Van Tol drew attention to the policy that had been followed by the Faculty during the 25 years prior to this; a policy of strengthening the limited bedding of design before the 1960s with more research and a better anchorage in the societal context. "At present, a part of these attainments threatens to be lost: from the inside, by putting (too) much weight on design in a restricted sense, and from the outside, by ill-balanced financial cuts" (Van Tol 1986:106-7). A question that remained unanswered was the relation between the disposition of spatial designers, especially architects, to 'autonomous' thinking and acting on the one hand, and, on the other hand, to the often very cautious and open policy recommendations made by social scientists about future society, and the design and planning tasks involved.

In 1986, financial cuts and reductions in staff resulted in the fusion of ISO and the Centre for Architecture Research (established in 1964). The new institute was called the Research Institute for Urban Design, Urban Planning and Architecture (OSPA). Notwithstanding a considerable output, contract research, and participation in research schools outside the Faculty, such as Nethur and Trail and in the Faculty's own educational programme, the Faculty remained rather indifferent to this new institute. To conclude this section, the early retirement of Heimans in 1988 is also worth mentioning, because this resulted in the discipline of spatial planning becoming the responsibility of one Chair-holder: Drewe.

## The 1990s onwards

Beginning in the mid-1980s, responsibilities for urban regeneration and extension were split up over different Ministries and other organisations (Hulsbergen & Stouten, 2001), which resulted in the spatial policies in the Netherlands becoming more and more disconnected. During the 90s, the complicated societal changes of the time forced the universities, Delft University of Technology included, to adapt to the new 'market' and 'management' orientations, and to the accompanying financial cuts. In the wake of these changes, the educational and research context of Spatial Planning within the Faculty began to change drastically. There were two obvious developments that affected research. The first was the dismantling of the Research Institute for Urban Design, Urban Planning and Architecture, the institute where the Faculty research on urban planning and architecture was clustered. The researchers were transferred to Chairs in Architecture and to the Urbanism Chair of Spatial Planning. The second was the organisation of research programmes throughout the whole Faculty. The aim behind this reorganisation was not only to increase research 'output', but also to force the Chairs to develop a clear vision on the relation between design and research. The policy change took place from 1997 onwards, with the appearance of the first Scientific Report [*Wetenschappelijk Verslag*] (Faculteit Bouwkunde, 1997).

An underlying problem that had to be coped with was that many 'designers' mistrusted architectural and urbanism research with a societal orientation. In their opinion, such research was out of place in the Faculty (Den Draak, 2001). An often-heard sentiment was that the Faculty needed 'its own' design research methodology. A general overview of the range of views and discourses can be found in the proceedings of the international conference *Research by Design* in 2000 (Langenhuizen, Van Ouwerkerk & Rosemann, 2001) and in the book *Ways to Study and Research Urban, Architectural and Technical Design* (De Jong & Van der Voordt, 2002). The outcome of reorganising the research was that, wherever possible, institutional research was discontinued or replaced by PhD candidates. Another effect that can be considered as a gain was that more staff members were encouraged, or forced, to produce research. This could not be said of the gradual phasing out of scientific research education, which had a detrimental effect. By successively reorganising the study programmes, less research was done by the students themselves during the Bachelor stage and early in the Masters courses. The new syllabuses suffered from an over-emphasis on form and composition and from the urge to 'free' the research from its socio-scientific orientation to 'proper architectural' research (which nevertheless still had to cope with the multitude of 'ways to study'). At the beginning of the 21st century, the research undertaken by students started to be channelled into so-called 'post-graduate study laboratories' that were linked to the research programmes of the Faculty departments. However, the problem that had to be tackled was that the majority of the students lacked the basic research skills needed.

To offer students design research opportunities during their final study project, the Chair of Spatial Planning devised the design studio 'The Network City VROM'. Co-financed by the Ministry of Housing, Spatial Planning and Environment [*Ministerie van Volkshuisvesting, Ruimtelijke Ordening en Milieubeheer*], this studio ran from 1997 to 2003 (http://www.networkcity.bk.tudelft.nl). For its own research, the Chair initiated the 'Network Cities' programme, which, in 2003, became part of the TU Delft spearhead: the Centre of Sustainable Urban Areas (http://www.sua.tudelft.nl). The new information and communication technologies (ICTs) and their uncertain and complex relations with spatial developments played a major role. The staff members focused their contributions on the simple programming research undertaken during the Bachelor semester course 'City and Housing', and on 'Design Strategies for Urban Problem Areas' given in the final semester of the Bachelor course. For the Masters course, contributions were organised during the 'Design and Strategy' semester in co-operation with colleagues of the Urban Design Chairs. These included lectures on the society-space relationship, the cyclical plan and systematic development, and the argumentation and justification of designs. In the final year of the Masters course, they co-supervised students' laboratory studies.

## The present

In the near future, the Chair of Spatial Planning [*Ruimtelijke Planning (RP)*] is being merged with that of Urban Renewal and Management [*Stedelijk Beheer en Stadsvernieuwing (SBS)*] under two Chairs: the nucleus Chair of RPS and the Practice Chair of RPS. The so-called *nucleus* Chair of Spatial Planning and Strategy [*Ruimtelijke Planning en Strategie (RPS)*] will have a full professor and staff, who will be responsible for long-term education and research programmes. The so-called *Practice* Chair RPS, as the name suggests, is meant to attract practitioners to take on a temporary position, as a Professor by Special Appointment, to strengthen awareness of changes in the profession, as practiced outside the university.

The Chairs of both Spatial Planning and Urban Renewal and Management have already reorganised and renewed their syllabuses within the framework of the educational transformation of the Faculty, and individual staff members of both Chairs already co-operate in several parts of the Bachelor and Masters courses. It is essential for the position of Urbanism (Urban Planning and Design) in the Faculty of Architecture that these processes of change continue under the new Chair of Spatial Planning and Strategy. Spatial planning is a wide and inspiring field of study. It aims at aspects and dimensions that are only partly known at the moment, such as the effects of new information and communication technologies currently being scientifically developed within the discipline. The overall question is not only 'what is the future of urbanism?', but mainly 'what sort of urbanism is relevant for the 21st century?' (Drewe, 2005). Spatial planning in Delft concerns all spatial scales: the urban object, neighbourhood, district, city, or region, and the international or even global scales. Moreover, spatial questions in urban planning and design are not limited to two dimensions, but to four (the three dimensions of space, and the fourth one, of time). Demographic developments and need changes have always been objects of research. New technologies are also a necessary field of study. Their spatial consequences are unclear at first, but from many past examples, we know that once a technology is widely adopted, its spatial impacts can be enormous. The changes in societal policies also need attention, as do the ways actors behave and safeguard their interests, not forgetting vulnerable and deprived groups affected by those transformations.

In the courses offered at Delft, spatial planning is treated more as a method or a way of thinking than as an obvious and final application, so the spatial planners could meet the urban designers in their ambition to make designs that go beyond the probable development, thereby making it possible to create an improbable (unexpected) but desirable future. The spatial planning contribution consists of an *ex post* evaluation of the built-up environment as well as an *ex ante* evaluation of plans and designs. Spatial planning in education also deals with the different roles played by urban planners and designers, and their importance; roles such as designer/visionary, political advisor and mediator, urban manager, scientist (European Council of Town Planners, 2003).

Research is more complex. The research carried out by the Spatial Planning staff is concentrated into the 'Network Cities' programme, which is has been included as part of the University spearhead Centre for Sustainable Urban Areas activities for the period 2002-2007. To disseminate Spatial Planning research, and to contribute to the development of bodies of empirical and practical scientific knowledge, the Chair started its own series of publications, the Delft Science Planning (DSP) Series, in 2004. Research conducted by Urban Renewal and Management falls under several programmes: Globalisation; Urban Form Governance, which is also used as the basis for ALFA-IBIS, an international European co-financed research programme (Drewe is one of the participants); and Space Lab. Although researchers of both Chairs communicate with each other on a personal basis, there is hardly any co-operation in terms of research programmes. Consequently, with the merging of the Chairs of Spatial Planning and Urban Renewal and Management, the relation between the research programmes of the new *nucleus* and *Practice* Chairs will have to be worked out. In terms of people, i.e. the researchers, there will be many changes in the coming years, and retirement will be a regular phenomenon. The task for the future will certainly be one of ensuring the continuance of academic research on subjects that have a lasting value for the discipline and the Faculty, and for society as a whole. Ensuring another kind of continuation: space for more fundamental research that is aimed at exploring and defining trends in society and technology where the causal relations with space are often unclear, though, potentially always important, is another large area of responsibility.

## References

Angenot, L.H.J., 1963, *De uitbreidende en uitdijende stad, Inaugural Lecture*, Technische Hogeschool Delft, Samsom, Alphen aan den Rijn

Angenot, L.H.J., 1971, De voorwaardelijkheid van de ruimtelijke ordening, Valedictory Lecture, Technische Hogeschool Delft, *Stedebouw en Volkshuisvesting*, January 1972

De Boer, N.A., 1990, *Stedebouwkundige Planvorming*, Internal publication Faculty of Architecture, Technische Universiteit Delft, Delft

De Casseres, J.M., 1926, *Stedebouw*, Van Looy, Amsterdam

De Jong, T.M. & D.J.M. van der Voordt (eds), 2002, *Ways to Study and Research Urban, Architectural and Technical Design*, Delft University Press Science, Delft

Den Draak, J., 2001, Onderzoek en (stede) bouwkunde, een moeizame relatie; In: Ter Heide, H., M. Bierman & A. Bours, 2001, *Even kennis maken. Een halve eeuw ervaring met onderzoek voor ruimtelijk beleid*, Siswo, Amsterdam, pp. 107-123

De Smidt, M., 1986, Een geografisch perspectief op onderzoek voor de ruimtelijke ordening; In: Van Dam J.B.Th. & L. Verhaar (eds), 1986, *Maatschappijwetenschappen en ruimtelijke ordening*, Rijksplanologische Dienst, Den Haag, pp. 9-24

Drewe, P., 1974, *Stedebouwkundig planologisch onderzoek, waarvoor?*, Inaugural Lecture, Technische Hogeschool Delft, Delftse Universitaire Pers, Delft

Drewe, P., 2005, Welke stedebouwkunde van de toekomst?, *Atlantis*, 16.2, pp. 39-43

European Council of Town Planners, 2003, *The New Charter of Athens 2003, The European Council of Town Planners' Vision for Cities in the 21st Century*, Lisbon, http://www.ceu-ectp.org/e/athens

Faculteit Bouwkunde, 1997, *Wetenschappelijk Verslag 1997*, Faculteit Bouwkunde, Technische Universiteit Delft (http://www.bk.tudelft.nl/scripts/inc/onderzoek)

Faludi, A., 1975, *Planningtheorie en planningonderwijs*, Inaugural Lecture, Technische Hogeschool Delft, Delftse Universitaire Pers, Delft

Faludi, A., 1978, Planologie en wetenschapsbeoefening, Inaugural Lecture, Universiteit van Amsterdam; *Rooilijn*, 11 (extra issue), October

Geddes, P., 1915, *Cities in Evolution*, London

Heimans, A., 1970, *Het ruimtelijk milieu, aktuele kritiek en onderzoektaken*, Public lecture, Technische Hogeschool Delft, Waltman, Delft

Hulsbergen, E.D. & P.L.M. Stouten, 2001, Urban renewal and Regeneration in the Netherlands: Integration Lost or Subordinate, *City Journal*, 5, pp. 325-337

Kruijt, C.S., 1961, *De verstedelijking van Nederland*, Inaugural Lecture, Technische Hogeschool Delft, Van Gorcum, Assen

Langenhuizen, A., M. van Ouwerkerk & J. Rosemann (eds), 2001, *Research by Design*, Conference Proceedings B, Faculty of Architecture, Delft University of Technology, Delft

Sitte, C., 1889, *Der Städtebau nach seinen künstlerischen Grundsätzen*, Wien, Verlag Karl Graeser; Dutch translation: Van der Woud, A.,1991, De stedebouw volgens zijn artistieke grondbeginselen, Uitgeverij 010, Rotterdam

Van der Cammen, H., 1979, *De binnenkant van de planologie*, Coutinho, Muiderberg

Van der Cammen, H. & L. de Klerk, 2003, *Ruimtelijke ordening. Van grachtengordel tot Vinex-wijk*, Het Spectrum, Utrecht

Van Lohuizen, Th.K., 1948, *De eenheid van het stedebouwkundige werk*, Inaugural Lecture, Technische Hogeschool Delft, Delft

Van Tol, J., 1986, Onderzoek en ontwerp, een spanningsveld?; In: Van Hoogdalem, H., D.J.M. van der Voordt & H.B.R. van Wegen (eds), *Tussen pen en potlood*, Delftse Universitaire Pers, pp. 95-107

# Part I

## Societal Changes and their Effects on the Use of Space

# Introduction to Part I

Part I: *Societal Changes and their Effects on the Use of Space* departs from the societal demand side of spatial and urban planning and design tasks. The changes in society have many faces. Improved living conditions help people to live longer. Individualisation, international migration, globalisation and the growing market orientation of government policy are influencing people's and organisations' opportunities on both neighbourhood and regional levels; and because of the increased cultural diversity, there is a need to reflect on society's social and spatial standards. The contributors to this part of the book reflect on particular aspects of these societal changes.

In the Chapter on *The Mobility of Older People and Urban Planning,* Mart Tacken discusses the changing and decreasing mobility of people aged 65 and over. The conclusions show how important accessible facilities and services are to this age-group from the point of view of distance, means of transport, and new information and communication technologies. Urban and regional planning needs to focus its knowledge and instruments more on improving the quality of these aspects of the living environment for the elderly.

In *Vulnerability and Deprivation,* Edward Hulsbergen looks at how urban problems are defined, and comes to the conclusion that much more attention should be paid to the careful aggregation and disaggregation of social and spatial data when identifying urban problem areas. He argues that, when the plans are defined and developed, and also at their realisation and evaluation stages, urban policy, planning and design should include vulnerability (dependencies) and deprivation (shortages) as key concepts.

Herman Rosenboom, in his chapter on *The Randstad: Its Position and Environment,* discusses the relation between the economy and spatial planning. To maintain a competitive position within the Netherlands, regional spatial policies in the last few decades have focused on the changing economy. For the last twenty years, the national Spatial Development Memorandums and related documents have also focused on the Netherlands' competitive position within Europe.

The subject of Luuk Boelens' chapter on *Streaming Spatial Planning: Technological Changes and their Impact on Space,* is the rapid development of information and communication technologies and their societal significance in relation to spatial developments. He comes to the conclusion that the rapid development of these technologies and the corresponding increase of possibilities for using them is the prelude to unforeseeably dynamic futures for which new concepts of (spatial) planning will be required.

In *The Impact of Spatial Logistics in the Cross-border Spatial Corridor of Flushing-Terneuzen-Gent,* Georges Allaert focuses on cross-border developments in Europe, and in particular on the common interests and cross-border initiatives of Belgian and Dutch entrepreneurs on both sides of the Scheldt River. To improve cross-border cooperation, he recommends new ways of planning, land use and knowledge transfer. The rise and fall of the International Scheldt University as a cross-border education initiative is a good illustration of the difficulties that have to be overcome.

Part I ends with Jan Goedman's chapter *Mixed Scanning and other Issues related to Development Planning in the Netherlands*, on the changes that have taken place in planning procedures and methods. Looking back, he clarifies the present debates on plans, parties and projects in the Netherlands, the relevance of spatial design in connection with spatial research and development, and indicates ways of facilitating tacit, explicit and scientific planning knowledge. He concludes that it must be possible to reconcile spatial research, design and development.

# 1 The Mobility of Older People and Urban Planning

Mart Tacken [1]

## Introduction

In the 1990s, as member of the Italian National Research Council, Beguinot started a discussion on a new charter for urbanism. This was necessitated by new developments, especially those in technology. The third statement in this charter, known as the Charter of Megaride 1994 was that: "The city of the future will have to guarantee to each citizen maximum access to places, services and information – a city where each different group finds conditions that respond to its particular needs, which it must be able to express freely" (Carta di Megaride 94, 1994:115). *The New Charter of Athens 2003* presents the vision of the European Council of Town Planners for cities in the 21st century. This Charter stresses more than before the ability of the planning profession to take a range of issues into account and translate them into spatial terms. Social developments occupy an important position in these issues. Human activities are widely spread and cities, large and small, are linked to an urban continuum, in a complex new network. This connected city has to be responsive "to the interests of the society as a whole, whilst having regard to the needs, rights and duties of various cultural groups and of individual citizens" (European Council of Town Planners, 2003:4). The ageing population is mentioned as one of the target groups of town planning. As stressed in both charters, explaining human behaviour and formulating guidelines for planning and design are primary goals in urban and regional planning research. The Spatial Planning group, and related research groups, such as the former Research Institute for Urban Design, Urban Planning and Architecture (OSPA), have focused their research on this interrelation between the individual and societal needs of specific groups of the population, and the spatial realisation of these needs. The European project MOBILATE (sponsored by the European Commission, Project QLRT-1999-02236) was carried out within this frame of reference. The main goal of this project was to improve mobility outside the home in later life.

European and North American societies are greying strongly, and this tendency will continue. The percentage in these populations of people aged 65+ years and older will increase from 15% in 2000 to 23% in 2030 in Europe and from 12% to 20% in North America, respectively (average variant set by the United Nations Economic Commission for Europe; UNECE, 2004). In both continents, older inhabitants will become a considerable part of the population, and hence will become important in terms of policy-making and marketing. Society has to consider how to offer this group the possibilities to live as independently as they themselves want, and to build spatial facilities that also meet the needs of older people. Mobility is an important aspect of their quality of life (OECD, 2001) and the environment must facilitate this mobility. A basic condition is that important functions should remain accessible. For a population whose physical mobility and financial means is decreasing, it is even more critical to find appropriate alternatives.

1 The research group consisted of Fiorella Marcellini, Heidrun Mollenkopf, Isto Ruoppila, Zsuzsa Széman and Mart Tacken.

Largely thanks to the car, activities and services in today's society are located, and take place, on a much larger scale than they did some decades ago. Garreau (1991) described this process of sprawl into edge cities for the United States, and Brand (2003) conducted similar research for his PhD Thesis on the development of the urban field in the Netherlands. Suburbanisation in most European cities has been enabled by the large-scale introduction of the car, which has improved accessibility in general. Recently, information and communication technology (ICT) has added new conditions for urban sprawl and for the relocation of a number of facilities. On the local scale, in particular, urban functions strongly related to the use of information technology are disappearing from the central locations. Branch offices of banks, post offices, insurance companies or municipal offices have either disappeared or have moved to central offices, or to offices located on the urban fringe at cheaper locations accessible by car. For a greying population, this development may create access problems to essential functions. Decreasing physical mobility and less money to spend reduce the available alternatives for older people. In the present older generation, many women have to manage without a car, and many older men and women consider that using a car takes too much effort, skill and money. They are less familiar with the use of computers and other new technologies that could offer them alternatives.

In this chapter we stress the essential role of spatial planning in enhancing and conditioning the outdoor mobility of older people. We describe the main insights and hindrances conditioned by the outdoor environment in relation to the out-of-home activities of older people. The MOBILATE 2000 data offer the opportunity of analysing the relationship between the main factors: personal and social characteristics, external conditions of the social network, transport facilities and spatial characteristics typified by terms such as access or accessibility, location, spatial scale and distance. We build up an explanatory model of this relationship and derive some guidelines to help urban and spatial planners improve the external conditions for mobility. After a short methodological paragraph on the MOBILATE project data, the insights available into the relationship between old age and spatial context will be described. The next part will present information on how older people in five European countries perceive certain aspects of their living environment. After this, the environmental conditions for their outdoor mobility will be described, followed by conclusions and some guidelines for spatial planners on how to create better conditions for outdoor mobility.

## MOBILATE data as a source for analysis

For the European project MOBILATE, 3950 people aged 55 years and older were interviewed. This survey took place in five countries (Finland, Germany (eastern and western), Hungary, Italy and the Netherlands) in 2000. The national samples were drawn at random from the municipality population registers in each of these countries. To analyse and gain an understanding of these samples, the total sample of 3,950 men and women aged 55 years or older was then stratified by gender and age (55-74 years; 75 years or older) in urban and in rural areas. The meaning of 'urban' and 'rural' differed somewhat from country to country. In Finland and Hungary, a 'rural' area is real countryside, but, in the Netherlands, for instance, 'rural' can more accurately be understood as a 'non-urban' area. Cities, middle-sized, compared with the other cities in that country, were considered as urban areas. Again, there are great differences between the extensive Finnish city, compared with, for instance, the more compact German city. To correct for oversampling in favour of older people and males, statistics of descriptive and comparative analyses were weighted by the share of the respective age and gender groups in the selected regions.

About 55% of the net sample of 7200 eligible participants was interviewed by trained, experienced personnel. In most countries (with the exception of eastern Germany), the net response rate was higher in the rural (58%) than in the urban areas (about 52%). The main reasons given for not taking part were: refusing without giving detailed reasons (22.6%); having no time (6.2%); unavailable when the interviewer called (5.4%). Only a minor percentage (5.1%) of the possible participants mentioned health problems as a reason for not being interviewed. In the sample achieved, the ages of respondents ranged from 55 to 98 years. Standardized questionnaires were used to assess essential features of the community, various kinds of activities, and mobility. The assessment also included demographic and health aspects, social networks, and personality measures. Internationally recognised measures were used, wherever available. Several parts of the questionnaire were also directed towards the spatial context. For instance, spatial elements have to be described in answering the questions concerning the residential area, the conditions of the location for outdoor mobility, the accessibility of services and the transport. Many questions were focused on personal experiences with the external conditions. The interviewees were also asked to keep a diary for two days in which they recorded all their movements outside the home. This last approach helped gain insight into the range of this mobility and its limiting conditions.

## Urban planning conditions required for the mobility of older people

In 1984, Altman, Lawton and Wohlwill published a book entitled *Elderly People and the Environment.* In this book, they made an overview of the available knowledge on the interrelation between the spatial setting and the behaviour of older people. They stress the importance of focusing on specific places, and of designing, or creating and shaping, these settings for older people. In many ways, it is urban and regional planning that influence the outdoor behaviour of older people.

For elderly people, decreasing physical mobility usually results in restricted mobility. "Land-use planning that aims to reduce the amount of travel needed to access services, facilities and social networks is of particular benefit to older people" (OECD, 2001:105). This concerns the location of dwellings and facilities, and also the distances to facilities that people want to use. Another relevant factor is the detailed local road layout and network.

The residential environment is very important. It can be assumed that spatial circumstances influence peoples' behaviour on different scales. The environment closest to people is their dwelling and its direct surroundings. Living in an apartment building without lifts can create barriers that prevent residents from going out. Climbing stairs can be a real hindrance for the elderly and disabled people. The location and characteristics of the dwelling create the first conditions for making the outdoor environment accessible. In his book *Pattern Language*, Alexander (1976) put forward a rather detailed design proposal on how to make a layout for a community containing older people, which makes allowances for the needs of these older people and the short distances they can cover. Accessibility is the main concept in the extended spatial environment. The residential environment has to be adapted to older users. "This concerns the accessibility of facilities such as public transport, shops and medical care and, increasingly, safety" (VROMraad, 1998:21). Two aspects condition this accessibility: the distance to the necessary or desired facilities and the transport system available. These two are interrelated. A good transport system offers an alternative for nearby facilities. For older people, the quality and layout of

the infrastructure is also relevant (Davidse, 2002). In a research project conducted by the Vrije Universiteit Amsterdam (Raaijmakers, Van der Beek & Rohde, 1996), the concept of resident-friendly areas has been used. This concept is based on the need to integrate the residential and living circumstances of older people. The weaknesses and opportunities offered by an area will be used as a starting point to develop a strategic spatial policy for the settlement of older people.
It is suggested that good transport alternatives and land-use planning facilitate walking as a mode of transport. That makes the decision to stop driving easier (OECD, 2001). Finally, this OECD report concludes that travel needs to continue after retirement, but that these needs may change, because elderly people are likely to experience problems with walking and public transport before they experience difficulties with driving. Rosenbloom (2003) tries to debunk some of the myths surrounding the travel needs of the elderly. The first one concerns the myth that older people first lose their ability to drive, then their ability to use public transport and finally their walking ability; special transport is then their last option. However, the order of this list is often incorrect. Driving is often the easiest physical task; easier than using public transport or special transport, which is easier than walking.

The hindrances experienced by elderly people are partly related to their own physical and mental condition, but these are also evoked or conditioned by external circumstances in the built-up environment. Rosenbloom (2003) stresses the importance of good urban planning in order to guarantee good accessibility to facilities for people with reduced mobility. Suburbanisation of the elderly is paralleled by the same movement in the United States. People nowadays live in built-up environments, where destinations are some distance away; 'discontinuous space' and the space between origin and destination must be connected by transport (Lavery, Knox & McKenna, 1998). This may create a 'mobility gap', especially for those who are less mobile, as are quite a number of elderly people. The increasing effects of information and communication technology, which broaden basic urban facilities, will enforce this tendency. Several aspects of peoples' spatial context and some of those that came to light from the 'trip' diary were used as research topics in the MOBILATE survey. What have we learned about these aspects as essential parts of the outdoor life of elderly people? First we will describe and explain the role of the spatial environment. Then we will start with the environment closest to people, that of the dwelling and its direct surroundings. Then we will examine the accessibility of necessary services and facilities, followed by a look at the two important conditions for accessibility: infrastructure and the transport system. This part will be concluded with an analysis of how these aspects can be integrated.

## Residential environment: dwelling and neighbourhood

Some aspects of the residential environment are relevant for the outdoor mobility of older people. Interviewees were asked: 'How important is being out of home for you?' Illustration 1.1 shows that, in both parts of Germany and in the Netherlands, country people place more importance on to go out of home than urban dwellers, but the opposite was found in Hungary and Italy. Can this be explained by the role of the larger family in countries such as Hungary and Italy, and the accessibility of services and facilities in German and Dutch urban complexes? The differences are big enough to justify further investigation.

Answers to the question: 'What is important for you in the area where you live?' give detailed indications of the role of the spatial context. In Illustration 1.2, these items have been listed in order of the importance attached to them. At the top of this list, interviewees chose 'a clean and quiet residential

| | urban | | rural | | t |
|---|---|---|---|---|---|
| | mean | std.dev | mean | std.dev | |
| Finland | 8,9 | 1,2 | 8,8 | 1,6 | 0,67 |
| Eastern Germany | 7,4 | 2,4 | 7,9 | 2,3 | -3,02* |
| Western Germany | 6,9 | 2,7 | 7,4 | 2,6 | -2,26* |
| Hungary | 7,3 | 2,5 | 6,1 | 2,6 | 5,71* |
| Italy | 7,4 | 2,4 | 6,7 | 2,7 | 3,26* |
| The Netherlands | 7,8 | 1,7 | 8,1 | 1,3 | -2,31 |

**Illustration 1.1:** **The average score to the question: 'How important is it being out?' (0 = not important, 10 = very important). (source: MOBILATE Survey 2000, N=3950, weighted data. Scale from 0 (not important at all) to 10 (very important). *t-test p<0.01)**

| | Finland | Germany (eastern) | Germany (western) | Hungary | Italy | Netherlands |
|---|---|---|---|---|---|---|
| A clean environment | 98 | 90 | 94 | 92 | 93 | 91 |
| A quiet residential area | 93 | 92 | 91 | 96 | 89 | 79 |
| Good medical care nearby | 78 | 92 | 94 | 93 | 96 | 87 |
| Easy access to shops/services | 79 | 86 | 89 | 93 | 84 | 85 |
| *A pleasant neighbourhood* | *90* | *52* | *69* | *92* | *83* | *88* |
| *Living close to friends/relatives* | *74* | *76* | *85* | *81* | *83* | *61* |
| *Living in an green area on the urban fringe* | *85* | *73* | *71* | *76* | *76* | *84* |
| *Well connected* | *65* | *64* | *76* | *79* | *79* | *81* |
| Cultural opportunities nearby | 33 | 40 | 18 | 24 | 26 | 37 |
| Living where 'the action is' | 33 | 6 | 9 | 19 | 18 | 43 |
| **Total N=** | **610** | **738** | **746** | **600** | **600** | **534** |

**Illustration 1.2:** **The features considered to be important in a residential area (listed in order of importance, in percentages of the total sample). (source: MOBILATE Survey 2000, N=3950, weighted data)**

area' as the feature of the living environment that they most wanted. 'Good medical care' and 'accessible shopping facilities' were also given a high ranking in all countries. In contrast to this, 'a vivid environment' and 'nearby cultural facilities' was given low rankings in all countries. Older people seek rest and peace in preference to a vivid environment, full of action. Between these two extremes, the importance given to the various items on the list seems to vary according to the cultural and spatial differences between the countries included in the survey, but access to facilities was a relevant item everywhere.

## The accessibility of services and facilities

Interviewees were given a list of 15 services and facilities and were asked to put them in order of availability (accessible within 15 minutes) with respect to their own homes. The outcome is that, in both urban and rural areas, a 'cemetery' and 'library' are given as being the least available facilities. More surprising is that, in Finnish urban areas, a bank is only available to 40-60% of the respondents, as is a pharmacy and a bank in rural eastern Germany. However, from the approach used here, no conclusions can be drawn regarding the lack of important services within acceptable distances.

In looking at accessibility, a significant factor is the means of transport chosen. In Dutch urban areas, 71% of the respondents visit important services on foot, and an additional 13% by bicycle. This shows the importance of slow modes of transport in the Netherlands: only 3% chooses the bus and only 5% (driving) their car. In the rural areas, however, these percentages are, 65% (on foot), 10% (by bicycle), 2% (by bus) and 16% (driving their car), respectively. So, apparently, the longer distances cause a shift to using their cars. Access to local services is the worst in rural areas, especially in Hungary, where 22% of the people interviewed even have difficulty in getting to a food retailer. People often mentioned health problems as their main reason for not being able to get to places, but in Hungary, 26% also gave long distances as a reason.

## Infrastructure and spatial organisation

Access to facilities depends on distance and on the available infrastructure. Although these facts are unavailable, we do have information on how transport and traffic are perceived. Respondents were asked to what extent they agreed with a list of statements about today's traffic. These scores have been ordered according to the average percentage of agreement with each statement. Illustration 1.3 shows that these older people are not impressed by the behaviour of others: they expect people to offer them a seat more often. The first six statements are all related to the behaviour of people as fellow road-users. The second statement, about cycle lanes, has to be combined with the nuisance caused by a mixture of pedestrians and cyclists on the pavements, as mentioned in the third statement. Cyclists need to take older pedestrians into account more, and, similarly, car drivers who approach pedestrian crossings too fast or drive too close to the sidewalks. The survey results show that people older than 75 agree more often with negative statements. This can be seen most clearly in the statements at the bottom of the list, which are focused on the use of public transport. However, only a small proportion of the elderly admit to feeling helpless or disadvantaged.

| | Urban | | | | Rural | | | |
|---|---|---|---|---|---|---|---|---|
| | 55-74 | | 75+ | | 55-74 | | 75+ | |
| | male | female | male | female | male | female | male | female |
| Too few people offer their seat in the bus to a person who needs to sit down | 47 | 57 | 55 | 59 | 51 | 53 | 51 | 55 |
| There are too few bicycle lanes | 52 | 53 | 57 | 64 | 46 | 41 | 46 | 48 |
| Children and adolescents should be forbidden to ride their bicycles on the pavement | 55 | 52 | 60 | 48 | 47 | 52 | 50 | 41 |
| Many cars and motorcycles drive too fast when approaching pedestrian crossings | 56 | 56 | 56 | 49 | 47 | 39 | 42 | 43 |
| Many cars and motorcycles drive too near the sidewalks | 31 | 39 | 41 | 47 | 38 | 39 | 47 | 50 |
| Cars and motorcycles drive so fast that you can only see them too late | 32 | 41 | 44 | 41 | 38 | 40 | 43 | 47 |
| It very often happens that the traffic light has already turned red when I am still only half way across the road | 27 | 32 | 26 | 25 | 45 | 44 | 44 | 46 |
| Buses run too infrequently at certain times of the day | 27 | 26 | 35 | 30 | 27 | 26 | 32 | 30 |
| The sidewalks are often so narrow that you have to step into the street to make way for other pedestrians | 24 | 30 | 25 | 25 | 26 | 28 | 29 | 28 |
| As an elderly person, you feel disadvantaged in today's traffic | 29 | 30 | 36 | 35 | 20 | 21 | 18 | 21 |
| The buses start too quickly and jerkily, so you get thrown about | 16 | 26 | 30 | 41 | 14 | 25 | 23 | 35 |
| I have difficulty getting in and out of the bus because of the high steps | 15 | 23 | 25 | 32 | 15 | 20 | 24 | 27 |
| Traffic is sometimes so busy that you don't dare to go out onto the street | 18 | 25 | 25 | 35 | 16 | 16 | 16 | 21 |
| The automatic closing and opening of bus doors is poorly installed | 12 | 17 | 19 | 26 | 10 | 17 | 20 | 30 |
| Nowadays, I often feel helpless in traffic | 14 | 20 | 21 | 26 | 9 | 10 | 16 | 16 |

**Illustration 1.3: The percentages of the response 'true' to general statements about current traffic conditions. 1= true, 2 = partly true, 3 = not true. (source: MOBILATE Survey 2000 N=3950, not-weighted data)**

The responses to the question 'Do you agree with some improvements of the traffic situation', give further information on how elderly people rate proposed general improvements in infrastructure (Ill. 1.4). Specific spatial items hardly feature in this list, and little comes to light regarding the role of spatial planning in creating some of the problems encountered. The respondents mostly agree with statements geared towards improving the behaviour of other participants in traffic or towards making it cheaper for them to buy mobility. The statements geared at improving the physical framework of infrastructure, such as placing more street benches, separating slow-moving traffic from faster traffic by constructing more cycle paths and broader sidewalks, get lower rankings. Both tables give indications of how spatial planning, especially that concerning traffic systems, could improve older peoples' quality of life. Spatial planning can improve accessibility by creating: shorter distances to facilities and services, safer routes, better connections, more parking places, better and more pedestrian crossings.

## The transport system

The transport system is another aspect of spatial planning. The degree of accessibility of facilities is conditioned by distances and by either public transport connections, or the form of private transport that is available. Taking the sample as a whole, 45% of all trips (as recorded in the 'trip' diaries) were made – at least partly – on foot and 10% by bicycle. 28% of the trips were made by car (as driver), and an additional 11% (as passenger). In 8% of the trips, public transport was used. 43% of all trips were made within a distance of 1 km and 68% within the respondent's own village or urban area (a distance of less than 3 kms). The most frequently used means of transport, especially for the oldest age group, are slow modes. However, this finding should not be taken to mean that older people go more on foot, but rather that because the other modes of transport have too many disadvantages to be used, trips on foot are seen as the only remaining alternative. As is shown in Illustration 1.5, older, less mobile pedestrians try to avoid specific situations. We have to realise that they do this in their most common mode of walking.

From Illustration 1.5, it is clear that older pedestrians try to avoid more or less extreme conditions: bad weather, darkness or busy traffic. This avoidance behaviour is most pronounced with respect to people with poor health. They have more problems negotiating bad roads and rush hours and coping with darkness. Other research shows that older road-users are more often victim of traffic accidents, but this can be partly explained by their greater vulnerability of the frail, physical condition and partly by the vulnerability of pedestrians (Insurance Institute for Highway Safety, 2001; OECD, 2001; SWOV, 2001). Car drivers are less vulnerable. Twenty-four percent of the car drivers, the self-confident drivers, do not avoid any of the situations mentioned in Illustration 1.6. Poor weather is the situation that is avoided most, followed by rush hours, darkness, unknown routes and unfamiliar areas.

Illustrations 1.5 and 1.6 show that road-users avoid several situations. This depends partly on how they use the roads. Pedestrians, for instance, are vulnerable to bad weather in different ways from car drivers. Good spatial planning can improve some of these situations, or at least reduce the negative effects. The effects of this avoidance behaviour can be seen in some of the trips recorded in the diaries. For instance, it can be seen that elderly people avoid being away from home in the evening or at night. They also avoid the busy times, mostly making their outdoor trips between peak hours. The destination of between 59 and 83% of the trips is somewhere within the respondent's own urban area or village. Most of these are short-distance trips of less than 1 km.

| Statements about improving traffic systems | In order of personal importance |
|---|---|
| More road courtesy | 55 |
| Improved economic resources for the elderly | 49 |
| More road (traffic) safety | 47 |
| Cheaper tickets for public transport | 46 |
| More people to help or accompany frail people | 42 |
| Subsidised taxi services for people with walking difficulties | 40 |
| More street benches | 39 |
| Bus designs that take the needs of the elderly into account | 38 |
| More closed circuit cameras in public places, streets, and underpasses | 35 |
| More surveillance personnel in public places, streets and underpasses | 35 |
| More bicycle paths | 32 |
| More shops and services within easy reach | 32 |
| Shops and public buildings that can be entered more easily (i.e. no steps) | 31 |
| Timetables that are coordinated better | 30 |
| Better bus/tram/train connections | 29 |
| Clearer and easier information about public transport | 28 |
| Longer crossing-cycles at traffic lights | 28 |
| Car designs that take the needs of the elderly into account | 27 |
| More pedestrian crossings (with islands) | 27 |
| Special parking places for people with walking difficulties | 26 |
| More parking places at bus stops and railway stations | 24 |
| Tram designs that take the needs of the elderly into account | 18 |
| More personnel at bus stops and rail stations | 17 |
| Bicycle designs that take the needs of the elderly into account | 16 |
| Shorter distances between bus and tram stops | 16 |
| Taxi sharing | 16 |
| Meeting-points e.g. at large bus stations | 11 |
| Women's taxi | 4 |

**Illustration 1.4: Important improvements in the traffic system (in %). (source: MOBILATE survey 2000, N=3720, weighted data)**

## Conclusions

For elderly people, urban and regional planning is important for improving the quality of their living environment. The spatial environment, however, has no coercive power. The spatial context offers good or bad conditions for living and/or other activities, but people have to make choices. They can choose for specific locations as a place to live or as the site of specific (occupational or recreational) activities, or they can adapt their spatial behaviour to the external conditions with no prior knowledge of the restrictions involved. The accessibility of facilities is just one of the factors needed for high-quality living conditions. People learn to cope with bad conditions. As the situation gradually deteriorates, they get used to them. Particularly in the countryside, people choose to live there for other reasons, despite the lower level of facilities available, and some families remain there for generations.

The use of a car is one of the ways to solve these problems. A car reduces distances. A decreasing health or a lower financial budget are reasons giving up driving a car. Health and financial resources are at risk for people of high age. Public transport could be an alternative for problems with private transport facilities, but this transport mode is sensitive for distances to and from stops, frequency, comfort, etc. New special transport services (or paratransit) could offer alternatives and contribute to more sustainable and liveable cities, but some specific characteristics as advance ordering, delay, unreliable departure and arrival times, create a high threshold. These services have to make a good start in order to convince older people that these systems offer good and accessible alternatives for regular public transport.

The findings of this MOBILATE project confirm Rosenbloom's (2003) statement about 'the debunking of the myth' that public transport, or special modes derived of public transport are a last solution for elderly people. Our figures show that driving, or using the car as a passenger, offers better, more chosen, solutions for mobility problems, and that, wherever possible, elderly people prefer private modes of transport, such as walking or cycling, to public transport, which they find too problematic.
Spatial planning can contribute to a better programming of the residential environment and the space needed by older people for their activities. Accessibility is the main concept, as it is an important condition for independence and flexibility, and, as such, is important for the quality of life. This concerns not only the distance to the destinations that older people want to reach, but also the conditions that will be encountered on these trips: road conditions, public safety, traffic safety, the availability of benches, broad sidewalks, well-placed croswalks, etc. As the spatial context influences accessibility, local services and facilities can be improved by taking this aspect into account at the planning and design phase. By so doing, much better design and planning solutions can be found for the needs of older people than by searching, at a later date, for compromises in the transport or care systems.

In his book *Pattern Language*, Alexander (1976) presents examples of possible solutions, using patterns. Older people's range of action is restricted, and the present generation of elderly people is rather attached to their home. So, for the essential activities of daily life, short-distance pedestrian trips form the best pattern to work with. In a study conducted by the Vrije Universiteit Amsterdam this has been translated into zones, each containing facilities that are easily accessible for elderly people (Raaijmakers *et al.*, 1996).

Urban planning can also improve public safety by eliminating places that the elderly view as risky and unfamiliar, and therefore unsafe, using, as indispensable tools, good routing, street lights, footpaths and sidewalks, etc. This project has also indicated, indirectly, where other improvements can be made. A good layout and well-located crossings can help to reduce car speeds before they approach pedestrian crossings. This can be achieved much more effectively by altering the layout of the road than by using

| Situation avoided completely, or wherever possible | % of pedestrians, with no walking difficulty | % of pedestrians, with poor or very poor mobility |
|---|---|---|
| Crossing the road at dusk or at dark | 48 | 78 |
| Bad roads | 47 | 78 |
| Walking in the rush hour | 42 | 74 |
| Crossing a road without a pedestrian crossing | 44 | 70 |
| Walking along a busy road which does not have sidewalks | 56 | 76 |
| Going to unfamiliar places | 52 | 80 |
| Busy traffic | 51 | 77 |
| Poor weather conditions | 62 | 85 |
| Insufficiently lit streets | 60 | 83 |
| **total N =** | **3728** | **542** |

**Illustration 1.5: Specific situations frequently avoided by pedestrians (in %). (source: MOBILATE Survey 2000, n=3950, weighted data)**

| | Finland | | Germany eastern | | Germany western | | Italy | | The Netherlands | |
|---|---|---|---|---|---|---|---|---|---|---|
| | urban | rural | urban | rural | urban | rural | urban | rural | urban | rural |
| Driving at dusk or at night | 55 | 44 | 35 | 38 | 38 | 39 | 56 | 45 | 30 | 31 |
| Driving on bad roads | 38 | 37 | 37 | 39 | 50 | 43 | 75 | 62 | 41 | 28 |
| Complicated junctions | 12 | 27 | 15 | 18 | 14 | 22 | 29 | 33 | 16 | 16 |
| Driving during rush hours | 41 | 49 | 29 | 27 | 39 | 33 | 41 | 33 | 41 | 36 |
| Busy roads | 20 | 22 | 12 | 23 | 23 | 30 | 30 | 46 | 28 | 24 |
| Driving long distances | 32 | 33 | 24 | 32 | 35 | 46 | 50 | 62 | 43 | 46 |
| Driving along unknown routes or in strange areas | 36 | 43 | 27 | 39 | 30 | 46 | 46 | 44 | 38 | 35 |
| Driving on motorways | 5 | 16 | 6 | 19 | 14 | 37 | 35 | 54 | 8 | 13 |
| Driving under poor weather conditions | 63 | 64 | 50 | 41 | 55 | 45 | 61 | 56 | 40 | 39 |

**Illustration 1.6: Specific situations avoided by car-drivers, by country and area (in %). (source: MOBILATE Survey 2000, N=1646, weighted data)**

traffic signs. Not only can well-located services and homes for the elderly improve traffic safety, but they can also take more into account the preferences of older people. The MOBILATE findings show that the elderly prefer quiet and peaceful residential environments to more lively, noisier ones.

The developments in recent years in information and communication technology are also contributing to improved accessibility. Technology can be helpful in facilitating mobility by making transport systems and road conditions more sophisticated. Communication technology can be an alternative for actually going to places. It can be used for contacting people, Internet banking, teleshopping, etc. While elderly people are not the largest group of users, our research clearly shows that older users are comfortable with this method of communication.

Spatial planning can create conditions that influence how older people can function in the public space, in the vicinity of their homes. Spatial planning can help to make that a self-assured, pleasurable experience, or one so fraught with problems that the elderly become house-bound. For spatial planning and design to respond to this, the basic input required is an awareness and knowledge of how elderly people behave, out of home. The knowledge and the planning instruments are available; what is needed is for urban planners to increase their awareness and then respond.

## References


Alexander, C., 1976, *A Pattern Language*, Oxford University Press, Oxford

Altman, I., M. Lawton & J. Wohlwill (eds), 1984, *Elderly People and the Environment Human Behaviour and Environment*, Plenum Press, New York

Brand, A., 2003, *Het stedelijk veld in opkomst. De transformatie van de stad in Nederland gedurende de tweede helft van de twintigste eeuw*, PhD Thesis, Faculty of Behavioural Sciences, Universiteit van Amsterdam, Amsterdam

Carta di Megaride 94, Citta della pace – Citta della scienza, 1994, Di.Pi.S.T. Universita studi di Napoli "Frederico II", Naples

Davidse, R., 2002, Weginrichting met oog voor oudere verkeersdeelnemers, Gerõn, *Tijdschrift voor ouder worden en maatschappij*, 4, nr. 2

European Council of Town Planners, 2003, *The New Charter of Athens 2003, The European Council of Town Planners' Vision for Cities in the 21st Century*, Lisbon, http://www.ceu-ectp.org/e/athens

Garreau, J., 1991, *Edge City, life on the new frontier*, Doubleday, New York

Insurance Institute for Highway Safety, 2001, *Status report, special issue: older drivers*, vol.36, no.8, September (http://www.highwaysafety.org)

Lavery, I., J. Knox & O. McKenna, 1998, *Towards the identification and evaluation of barriers to travel on public transport by mobility impaired people*, WCTR paper July 1998, Antwerp

OECD, 2001, *Ageing and Transport. Mobility needs and safety issues*, OECD, Paris

Raaijmakers, A., M. van der Beek & I. Rohde, 1996, *Woonomstandigheden van ouderen in Overvecht verkend. Een GIS-zoneringsonderzoek en marktonderzoek*, Stichting voor Toegepaste Gerontologie, Vrije Universiteit Amsterdam, Amsterdam

Rosenbloom, S., 2003, *The mobility needs of older Americans: implications for transportation reauthorisation*, The Brookings Institution Series on Transformation Reform, Centre on urban and metropolitan policy, Washington, http://www.brookings.edu/urban

SWOV, 2001, Stand van zaken in de verkeersveiligheid, *SWOVschrift*, 85, March (http://www.swov.nl/nl/swovschrift/stand_van_zaken_in_de_verkeersveiligheid.htm)

UNECE, 2004, http://esa.un/unpp/p2k0data.asp

VROM-raad, 1998, *Advies over wonen met zorg*, Advies 002, Den Haag

# 2 Vulnerability and Deprivation

« Le problème n'est pas de savoir si ça va aller mieux demain, mais de connaître le prix économique et social que nous allons payer pour nous y rendre. »
Alejandro Rada-Donath (2004:37)

Edward Hulsbergen

## Introduction

In urban planning and design (urbanism) the concepts vulnerability and deprivation are basically about the inclusion and exclusion of urban residents and users of urbanised areas. Firstly, this has to do with whether they are included or excluded in the problem definition of urban transformation. Secondly, it has to do with whether they are included or excluded in making the proposals, implementing the plans, and evaluating the effects.

Many publications on the topics of poverty, deprivation and vulnerability refer to Chambers (1989:1), who stated in a rather pointed fashion that "vulnerability is not the same as poverty (…) not lack or want, but defencelessness, insecurity and exposure to risk, shocks and stress", and, as such, should be understood as being the opposite of security. Vulnerability "has remained curiously neglected in analysis and policy, perhaps because of its confusion with poverty". The view held here is that since the 1980s, this has been true not only of vulnerability, but also of deprivation.

Nowadays, the terms 'vulnerability' and 'deprivation' can be found in a variety of disciplines and fields. There seems to be an often unrelated tendency to associate these concepts with local policies (see: the *vulnerable neighbourhood* in Van der Pennen, Veldheer, Ter Borg, Kunst, Boelhouwer & Knol, 1998), housing and control by residents (Van de Lindt, 2002), or with social exclusion. De Haan (2003) defines vulnerability as a specification of deprivation within the framework of social exclusion. Whelan and Maître (2004:29), in their comparisons of European countries, also use vulnerability and deprivation as an indicator of social exclusion, but they understand it "as the outcome of a process in which the accumulation and erosion of resources over time interacts with variability in the demands with which households must cope". The terms can also be encountered in relation to education, work and health. An additional contribution to the debate comes from research on climate and environment, where O'Brien, Eriksen, Schjolden and Nygaard (2004) perceive two approaches to vulnerability: the 'end point approach' which views "vulnerability as a general characteristic generated by multiple factors and processes" and the other which takes 'vulnerability as a starting point', maintaining that it 'determines adaptive capacity'. It is accepted that poverty, exclusion, vulnerability and deprivation are multidimensional and interrelated.

One might ponder on the relevance of the above for urban planning and design, because, in the Netherlands, in both scientific fields and in practice, 'vulnerability' is usually seen as a point of attention for the socio-cultural, and 'deprivation' for the spatial-economic sectors. The view held here is that this fuels a misunderstanding about urban processes; it restricts outlooks on urban renewal and regeneration, and contributes to neglecting societal needs. Urban planners and designers, therefore, need to know about vulnerability and deprivation and participate in the debate, and then include these aspects in their theories and instruments.

This chapter focuses on defining vulnerability (dependencies) and deprivation (shortages in supply). The objects of study may be areas, groups or individuals. It is maintained that the necessary point of departure is to keep a clear distinction between the two, but that, for an integrated approach to urban problems, the two concepts need to be related from the start. Their true relevance for theory, application and policy lies in combining the two approaches, and in carefully considering them with regard to spatial and social (dis)aggregation.

## Urban complexity

Today's urbanised areas, our towns and cities, are characterised by complexity, not only in their functioning but also in their development through time. At the beginning of the 1960s Jane Jacobs used the now widely known term 'organised complexity'. Urban development, however, should not be understood as simply the re-organisation of specific components of the urban space. Urban networks, and artefacts and their total uses, in fact surpass present-day urban theory. At best, defining urban problems is a way of coming to grips with urban complexity; though more often than not, it seems to be mainly the short-term problems that are defined. Although there is a more or less common agreement on which components can usefully be distinguished in the urban debate – physical-spatial, social, economic, cultural, ecological and administrative – integrating these components into a valid definition of urban complexity is a difficult task. In reality, scientific and policy viewpoints are strongly influenced by disciplinary and sectorial divisions, which seem to be based on "the premise that particular aspects of human activity can be isolated for separate (...) individual professional treatment" (Moser, 1997:46). Despite attempts to bridge these divisions, periodic reorganisations within the various sectors, and good intentions, real integration is usually far off. This can be, for instance, illustrated by a quick scan of policy-oriented research and policy documents on segregation, the impact of new technologies, unemployment, economy and sustainability. Even reports on the combined efforts of a variety of scientific, policy and practice-oriented disciplines, from very influential sources, reveal the problem of lack of integration (Roberts & Sykes, 2000; Paddington, 2003).

There appears to be no simple answer to the question: which societal developments are relevant for urban planning and design, and, in particular, for spatial planning? An ongoing field of study and speculation has made 'trends' the subject of its investigation. Institutions and scientific disciplines make demographic, spatial, economic and cultural descriptions and from these, interpretations; they identify slow and fast trends, give advice, and, as such, help to construct an image of the world we live in, both globally and locally. It is a general academic interest to study change; to find out what is going on, and how to look at the near, and more distant, future. What is happening now, what should we do next, what could be beneficial, where should we invest? Alternatively, the intention might be to avoid unwanted developments and form ideas and options for coping with uncertainties.

In this chapter, no attempt is made to give an overview of the dynamics of society as a whole nor to select the most dramatic dynamics (such as social safety, global population growth, pollution, and the world wide web). Statistics Netherlands [*Centraal Bureau voor de Statistiek (CBS)*] publications offer abundant data; planning offices like the Netherlands Institute for Spatial Research [*Ruimtelijk Planbureau (RPB)*] and other institutions like the Social and Cultural Planning Board [*Sociaal en Cultureel Planbureau (SCP)*] and the Society and Business Foundation [*Stichting Maatschappij en Onderneming (SMO)*] construct lines of development, define short and long-term cycles and give advice. Trend-watchers share their views on fashion, business opportunities, housing and other markets. The focus in this chapter is on the places and people that, often unintentionally, are 'forgotten' in the turmoil of societal transformation; Meltzer (1984) talks about the hidden modernisation tax. The local level is important, because that is where households conduct their daily life. It is there that the effects of macro processes (international, national, urban) can be observed in detail, and where pilot projects and living experiments can be carried out, to use as tools for improving living conditions. If there are no valid ideas on how to approach and reduce real-life problems for households, then what hope is there for a better future in general, micro and macro? The view held here is that the concepts 'vulnerability' and 'deprivation' are crucial. Vulnerability indicates different forms of dependency (social, economic and, physical), while deprivation points to a variety of shortages in the living environment. In recent Dutch policy discussions, vulnerable groups have often been simplistically defined as old, handicapped, socially isolated, unhealthy and/or allochtone. It seems as though deprivation is viewed as being synonymous with segregation, or, simply, homelessness. As much of the confusion might have its origin in (dis)aggregation fallacies, we start with this subject, as a prelude to the main subject of this chapter: vulnerability and deprivation.

## Aggregation and disaggregation

The verb 'to aggregate' and the nouns 'aggregation', and the opposite 'disaggregation', are derived from the Latin *aggregare, aggregatum* which mean to bring together, as a flock. An aggregate is a total, a collection in the form of a mass. 'Aggregation' describes the state of being collected and 'to aggregate', the act of bringing something together. The opposite meaning, 'disaggregation' describes the state of acting as an individual or independent entity, and 'to disaggregate', the act of allowing the various individuals of elements to operate as individuals or separate parts, each with their own set of specific characteristics.

In the social sciences, there is a long history of methodological and theoretical discussions on (dis)aggregation (cf. Scheuch, 1966; Hannan, 1971). A particular source of awareness has been the consequences of having many separate units in empirical research, thereby losing sight of the interdependent, often implicit, relationships between theoretical formulations, research designs, and data elaboration methods; of being able to draw comprehensive conclusions to explain real-life situations, that draw on the expertise of all the disciplines relevant to the research topic, not just one or two.

There are all kinds of fallacies about aggregation and disaggregation. Basically, these are the ones that generalise a specific relation (or calculated correlation) perceived on one level of (dis)aggregation as being the same as that perceived on another level. In other words, they treat units of analyses as interchangeable manifestations of the same order. A good example is the relation between the quality of the built environment and the composition of the population. Even if statistical correlations can be found between housing quality, unemployment and the presence of ethnic minorities, the inference that the presence of an ethnic minority in a particular district is indicative of an urban problem area, first needs to be proved at the level of the district itself. A second example can be found in the regeneration of districts. Even if the urban structure is improved, and the composition of dwellings changed by demolishing housing blocks in a specific district, this is no guaranteed solution for urban decay in a 'problem area', despite the current general view that urban decay in such areas can only be stopped by replacement development.

Though there is a sharp distinction between aggregation and its opposite, disaggregation, they are interdependent; the one having no meaning without the other. Aggregated data cannot be made unless there are individual data, and in order to make a broader picture, individual data have to be aggregated. Aggregation ought to be carried out consciously, as a stage in a process of generalisation, or to connect different aspects in order to gain knowledge or take action.

Information on the individual level is not collected just for the sake of it. The relevance of such activity can often be explained by expectations, which in turn are based on earlier (aggregated, integrated) experience. Even if the aim is to collect non-biased information, there is always a certain starting point. Language, perception and notation are also the products of scientific and other cultures, all of them also subject to earlier aggregation. Any categorisation is a form of aggregation. It can broaden people's views, uncovering unthought-of dimensions, and it can stimulate theory and ideology. Disaggregation satisfies attention to the specific, the characteristic; it clarifies the unique, cautions against unjust deduction, and is the critical companion of careful generalisation. Dealing with (dis)aggregation is an art; a responsibility not to be taken to lightly. From an evolutionary perspective, it is interesting that, even though our species is inclined towards making quick generalisations, it also spends time caring for individual members of its species (e.g. De Waal, 1996).

Data – disaggregated and aggregated – are tools for defining situations and for making the distinction between (a) a perceived situation, (b) a problem situation and (c) an actual problem that has to be solved (see Box 2.1). Municipal statistics on the quality of housing, amenities, infrastructure, unemployment, and the lack of safety, may give the impression that an urban area (district or neighbourhood) needs regeneration. However, at what point can these data be considered worrying enough with respect to vulnerable and deprived households to justify defining the area as a regeneration problem?

> In *100 imagenes de un mundo desigual*, Sutcliffe (2001) describes a number of inequalities – production, income, health, trade, investments, debts, repression, gender, discrimination, hunger, etc. – in the world, between countries or regions. For this, he uses many international sources of available data. The data are visually presented in diagrams and maps, followed by a brief comment. The intention is to give a sort of overview of the world we live in, and to stimulate thinking, discussion and debate about the causes, consequences and possible approaches of inequality. The aggregate data prove to be of great help in giving general pictures. Such pictures could not be made without data aggregation. The author uses these sources to underline his views that a number of inequalities are becoming growing and worsening problems that should be placed higher on the world agenda. The book does not give precise answers on how to define and solve these problems. This is understandable, as the information is not specific enough, certainly not for approaches on the spatial scale on which people live their lives. It tells which aspects or variables could be used as components for defining the problems that need to be solved.

**Box 2.1: Descriptions of inequalities**

## Defining urban problems

Given the complexity of urban areas, and the fact that fundamental research needs to operate within clear frameworks, and that sectors and short-term successes are indispensable for policy and public administration, where then are the opportunities for a workable and responsible line of thinking? At the heart of the question is the importance of defining a problem well. What, actually, is the problem to be approached and if possible solved? How can we be certain that one problem definition is preferable to another? Which problem definition justifies leaving trusted, well-trodden paths? When is an urban area a problem area that must be dealt with? How can we avoid being blinded by mesmerising architecture and other diversions on the supply side? How can we ensure that we do not neglect those urban needs that are not explicitly mentioned in the problem definitions? Even if it were true that an incorrect problem statement could lead to an adequate approach, a 'good' problem definition is preferable. There is no simple answer to these questions. However, a few considerations should be borne in mind: it must be possible to conduct research on a problem definition; it must be possible to underpin the validity of the definition; defining a problem is more a process than a one-way affair; the point of departure, and who takes this position, must be made explicit; in defining the problem, care should be taken not to stigmatise a group of people or an area; people's characteristics must be separated from the characteristics of the areas in which they live, being especially alert to avoid (dis)aggregation fallacies; before using them, it is best to review potential approach directions critically and also the routes taken to find solutions (Drewe & Hulsbergen, 1987; Hulsbergen, 1992:49, 51).

Common policy and practice in solving urban problems are usually based on aggregated information taken from data ranging from existing statistics to newly collected information. The aggregated information is studied and (re)organised. The elements in this information are then used for constructing a problem definition, which is the step towards a more or less integrated approach to the urban problem. Even if data are only available on the 'individual' and 'household' level, their elaboration and interpretation are focused on categories of considerable size, aggregated in order to compare them with existing aggregated data. A problem definition built up in this way is assumed to be representative, and thus the best definition possible. This is not an incorrect way of using information as such. Moreover, there is often little choice, as data are not always available on the household level, nor for similar households and areas. The important thing is to establish whether the aggregation that *is* available is sufficient to deal with the complexities of urban regeneration. In defining urban problems of this kind, our observation is that the needs of the most vulnerable and deprived residents tend to disappear underground, so to speak, so that urban 'sores' continue to fester, both in the improved areas and in other parts of the urban region as well. In recent decades, we have examined the relevance and operational contents of different concepts, within a multidimensional framework with different data bases. In our views concerning the urban debate, 'vulnerability' and 'deprivation', and the relation between the two, have become crucial development concepts.

## Vulnerability and deprivation as development concepts

Cities and other urban environments are typical attributes of human existence, because they facilitate human needs for (re)producing their cultures and themselves, as a species. Cities are also places with great welfare differences between groups and households, which may take the form of intentional or incidental marginalisation. Urban environments are complex, or rather complicated, spatial-functional and social systems. They are complicated in the sense that it is hard to identify easy, cause-effect relations. No matter what policy-makers may think of the proposed outcomes, we should be pleased if we are able to define the (cor)relations in such a way that regeneration activities are likely to have the desired result. Urban regeneration is a very complex enterprise: it has to deal with physical–spatial as well as social, economic, cultural, ecological and administrative aspects, and to find more and more sustainable solutions within each of them.

It is important to have a clear idea of the desired effects of the approach used to identify and define urban problems. The basic questions here are: Who defines what the effect should be? Who dominates the processes, and whose interests are directly linked to the effects being pursued? The next basic question is: How are the problems of vulnerable and deprived households and groups being identified and defined within the regeneration approach?

This author's empirical research on 'vulnerability' and 'deprivation' began as a sequel to the research project 'Segregation in Rotterdam' during the second half of the 1970s (several reports on this project were published in 1979, see Mik, 1987 and Drewe & Hulsbergen, 1986). In testing segregation hypotheses with data bases such as existing municipal statistical data and newly collected data from a survey on individuals and households, it was clear that calculating the correlations in the way that had been generally accepted up to that point, using mainly aggregated data, would not apply any more. The research outcomes criticised existing policies, which focused solely on reducing segregation and neglected the complexity of the problems in the neighbourhoods and districts. Despite that, the so-called problem areas and problem groups in major Dutch cities are nowadays again being defined in terms of ethnic concentration and spatial decay, although somewhat less simplistically than in the 70s. This is also happening elsewhere in Europe (e.g. Andersen, 2003). Following the Rotterdam Segregation project, our focus of research shifted towards solving methodological questions, and investigating 'vulnerability' and 'deprivation' (see Ill. 2.1).

In re-working the results of the 'Living Conditions Survey' (Hulsbergen & Drewe, 1984), we focused on methodology (on how to process information from a large, multivariate, categorical data base) and on hypotheses concerning the relation between vulnerability, socio-economic status, life-cycle phases, housing quality, employment and urban environment, all of which were defined in terms of a number of operational variables (based on response scores). In a detailed step-by-step elaboration, data sets were made and the (canonical) correlations between the main variables were established. The groups of operational variables in the original scores were systematically reduced to remove redundant scores, and to construct new, operational, variables based on a combination of scores in order to operationalise the main variables in a meaningful way. We also tried to quantify (multivariate) vulnerability and deprivation in different ways. In our present definitions (see below), the percentage of vulnerable respondents amounted to 7.4% and deprived respondents to 21.9%.

The 'Marginality in Managua' research project (Drewe & Hulsbergen, 1987), which was a joint project between the municipality of Managua and the Chair of Spatial Planning, focused on two aspects: policy choices for the city, and developing the methodology further, especially multivariate data processing. The percentages for 'vulnerable and deprived' in our present definition are: 32% for dwellings, and 24% for people. We also quantified the marginal position for all kinds of living conditions. Under the heading 'multiple marginal positions' the percentages were 17% for dwellings and 13% for people.
The municipality's research project on unemployment in Rotterdam originated in the policy request for detailed insight into the problems and the potential measures that should be taken (Feddema & Hulsbergen, 1991). Our contribution was part of the questionnaire, data elaboration and an improved

| | No. of respondents in the data base | No. of initial scores | Main variables (no. of variables in the multivariate analyses)* |
|---|---|---|---|
| **Living conditions Survey** (Hulsbergen & Drewe, 1984) | 4693<br><br>secondary analyses of 1974 survey data, national sample | 559<br><br>Testing the (one- or multi-) dimensionality of the data, and the multivariate data elaboration programmes | Deprivation (dwelling and district) (12)<br>Living environment (larger scale) (3)<br><br>Vulnerability (17)<br>Socio-economic status (3)<br>Phase in life cycle (5) |
| **Urban marginality** (Drewe & Hulsbergen, 1987) | 568<br><br>analysis of survey data of selected districts in Managua | 64<br><br>Testing the multivariate nature of marginality | Deprivation (2)<br>Name of district, representing living environment (1)<br><br>Vulnerability (4)<br>Socio-economic status (3)<br>Phase in life cycle (5)<br>Origin (3) |
| **Urban unemployment** (Feddema & Hulsbergen 1991) | 598<br><br>contribution to construction of questionnaire, analysis of survey data of sample of unemployed in Rotterdam | 489<br><br>testing the multivariate nature of unemployment, and the existing (simplistic) unemployment categorisation | Dwelling (10)<br>Neighbourhood/district (8)<br><br>Unemployment position (8)<br>Socio-economic position (10)<br>Level of income (2)<br>Phase in life cycle (6)<br>Attitude towards work and change (11)<br>Wishes, future expectancies (7)<br>Vulnerability (16)<br>Origin (3) |
| **Cochabamba** (Ledo, 2002) | 1988: 10,250 in 2,313 households<br><br>1996: 2,374 in 532 households | 1988: 124 questions<br><br>1996: 80 questions | Living conditions – deprivation (housing, drinking water, sewerage, electricity) (11)<br><br>Socio-economic status (4)<br>Socio-spatial status (3)<br>Position in life-cycle, household (6)<br>Vulnerability (4) |

**Illustration 2.1: Short overview of research projects, data bases and main variables**

* The variables in the multivariate analyses can be initial or newly constructed scores. New variables are always based on existing material, e.g. to diminish overlap, compensate for missing information, in order to have a workable and relevant data base. The aim of operationalising the main (theoretical) variables was to construct meaningful score profiles on the individual level of the respondent in the data base.

(and more detailed) typology of the unemployed that departed from a broad framework by including vulnerability and spatial aspects. To give an idea of the percentages obtained in relation to vulnerability, here are a few variables: 'social network': nobody else 27.4%, partner/household 21.1% – so half of the sample had no social network outside the home; 'living in a neighbourhood where many people are unemployed': 43%; 'health': health problems, some of them severe 24.9%. Variables and percentages related to deprivation: 'more than 3 deficits in the dwelling': 45.7%; 'less than 5 local amenities': 17.6%;,and 'no more than 6 local amenities': 47.3%.

Based on the multivariate calculations and tests carried out on the hypotheses in the research projects mentioned above, we arrived at the following definitions of vulnerability and deprivation (Hulsbergen, 1992:37).

*Vulnerability*: Vulnerability is defined as a state of dependency, characterised by a variety of manifestations. These manifestations can be categorised under: social contacts, the social network, how one provides for one's household, the time spent on all kinds of activities (including leisure); social (societal) participation; aspirations and future expectancies; knowledge of societal developments and the use of societal amenities (services); as well as health and handicaps. We call someone vulnerable when changes in their living conditions always have an enormous influence on them. Such changes mostly have a negative effect on living conditions in general, and the person or household can do little to improve the situation (apart from trying to cope with it). Put in another way: 'vulnerable' means that the system that creates the changes cannot be used to improve one's own situation.

*Deprivation*: Deprivation is about the form and uses of available space. The importance of the spatial environment and its form is that they create the material conditions for either facilitating or restricting one's life. In other words, deprivation concerns the shortages experienced in providing for one's household brought about by the physical space. Just as vulnerability is about the social constraints, so deprivation is about the spatial-physical constraints.[1] Deprivation is relevant on different spatial scales, ranging from the home, neighbourhood and district to the urban and regional level. It includes the quality and quantity of housing, and accessibility to all kinds of relevant amenities both near the home and further away, as well as to suitable employment, and to modern information and communication technologies.

However, the story does not end here. In the research project on urbanisation processes, with its focus on poverty in Cochabamba, Bolivia, Ledo (2002) also analysed municipal data bases of surveys held in 1988 and 1996. One of the aims was to test a model consisting of spatial units (living conditions, such as housing and amenities) and social units (socio-economic and socio-spatial status, position in life cycle, vulnerability), along the lines of our earlier experiences. With these data, it was possible to test vulnerability and deprivation at a specific moment, and, in addition, to focus on changes in time by comparing the data and the structure of these data with the two earlier data bases. The outcomes of the multivariate analyses were remarkable (*ibid.*: chapter 4). For both years, the data revealed two-dimensional (comparable) structures. The first dimension differentiated households in terms of the

---

1 Peters & De Wilde (2004), in an interesting article on mobility, point at the relevance of spatial form and the rules on the use of form; rules aimed at clear and safe use for all participants. The authors do not make the comparison between traffic and social movements, but this is a small step. Their concern is designing mobility space which is useable and safe for both cyclists and car drivers. In our terms, cyclists are vulnerable compared with car-users, because they are often hindered by insufficient infrastructure, as they have to share it with motorists.

(high to low) quality of the environment (with respect to neighbourhood location, sewage system, drinking water, housing quality); the second dimension differentiated households in terms of (high to low) vulnerability and inequalities (in terms of source of income, age, labour insecurity, social stratification, house ownership). Though the structure of the two-dimensional model was the same for both 1988 and 1996, the shifts in the dominant operational variables (which determine these dimensions) revealed that the main aspects of the living conditions were worsening: there was more discrimination with respect to water and unemployment in 1996 than there was in 1988. Ledo quantified the databases within the framework of these two dimensions, which she called "the multi-dimensional face of poverty, vulnerability, deprivation and social inequality in Cochabamba" (2002:172). See Illustration 2.2 for the household distribution based on the 1988 and 1996 representative samples.

The table not only shows the growth of the population, but also the absolute and relative changes in terms of deprivation and vulnerability. Though more households are less deprived in absolute quantities, in percentages, the situation in 1996 is worse than in 1988 in all respects. Notable is the absolute, as well as percentage growth of deprived and vulnerable households. The expected frequencies in 1996, based on the 1988 proportions, are not only telling with respect to what the frequencies should have been, *ceteris paribus*, but also make clear that deprivation and vulnerability are relevant criteria for measuring the social-spatial developments. When disaggregated into neighbourhoods and districts, the most affected areas are revealed.

It would be interesting to know which households have remained deprived and vulnerable (stayers), which became deprived and vulnerable (newcomers), and which ones have the potential to escape from vulnerability and deprivation. These last-mentioned households are especially interesting from the point of view of policy, because they can be used to evaluate the effects of measures and of developing new instruments. However, for that, one needs data bases that can be compared on the level of households.

| **Year of data base** | **1988** | | **1996** | | |
|---|---|---|---|---|---|
| **Percentages and frequencies** / **Population Categories** | **% in sample** | **Frequency generalised to total population** | **% in sample** | **Frequency generalised to total population** | **Expected frequency, compared to 1988** |
| No deprivation, no vulnerability ("low risk") | 26.7 | 19.477 | 23.6 | 26.861 | 30.485 |
| No deprivation, yes vulnerability ("recent poor") | 23.5 | 17.118 | 21.3 | 24.271 | 26.792 |
| Deprivation, no vulnerability ("inertial poverty") | 23.8 | 17.316 | 24.6 | 28.102 | 27.102 |
| Deprived and vulnerable ("high risk") | 26.0 | 18.960 | 30.5 | 34.821 | 29.676 |
| Total | 100.0 | 72.871 | 100.0 | 114.055 | 114.055 |

**Illustration 2.2: Deprivation and vulnerability in Cochabamba, 1988-1996. Based on: Ledo, 2002:172**

## Conclusions

The urban complexity of daily life is a complicated mix of social and spatial aspects. It is necessary to distinguish between the variables of these two aspects in order to understand what is going on. Though many social scientists (and policy-makers) view 'space' as a social construction, it seems to have been re-discovered as a (relatively) separate entity, requiring understanding as such, with its own place in theory and practice (Gieryn, 2000). This points to urban planning and design (urbanism) as having greater relevance for the development of urban science and practice than is usually perceived from either side.

Vulnerability should be distinguished from aspects such as (the traditional) social, economic, cultural variables within the context of social aspects; likewise, deprivation should be distinguished within the context of spatial aspects. In defining urban societal problems, it is crucial to combine vulnerability and deprivation, so that those who really need support (because they have no 'choice') in improving or maintaining their living conditions can be identified, and so that they can be used as criteria for measuring the effects of urban developments and policies. Above all, study directed towards vulnerability and deprivation is necessary for improving problem definitions and keeping policy and practice discussions active. As society changes, re-defining definitions will be an ongoing task. Moreover, practice demands a comprehensible reduction in the complexity of society's problems. For politics and policy, problem definitions need to be simple, finite, manageable and convenient (Edwards & Batley, 1978:246) in order to decide on general measures for areas and population groups, and to make concrete plans and designs. Decision-makers and researchers are stimulated to take action by 'signals' from society; even though it is characteristic of vulnerable and deprived households that they lack 'voice'.

All this is also relevant for the discussion on 'social innovation' in both the Netherlands (government, employers, unions, municipalities and social organisations) and in the European context (cf. URBAN initiatives for the urban environment; e.g. European Commission, 1999). Social innovation concerns the development of a dynamic, competitive knowledge economy and a well-organised, productive, participative population, which not only trusts the economy, politics, policy and society, but also cares about the restricted future perspectives of the vulnerable and deprived; those who find it difficult to hook onto the main streams of society.

A point of attention right now is the impact of new technologies. Though there is no direct relationship, and the consequences for urban planning and design are unclear, they have undoubtedly been changing the ways in which businesses operate and how people use urban space. These changes can be expected to continue in the future, and their impacts on the vulnerable and deprived should not to be underestimated as tools for improving living conditions and perspectives (Drewe, Fernández-Maldonado & Hulsbergen, 2003; see also: http://www.networkcity.bk.tudelft.nl/). On the local level, ICTs "can serve as tools for urban development, as they can be used to provide basic services such as education and health, as well as new impulse for business and commerce in a relatively cheap and flexible way", and also as potential sources of empowerment and bottom-up developments (Fernández-Maldonado, 2004:18). For this, ICTs are subject to urban design and planning.

Finally, it is the task of spatial planners and designers to raise the alarm in their field, if necessary, or at least to guard against plans and designs that undermine the already weak position of vulnerable and deprived households.

## References


Andersen, H.S., 2003, *Urban Sores: On the interaction between segregation, urban decay and deprived neighbourhoods*, Ashgate, Aldershot (UK). Book review: Hulsbergen E.D., 2004, Urban Studies, 41/7, pp. 1396-1398 (June)

Chambers, R., 1989, Vulnerability, coping and policy (Editorial Introduction); In: *IDS Bulletin, Vulnerability: How the Poor Cope*, 20/2, pp. 1-7

De Haan, A., 2003, *Social Exclusion: Enriching the Understanding of Deprivation*, Poverty Research Unit, University of Sussex

De Waal, F., 1996, *Good Natured: The Origins of Right and Wrong in Humans and Other Animals*, Harvard University Press, Harvard Mass.

Drewe, P. & E.D. Hulsbergen, 1986, *Segregation in Rotterdam and the aftermath, Recent developments concerning ethnic groups in the Netherlands*, Instituut voor Stedebouwkundig Onderzoek, Reeks Werkstukken, 9, Delft University of Technology, Delft

Drewe, P. & E.D.Hulsbergen, 1987, Marginality in Managua – a multivariate approach, *Angewandte Sozialforschung*, 1986/7, 14/ 2/3, pp. 201-210

Drewe, P., A.M. Fernández-Maldonado & E.D. Hulsbergen, 2003, Battling urban deprivation: ICT strategies in the Netherlands and Europe, *Journal of Urban Technology*, 10/1, pp. 23-37

Edwards, J. & R. Batley, 1978, *The Politics of Positive Discrimination. An Evaluation of the Urban Programme 1967-77*, Tavistock Publications, London

European Commission, 1999, *Guidelines for the New URBAN Community Initiative 2000-2006*, European Commission, Brussels

Feddema, R. & E.D. Hulsbergen, 1991, *Grootstedelijke werkloosheid. Ideologie en realiteit*, Publikatieburo Bouwkunde, Faculty of Architecture, Technische Universiteit Delft, Delft

Fernández-Maldonado, A.M., 2004, *ICT-related Transformations in Latin American Metropolises*, PhD Thesis, Design/Science/Planning Series, Delft University Press, Delft

Gieryn, T.F., 2000, A space for place in sociology, *American Review of Sociology*, 26, pp. 463-496

Hannan, S.E., 1971, *Aggregation and Disaggregation in Sociology*, Lexington Books, Toronto/London

Hulsbergen, E.D., 1992, *Positie en Ruimte. Kwetsbare groepen in de stad: denkbeelden en feiten*, PhD Thesis, Publikatieburo Bouwkunde, Faculty of Architecture, Technische Universiteit Delft, Delft

Hulsbergen, E.D. & P. Drewe, 1984, *Herbewerking leefsituatiesurvey 1974; een methode voor identificatie van probleemgroepen*, Instituut voor Stedebouwkundig onderzoek, Report no. 46, Technische Hogeschool Delft, Delft

Ledo, C., 2002, Urbanisation and Poverty in the Cities of the National Economic Corridor in Bolivia, Case Study: Cochabamba, Delft University Press Science, Delft.

Meltzer, J., 1984, *Metropolis to Metroplex, the Social and Spatial Planning of Cities*, John Hopkins University Press, Baltimore

Mik, G., 1987, *Segregatie in het grootstedelijk milieu. Theorie en Rotterdamse werkelijkheid*, PhD Thesis, Economisch-Geografisch Instituut, Rotterdam

Moser, C., 1997, Urban Social Policy and Poverty Reduction; In: Burgess, R., M. Carmona & T. Kolstee (eds), *The Challenge of Sustainable Cities*, Zed Books, pp. 44-61

O'Brien, K., S. Eriksen, A. Schjolden & L. Nygaard, 2004, *What's in a word? Conflicting interpretations of vulnerability in climate change research*, CICERO Working Paper 2004:04, Center for International Climate and Environmental Research, Oslo, Norway

Paddington, R. (ed.), 2003, *Handbook of Urban Studies*, Sage, London

Rada-Donath, A, 2004, interview, *Découvrir*, 25/6, pp. 36-38 (November/December)

Roberts, P. & H. Sykes (eds), 2003, *Urban Regeneration, A Handbook*, Sage, London

Scheuch, E.K., 1966, Cross National Comparisons Using Aggregate Data: Some Substantive and Methodological Problems; In: Merrith, R.L. & S. Rokkan (eds), 1966, *The Use of Quantitative Data in Cross-National Research*, Yale University Press, New Haven, Con., pp. 131-167

Sutcliffe, B., 2001, *100 imágenes de un mundo desigual*, Intermon Fundacion para el Tercer Mundo, Barcelona. English edition by Zed Books, London. The Netherlands edition by Novib, The Hague, 2002

Van de Lindt, M.C., 2002, *Kwetsbaarheid en maatschappelijke dynamiek*, Universiteit van Maastricht
Van der Pennen, A.W., V. Veldheer, E. ter Borg, M. Kunst, J. Boelhouwer & F.A. Knol, 1998, *Sociale vernieuwing: van plan naar praktijk: onderzoek naar de voorwaarden voor een effectief lokaal achterstandsbeleid*, SCP cahier 146, Social and Cultural Planning Office, Rijswijk (NL)
Whelan, T. & B. Maître, 2004, *Vulnerability and Multiple Deprivation Perspectives on Social Exclusion in Europe: A Latent Class Analysis*, EPAG Working Papers, 2004-52, University of Essex, Colchester (UK)

# 3 The Randstad: Its Position and Environment

Herman Rosenboom

## Introduction

In this chapter we investigate how the quality and potentials of the position and environment of the Randstad were evaluated in the period from 1980 to the present day.[1] To do this, we refer to a number of research papers and policy documents of that period, which, rather than giving an exhaustive overview, favour a more chronological insight into some important aspects of the period. So, to bring the dynamics behind these aspects to the fore, we divide the period into two, at the turn of each decade, comparing the Randstad of the 1980s with that of the 90s to the present day. An attention point for spatial planning and the spatial economy is the quality of a location's environment. In the Netherlands this was also an important issue in the 80s. Unlike earlier memoranda, the *Vierde Nota over de Ruimtelijke Ordening* [Fourth Memorandum on Spatial Planning] (Ministerie van Volkshuisvesting, Ruimtelijke Ordening en Milieubeheer, 1988) was oriented towards strengthening the favourable aspects. This document drew attention to the competition between cities and regions in Europe and throughout the world.

The Netherlands, and especially the Randstad region had to compete with other regions in Europe, such as the Rhine-Ruhr area, Paris, London, Frankfurt and the Antwerp-Brussels-Gent urban region. The north-western part of Europe, including the Netherlands, had long been considered an important economic location. The Netherlands, in particular, was considered important as an European transport and distribution centre. In addition, it had two strong trump cards, the seaport of Rotterdam and Schiphol Airport. However, various developments and changes in the economic and technological spheres, have threatened that strong position.

In the first place, the economic point of gravity in Europe has moved towards the south, and German reunion has opened up the possibility for developments in an easterly direction. The construction of high-speed links and the Channel tunnel have opened up further possibilities, which could force the Netherlands into a more and more peripheral position. The economic structure of the Netherlands has also changed; what was once an industrial economy is now becoming a service economy – so the spatial requirements have changed too.

This means that, for the Netherlands, and especially for the Randstad, to be able to compete in international economic development, the quality of the environment will be an important factor determining whether or not it will succeed. For it to be considered as an important economic location, both now and in the future, the Randstad has to be able to offer more than just an airport and a seaport. The task at hand is to raise environmental quality in the Netherlands in general, and in the Randstad in particular, so as to keep abreast of changes that take place in European spatial and economic structure.

---

1 See on the spatal policies of this period in the Netherlands also chapter 11 by Joost Schrijnen.

# The position of the Randstad 1980-1990

Due to the weak economy in the Netherlands at that time, much attention was given to the Randstad as a location for (international) economic activities. The *Rijks Planologische Dienst (RPD)* [National Spatial Planning Agency] arranged a meeting in 1984 to discuss this potential role for the Randstad with people involved in the development of the spatial structure. This meeting, held under the theme *De Randstad van binnen en buiten bekeken* [Looking at the Randstad from outside and inside], was focused on developing the spatial structure of the Randstad. Because the position of the Randstad as a potential location for economic activities was an important topic of discussion, we take this meeting as the starting point of our analysis here.

Drewe *et al.*'s research *De kwaliteit van het vestigingsmilieu ter discussie* [Discussing the quality of location factors] (Drewe, Guyt & Rosenboom, 1985) was presented at this meeting. This investigation looked at well-known alternative spatial developments, and their effects on the qualities of locations viewed against economic–technological scenarios. This was inspired by the long-term need to find new ideas for spatial development.

During this period, much attention was also given to 'large city' problems, and especially to those of Amsterdam, Utrecht, Rotterdam, and The Hague, the four largest, and most important cities of the Randstad. In spite of the substantial problems encountered by large cities, as economic centres, they have enormous overall potential (Sietsma, 1986). The city has played an important role in every economic change that has taken place. In the Middle Ages, the city played a key role in the shift from agricultural to trading societies. When these societies became industrialised, the city was again the centre of change. It led first to urbanisation and then to suburbanisation and decentralisation, and, in the process, to an ongoing division of labour and de-industrialisation, both nationally and internationally. Suburbanisation is the movement of city inhabitants to urban peripheries in search of better living conditions. Decentralisation is the process in which companies leave the city and settle in the suburbs and further away. The potential for cities then lay in shifting from an industrial to a service economy. A number of characteristics inherent in the urban environment favour such development: e.g., the availability of a wide range of goods and services, access to more advanced communication systems, a concentration of institutions of higher education to supply the skills needed for these changing economies, plentiful labour, which simultaneously produces consumers in search of an ever wider range of services, especially those in the sphere of leisure activities.

However, the living environment remains an important point of attention for all urban centres, because, to sustain a service economy located there, it is crucial for cities to invest in a good supply of housing and living environments that the highly-skilled labour on which the service sector depends will find attractive. It is essential that highly-skilled young professionals do not desert the cities later in their careers.

In the *De Randstad als vestigingsplaats voor internationaal georiënteerde bedrijvigheid* [The Randstad as a location for internationally oriented business] (Boeckhout, Verhoeff & Verster, 1987), the position of the Randstad is compared with other competing urban regions in Western Europe, such as London, Paris, Hamburg, Frankfurt, Munich and Brussels/Antwerp. Based on literature and on interviews with experts, information was gathered about the requirements that these competing locations had to fulfil, and the relative importance of each. These requirements were investigated for those functions of the service economy that have a strong influence on economic restructuring and growth, for instance, the requirements set for the location of company head offices and business services, research and development activities, and the production and distribution of high-quality technological products in these competing places. The same was then done for the Randstad. This study revealed that the Randstad is a potentially good location for a number of internationally oriented economic activities, of which the main one is distribution. It is also favourable for specific parts of company decision-making centres and business services.

On an international scale, a location's accessibility, and the attractiveness of the environment in which it is situated, are clearly very important, both nationally and regionally. The important factor here is infrastructure, and not just infrastructure in the narrow sense of the word – harbours, roads, airports, etc. – but all kinds of information networks related to them. By creating new infrastructure and improving existing systems, the Netherlands can maintain its position as a node in international networks. But to be competitive, it is important to create an international style. There must be top locations for international offices, high-quality housing and a higher level of socio-cultural and distribution facilities.

*De Vierde Nota over de Ruimtelijke Ordening* [The Fourth Memorandum on Spatial Planning] (Ministerie van Volkshuisvesting, Ruimtelijke Ordening en Milieubeheer, 1988) is oriented towards economic growth, on strengthening the advantages offered by the region and on utilising promising developments to the full. These aims are also included in the perspective for spatial development:
- Strengthening the economic and spatial advantages offered by the region and utilising promising developments to the full;
- Increasing spatial diversity;
- Strengthening and utilising the qualities offered by the different parts of the country.

In the Document, the spatial development perspective is viewed from seven different angles:
- The urban node;
- The inherent strength of a region;
- The urban ring of central Netherlands;
- The Randstad;
- The Dutch water 'land';
- Rural maintenance and renewal;
- Environmental policy.

The Randstad was seen as the region that would be most able to meet rising competition from European cities and those worldwide. The parts of the Randstad considered to have the best potential for doing this are the agglomerations of Amsterdam, Rotterdam and The Hague. To realise this aim, it was considered crucial for a number of redevelopment projects to be carried out. Besides having one or two top locations for offices, it was necessary to improve telecommunication connections, build high-class hotels and expensive housing to accommodate international personnel, and high-quality community facilities, and to improve tourist attractions, museums, public outdoor urban spaces, public transport and the motorway network. However, in trying to realise this list of aspirations, less attention was given to spatial imagination.

In the report *Grote steden grote kansen* [Large cities, large opportunities] (Externe Commissie Grote Stedenbeleid, 1989), the remark was made that, since the 80s, there has been a revival in the large urban regions. Although typical urban problems had not been eradicated, attention focused more on the potentials of these regions; for instance, on their economic significance as agglomerations providing one third of the employment and the GNP. An important force behind this trend was the internationalisation of the economy. National boundaries were becoming less important, and with this scale enlargement, activities were concentrated more and more into favourable locations, and to make these locations attractive, urban renewal was focused on improving the quality of the living and working environment.

To compete, as a city or region, especially for the location of (international) financial and business services, the quality of the office environment was considered very important. In the research *Naar hoogwaardige kantorenlocaties* [Towards high-quality office locations] (Bongenaar & Lie, 1991), the criteria that had to be satisfied for office locations to be highly valued were established on the basis of international office sites that had proved to be successful in other countries. The aim was to gather information to get some idea how to develop top locations in the Netherlands. A top location can be defined as a location in a metropolis with a highly developed urban economy and a strongly diversified business service economy which attracts a strong concentration of high-quality offices. These locations enable highly qualified, internationally oriented office activities in financial and business services, and in top management, to be carried out with optimal effectiveness. A high concentration of offices at a top location is evidence of its high user-value. This is manifested in good internal accessibility and in the location's high image. Many companies prefer to locate in cities that have a diversified service sector, and they are attracted to locate there by a number of closely related characteristics of that environment. These include the representativeness of the location; its status, sphere, and the level of amenities that it offers. Compared to countries elsewhere, the Netherlands did not at this time have a top location for the offices of international companies, because nowhere was the spatial concentration of offices dense enough. Office locations in the main agglomerations and elsewhere in the Randstad were very dispersed.

## Conclusions: Randstad 1980-1990

The Randstad, in particular, was seen as a suitable location for international economic activities. The large cities were viewed as having potential, especially in transforming their industrial economy into a service economy. However, alongside this drive for development, typical urban problems continued. Some problems had to be solved, though. One of the most important of these was the problem of improving the quality of the living environment. This meant improving the housing stock, raising the level of cultural amenities, and the quality of shopping facilities. Increasing accessibility was viewed as an especially important problem. There was no top location for offices, at this stage, and if they were to develop, there would only be space for one or two of them, because only one or two sites could satisfy the very special conditions needed for the development of a high quality international office location.

The policy formulated in the Fourth Memorandum incorporated the conclusions of the different research projects. Amsterdam, Rotterdam and The Hague were named as suitable urban centres for a concentration of multinational and foreign company offices. Two top office locations were envisaged. Expensive houses would be built, high quality shopping facilities, tourist attractions and good connections.

One of the most striking aspects of these plans is the continuing emphasis that was placed on the quality of the living environment. This was because housing and living conditions were viewed as important limiting factors of a service economy. Not surprisingly, therefore, this aspect was named in almost all the reports written during this period. As companies in the service economy have more freedom in where they choose to locate their offices, raising quality and increasing diversity in the housing market and living environment was viewed as being one of the ways of competing with other regions in attracting companies to locate there. This meant that the concepts for urbanisation had to be reconsidered and its main components – housing, workplaces, landscape, infrastructure and amenities – designed more integrally. Such developments should also be possible along the infrastructural flow to and from cities.

Making lists of targets and aims would not suffice; designs of possible futures should be presented. Decisions also had to be made about the actual location of a top international office park. However, for the Randstad, that proved a big problem, as there is no government authority at Randstad level that can make these decisions.

## The Randstad 1990-2005

In the 90s, developing the economic performance of Dutch cities and the Randstad, and how this compared with important urban centres elsewhere, was still a hot item on government agendas. Much attention was paid to the way in which world cities were developing because, largely due to the growth of business and financial services there, they were seen as the command centres of the world economy (Priemus, 1994). At the same time, powerful economic development was also taking place in the suburbs, because large cities are subject not only to centralising forces, but to decentralising ones as well. Cities became more closely linked and closer ties were established with urban centres further away, both inside the country and abroad. This development led to the formation of urban networks; a development that appeared to support economic growth and social cohesion. However, for this to happen, a number of conditions had to be fulfilled. In the first place, the nodes in the urban network had to be connected well with each other, and also with the all-important mainports in their logistical chains. In the second place, there had to be an innovative production environment characterised by task division and specialisation, to stimulate the dynamics of economic development. Research & Development centres were of great importance here, as were fruitful relations with high-quality research and education centres, such as universities. Within the network, there also needs to be functional division and specialisation.

In 1997, the *Nota Ruimte voor Economische Dynamiek* [Memorandum on Space for Economic Dynamics] (Ministerie van Economische Zaken, 1997) was published. This report was based on the premise that, within the spatial-economic structure of north-western Europe, the Netherlands can offer competitive environments for top office locations. It deals with the relation between economic development and spatial policy.

The most important challenges at that time were seen as:

- Developing the economic aspects of cities, and especially those in the Randstad;
- Keeping the expected increase in mobility under control;
- Being able to meet the different spatial demands.

The main spatial-economic structure of the Netherlands consists of urban agglomerations, (air)ports and transport axes. These main structures occupy roughly 40% of the surface area of the country. This same area is also home to 75% of the population; about 80% of the employment is concentrated here and it is the area where 80% of the GNP is generated. Both concentrating as well as de-concentrating tendencies are evident in the spatial-economic dynamics. Concentration has come about due to the influence of the large cities and the developments around Schiphol Airport. At the same time, there has been de-concentration due to the expansion of the core economic region to parts of Gelderland, Noord-Brabant, Limburg, Overijssel and Flevoland. The assumption in the Nota is that the important factors for attracting new economic activities will be:

- The presence of suppliers and knowledge centres;
- A favourable labour climate;
- Appropriate locations with enough space for expansion;
- Accessibility: easily accessibility to and from important economic centres and access routes to international infrastructure and mainports.

Developing the network economy was mentioned in this Memorandum as being an important target. This was based on the assumption that enlarging scale and specialisation would also increase the functional relations between and within companies. Chains and clusters are expressions of these kinds of inter-company functional relations, and such relations arise on different scale levels. For companies, these levels can be both local and global. The movement of economic growth along a spatial network is a complex process, because there are different types of concentrating and deconcentrating powers at work. The normal benefits to be gained from an agglomeration – a varied labour market, large numbers of consumers and a range of urban facilities – attract new economic activities. However, when, for example, relations between the companies in a cluster become so intensive that spatial proximity becomes an advantage, additional benefits may also arise.

The report entitled *Steden en Stadsgewesten* [Cities and Urban Regions] (Van der Vegt & Manshanden, 1996) is further evidence that economic growth, especially in the large cities, was still a matter for concern. After the suburbanisation period and the resulting degeneration of city centres, urban populations again began to increase slightly, although the pace of population growth in the suburbs was still at a somewhat higher level. That there has been positive development in the large cities of the Randstad can be partly explained by the compact city policy and partly by demographic trends in the foreign population. How can the higher economic growth in urban hinterlands be explained when there are only small differences between the economic structures of urban-core companies, compared with those in surrounding areas? In this report, the difference is explained as being due to the size of these cities, and thus to the benefits and disadvantages of agglomeration. Because of their size, large cities have an economic advantage, although the disadvantages increase even faster. Limits to physical growth start to become a problem, due to congestion and lack of space, and this has a negative effect on the urban environment. This also explains why the ratio between advantages and disadvantages is more equally balanced in smaller cities, thereby explaining their more favourable growth perspectives.

In the report *Ruimtelijk Economisch Beleid* [Spatial-Economic Policy] (Ministerie van Economische Zaken, 1999), the importance of having an excellent environment for business location is expressed. On the scale of the Randstad a further concentration of economic activities along the North Wing can be observed. A strong growth in employment opportunities was forecast here, due to the combination of the environment created by the metropolis of Amsterdam, the centrality of Utrecht and the strong growth potential of Schiphol. On this scale, though, the nuclear spatial-network structure of the Randstad concentration does not automatically mean that the urban centres will be strengthened, because it is mainly the regions in between these centres that are favoured, due to their easier accessibility.

In *De ruimte van Nederland: startnota ruimtelijke ordening* [Space in the Netherlands: An Initial Memorandum on Spatial Planning] (Ministerie van Volkshuisvesting, Ruimtelijke Ordening en Milieubeheer, 1999), a number of starting points were formulated. In this document, it was assumed that network cities would be created on a regional scale. 'Network cities' is the term used to describe a complex of different urban centres and nodes, which nevertheless form one housing and labour market, offering a comprehensive supply of living and working environments, and amenities, to satisfy the changing and varied requirements of people and companies. Three large network cities were identified in the Randstad: The North Wing, the South Wing and Utrecht and its hinterland. For the economies of the large urban regions, two developments are important (Van der Wouden & De Bruijne, 2001). The concentration and deconcentration of economic activities in the cities is the first of these. The second one has been the transformation of the industrial urban economy into a post-industrial urban economy. This transformation has significantly changed the pattern of labour requirements, in that it has reduced the need for low-skilled labour. Particularly the cities have lost industrial activities, capital and space-consuming industry. The position of these industrial activities has been taken up by companies in business and financial services attracted to city centres by the proximity of financial headquarters and markets.

The classic spatial pattern of centre and periphery was rapidly disappearing, to be replaced by a spatial pattern consisting of the different economic centres of the network city or urban network. Each network consisted of a number of highly concentrated (economically, demographically and recreationally) urban centres, connected with each other by infrastructure (see also chapter 12 by Remon Rooij). In particular, the new economic concentrations attracted the large-scale economic activities from the large cities. In recent decades, new concentrations have grown up at nodes in the infrastructure. These begin with an urban function (companies and a shopping centre locate there) and then other urban functions follow as growth continues. Where there was parity between the nodes, the term 'multi-centred urban region' was used (or multinodal region – see chapter 28 by Marc Jacobs).

Some of the developments have been very dynamic. In 2002, urban development changed drastically. The cities became the fastest growing areas of the country (Rienstra, 2002). The growth of 25 cities and their hinterlands involved in the 'Policy on Large Cities' was investigated. From 1997 onwards, these cities appeared to be recovering from a long slump in growth. The economic recovery was the most promising in the four largest cities, Utrecht, Amsterdam, The Hague and Rotterdam, and their hinterlands, and of these, the first three were found to be performing much better than Rotterdam. The growth in employment was even better. The growth in production and employment in these cities was particularly visible in the commercial service sector. The increase in commercial services, in fact, explains much of the success of large cities at this time, and not just in the Netherlands. Knowledge-based activities are best placed in urban regions, because spatial proximity facilitates face-to-face contact and knowledge exchange.

In the *Nota Ruimte: Ruimte voor ontwikkeling* [National Spatial Strategy: Space for Development] (Ministerie van Volkshuisvesting, Ruimtelijke Ordening en Milieubeheer, 2004), which was the report that replaced the Fifth Memorandum [*Vijfde nota over de Ruimtelijke Ordening*], the national spatial policy for cities and networks was oriented towards space for living, working and mobility, together with the amenities, green areas, recreation, sports and water facilities associated with them. One of the aims in this memorandum was to strengthen the international position of the Netherlands. To bundle economy, infrastructure and urbanisation together, the spatial strategy that was chosen was 'networks and cities'.

The Randstad with its three core economic regions: the North Wing, the South Wing and Utrecht and its hinterland was defined as a 'national urban network', as it was the largest in the Netherlands. The North Wing of this network has enormous economic dynamism and diversity. Its economic engine is centred on the headquarters of various companies, and on Schiphol Airport and the concentration of transport and logistic activities around it. An important economic hub of the South Wing is the harbour of Rotterdam, although Rotterdam's economic structure is rather limited. In this Memorandum, the location policy for companies and amenities, and the location policy for peripheral and large-scale retail enterprises was replaced by an integrated policy with a number of goals. These were:

- Finding possibilities for economic development;
- Encouraging proximity;
- Ensuring liveability.

In the document *Pieken in de Delta* [Peaks in the Delta] it was acknowledged that, if ranked internationally, the Dutch economy would score badly in comparison with competing countries (Ministerie van Economische Zaken, 2004). For growth to recover, the current bottlenecks in creating a good environment for business location would have to disappear. For this to happen, the employment market would have to improve and there would have to be an excellent environment for stimulating innovation. Physical factors affecting the location were also viewed as being significant here. To improve competitiveness, the following were seen as being crucial:

- Mainports that function well;
- Improving the accessibility of cities, work locations and mainports;
- A suitable supply of labour;
- Attractive housing and working environments.

Six regions were distinguished in this report, all of which were expected to add substance to the ambition of creating a dynamic, competitive Dutch economy. The three core economic regions were based on different economic profiles of the Randstad. Along the North Wing, with its favourable production structure and relatively high economic growth, the important foci for spatial development would be Schiphol, new industrial parks, and increasing accessibility by car. Along the South Wing where the share of employment in traditional sectors, such as the harbour and industry is on the decrease, new sectors have not developed fast enough to compensate for this. An important problem here is acknowledged to be the shortage of space for new economic activities. Other problems were identified as being accessibility, housing and the quality of the environment. However, the region is also seen as having a number of potential advantages, such as the high-speed rail link (HSL), the International Court of Law, and a concentration of well-known knowledge and educational institutions.

## Conclusions: Randstad 1990 -2005

As in the period 1980-1990, great attention was given to the relation between economic developments and spatial policy. Providing a competitive location for establishing the core departments of foreign companies in the Randstad was still a very important target. A remarkable change took place during this period. The economic growth of cities, which had lagged behind during the 80s and early 90s, was showing signs of recovery by the late 90s. In particular, the large cities were losing their industrial and space-intensive activities. An important explanation for the success of the large cities at this time was due to the emergence of high-quality commercial services. Meanwhile, government policy focused on the same aspects as in previous policies: i.e., 'space for industrial parks' and 'accessibility'. But, this led to a dual labour market which favoured highly-skilled jobs. So, although there was economic growth in the cities, many city dwellers, especially the low skilled, entered the ranks of the long-term unemployed and did not benefit at all from this growth. Much attention was given to improving the physical structure of urbanisation. In the 80s, the centre-periphery model used at that time had brought about a distinction between cities cores and suburbs. The concept for urbanisation was the compact city. By the 90s, however, planners were thinking in terms of urban networks, consisting of nodes and relations, and connections between those nodes.

These networks were identified on all levels of scale. Where there was more parity between the different nodes from which these urban networks were constructed, the term 'multi-centred urban region' was used instead. The Randstad was defined as a national urban network with three core economic regions: the North Wing with its relatively high economic growth; Utrecht and its hinterland; and the South Wing struggling with its more industrial economy, the seaport of Rotterdam and its associated industry. However, although the region has many potentials, it lacks a strategy good enough to channel them into economic growth. The most important challenges for spatial-economic development are being able to offer sufficient suitable top business locations, increasing accessibility and creating optimal liveability.

## Final Conclusions

In reviewing a period of over 20 years, certain points of attention come to the fore. As evidenced by the positive turn in the economic positions of the large cities during this period, it is clear that economic vitality can change very quickly. However, these large cities are still struggling with a lot of problems. Despite all the attention paid to research and policy-making, during this entire period, there has been a call for improving the housing and living environments. This still seems to be one of the weakest aspects of the region. Why is this, we may ask? In spatial-economic policy much attention is still being given to creating sufficient space for industrial parks.

However, if we look at the development of the economic structure, then we can observe that the service economy is becoming more and more important. These economic activities are more suited to an urban environment than industrial activities are. The need for face-to-face contact and proximity are important factors in the service sector. The shift to the service economy has also changed labour market requirements; highly skilled labour is now more in demand. In turn, this has influenced housing market requirements and the demand for additional amenities such as culture and recreation.

This must lead to a spatial policy that focuses more on an integrated approach to urban development and a policy that also pays attention to other factors, for instance housing, work opportunities and amenities for different population groups. More diversity is needed, and a broader range of choice in the working and urban and rural living environments. To further the service economy, in urban development more attention needs to be given to the location factors that this sort of economy requires. During the entire period reviewed here, the research undertaken at the time repeatedly resulted in the conclusion that the Randstad has the potential to compete internationally, on an economic level. However, at the same time, we see that the problems that were evident at the beginning of the period have still not been solved. We also see that much of the discussion about the Randstad has been about the *concept* of the Randstad. However, during this period, different concepts have been used: the compact city, wings, and now the network city. It is unclear how these changing outlooks have affected, and will affect, the development of the Randstad over time. Is it possible to make the step from the compact city to the network city without destroying existing investments? Which interventions needed for the compact city can be used for developing the network city? How long will the concept of the network city hold before it is superseded by new ideas?

We have seen that the Randstad does not yet operate as a single region; it has strongly differentiated northern and south wings, which are seen as two different urban systems. The question is: what added value are we striving to achieve in trying to bring the whole area together, under one urban system? Another difficulty with the Randstad is that it has no overall governing body responsible for its development. At present, the development of the Randstad is the combined responsibility of the national government, the provinces and the municipalities. If one institution could take over this responsibility, then a single structural plan could help to solve the on-going problems in a more integrated way. One of the main on-going problems for developing the area further is the limited housing market and the need to improve living conditions. Although, creating more diversity could provide an answer, identifying the problem is no guarantee of its solution. What is needed in the coming years is adequate research and design to develop solution directions ('what', 'where', and 'how') and effective *practical* measures for solving these housing and living problems.

## References


Boeckhout, I.J., B.M. Verhoeff & P. Verster, 1987, De Randstad als vestigingsplaats voor internationaal georiënteerde bedrijvigheid, *Economisch Statistische Berichten*, 72/3634 pp. 1138-1145

Bongenaar, A. & R.T. Lie, 1991, Naar hoogwaardige kantorenlocaties, *Economisch Statistische Berichten*, 76/3820, pp. 792-796

Drewe, P., P. Guyt & H. Rosenboom, 1985, De kwaliteit van het vestigingsmilieu ter discussie; In: *De Randstad van binnen en buiten bekeken, Studierapport Rijksplanologische Dienst*, Staatsdrukkerij (Sdu), Den Haag

Externe Commissie Grote Stedenbeleid, 1989, *Grote steden grote kansen*, Ministerie van Binnenlandse Zaken, Directie Interbestuurlijke Betrekkingen en Informatievoorziening, Den Haag

Ministerie van Economische Zaken, 1997, *Ruimte voor Economische Dynamiek*, Den Haag

Ministerie van Economische Zaken, 1999, *Ruimtelijk Economisch Beleid*, Den Haag

Ministerie van Economische Zaken, 2004, *Pieken in de Delta*, Den Haag

Ministerie van Volkshuisvesting, Ruimtelijke Ordening en Milieubeheer, 1988, *Vierde nota over de ruimtelijke ordening, deel a: beleidsvoornemens*, SDU uitgeverij, Den Haag

Ministerie van Volkshuisvesting, Ruimtelijke Ordening en Milieubeheer, 1999, *De ruimte van Nederland: startnota ruimtelijke ordening*, Sdu Uitgevers, Den Haag

Ministerie van Volkshuisvesting, Ruimtelijke Ordening en Milieubeheer, 2004, *Nota Ruimte: Ruimte voor ontwikkeling*, Ministerie van Volkshuisvesting, Ruimtelijke Ordening en Milieubeheer, Den Haag

Priemus, H., 1994, Economie van stedelijke netwerken, *Rooilijn*, No. 1, pp. 4-9

Rienstra, G.U., 2002, Agglomeratievoordelen zijn terug, *Economisch Statistische Berichten*, 87/4343, pp. 56-58

Sietsma, W., 1986, Kansen voor de stad, *Economisch Statistische Berichten*, 71/577, p. 997

Van der Vegt, C. & W.J.J. Manshanden, 1996, *Steden en Stadsgewesten*, Sdu Uitgevers, Den Haag

Van der Wouden, R. & E. de Bruijne, 2001, *De stad in de omtrek*, Sociaal en Cultureel Planbureau, Rijswijk

# 4 Streaming Spatial Planning: Technological Changes and their Impact on Space

Luuk Boelens

## Introduction

The spatial approach to studying and interpreting telematics is a recent development. About 15 years ago, after some initial futurological studies, spatial planners and researchers made the computer and telematics revolution a serious object of study. The overwhelming majority of studies, however, have examined the consequences of this technological revolution for the existing hierarchy in the global-city network – the layered structure of the telematics landscape – or they have included at least a few telematic experiments within the existing Cartesian understanding of time and space. The key shift, that of viewing Information and communication technology (ICT) more as an *enabling*, rather than a driving force, has yet to be made. When made, this shift will open the way for research and design that will allow a much broader approach to ICT.

## A brief history of telematics

Telematics has a long history. Its origins go right back to the invention of the abacus and the use of drums and smoke signals to convey messages. It moved forward with Charles Babbage's invention of the analogue computer in 1833, Samuel Finley Breese Morse's electric telegraph in 1837 and Arthur Graham Bell's telephone in 1876 (Pieterson, 1981). But the modern history of computers and telecommunications is generally agreed to date from the time of the Second World War. The War increased the demand for both reliable and efficient communication systems (preferably wireless) and calculating machines capable of cracking complicated military codes. The first electronic computers, such as the Colossus in Bletchley Park, London, and the ENIAC (Electrical Numerical Integrator and Calculator) at the University of Pennsylvania, date from this period. These were enormous machines, requiring a whole team of operators to run them (McCartney, 1999). Even then, they could only perform simple calculations. Worthless for everyday use, they only became interesting with the availability of applications, and the discovery of semiconductors. These were used first in transistors, and later in computer chips. Computers became smaller, faster and more efficient. But they only really took off after the introduction of the Personal Computer by IBM in 1981, followed by Steve Jobs' 'graphical user interface' in 1983 and Microsoft's 'multi-tasking environment' in 1984. The computer became a significant force in society. With Compaq's introduction of the laptop at the end of the 1980s, and Apple's at the beginning of the 1990s, computers then became mobile, too. Once 4-6 kilos in weight, modern laptops, such as the Apple Power Book, are now half the weight of the first models, a hundred times faster, and they now support numerous wireless applications as well (Linzmayer, 1999).

The history of (mobile) telecommunications and Internet has been just as dramatic. It, too, can be traced back to Morse's and Bell's inventions. The first applications were with the development of railways, regulated shipping and, later on, aviation. Again, though, it was not until after the Second World War that the telephone and television came within reach of the broad majority of the population. From 1957 onwards, the US defence industry, with its Advanced Research Projects Agency (ARPA) and a network that included universities (Castells, 2001), became an important driver for the next technological revolution. Related computer networks sprang up within and outside the United States, such as Telenet, Usenet, CSNet and Eunet, each of which was later connected to ARPANET. By 1984, the resulting 'backbone' consisted of 1000 hosts. Boosted by the popular e-mail program, it was a rapid evolution that grew from 5000 in 1986 to 28,000 in 1987. Developed as a facility for building protocols, it proved highly efficient for the exchange of personal information. With growing use, there was an increased demand for more bandwidth. This was first provided by UUNet in 1987, and later by other commercial Internet providers. These new backbones finally pushed ARPANET out of the market in 1990. At the same time, it became much easier to make web pages and establish links to existing pages, and this fed an exponential growth in use. By 1993, the Lycos search engine had indexed 800,000 web pages; Google now has more than ten thousand times as many, equivalent to eight billion web pages (http://nl.wikipedia.org/). The use of Internet has grown enormously, from about 20 million users in 1995 to more than a billion registered Internet users now (Mau, 2004:222). Moreover, UMTS frequencies and the bottom-up development of the WiFi network (and soon, maybe, Wi-MAX) are spawning wireless applications and driving up mobile useage, in line with the rapid growth of mobile telecommunications. The number of GSM subscribers in 2001 was estimated at about 600 million, worldwide (Katz & Aakhus, 2002). In the Netherlands alone, the number of subscribers rose from around 100,000 in 1995 to 13 million now, while, by 2003, the proportion of households with only mobile services and no land line had again doubled to 8% (Ministerie van Economische Zaken, 2004).

## The spatial planning of telematics?

Given this relatively short social history of the computer, Internet and GSM, serious research and thinking on the spatial impacts and requirements of telematics is also recent. It has only been identifiable since the beginning of the 1990s. Although several works had already appeared, including *The Rise of the Global Village* (Marshall McLuhan, 1964), *Die dritte industrielle Revolution* (Dieter Balkhausen, 1978) and *The Third Wave* (Alvin Toffler, 1980), these were mostly exercises in futurology, written before access to computers and telecommunications technology had become widespread. Many of these writers' predictions have simply not come about. We have not entered a paperless world; rather the opposite. Neither has the world become a 'global village'; some places have become closer, others have moved relatively further apart. Telematics has not replaced traditional face-to-face contacts and physical mobility, but has added another mode which operates alongside, or greatly enhances, traditional methods of movement and communication (Gillespie & Richardson, 2000). Studies conducted in the 1990s, and even in the early years of this century, have also failed to provide an adequate analysis of the new spatial reality. Their predictions were either too extravagant (Cairncross, 1997; 2003) or too traditional (Dreier *et al.*, 2004; Boddy & Parkinson, 2004), and many of the assumptions made at that time have since been proved false.

Distance and geography *do* still matter, although with an entirely different significance and scope than was traditionally supposed. Instead of places being exclusive and independent – like biotopes growing on a substrate of natural and cultural history – they have become more relational, acting as landing and departure points in a wider environment (Thrift & May, 2001). It is true that glass-fibre cables and Internet hubs tend to follow existing infrastructure and geographic hierarchies, thus strengthening existing centres and discrepancies (Graham & Marvin, 1996), but, with increasingly rapid and cheaper modes of transport, they have totally transformed the use and organisation of these spaces (its software and orgware). With the growth of wireless and mobile networks, the diffuse nature of which seems to be creating a more uncertain situation (De Wit, 2004), this change has become more obvious.

## Beyond the *idées fixes*

The huge growth in computer use and telecommunications has made a more complicated, varied and sometimes more paradoxical impact on development space than was assumed (Boelens, 1994). It seems sensible, therefore, to abandon the simplistic, one-dimensional *idées fixes* about this relationship, once and for all.

First, the world of computers, high mobility, Internet and telecommunications has not replaced the traditional world of travel, meeting people and communicating, but it has added another layer to it. It has created more diversity and much greater freedom of choice. We can communicate with people on the other side of the world as well as with personal friends, colleagues and acquaintances in our immediate environment; we can do our telebanking in the evening, and then queue up the next day to collect a permit for a residents' parking place. This implies that the traditional spatial order is still in place. All those sweeping claims about a truly global society, footloose businesses and the emergence of a 'kinetic mass' – with the inevitable consequences for urban design and spatial planning (Koolhaas, 1995:1248-1264) – should be taken with 'a pinch of salt'.

Secondly, it is too simple to conclude that progressive globalisation will itself rekindle an appreciation of the uniqueness of 'place'. 'Think global, act local' – or the opposite – is much too simple a reflection of the complex relations at play here. Moreover, it provides an alibi for those who take a far too global approach to datascaping and then fall back on traditional design, shaped by the landscape, urban planning and spatial contexts. As a result, the network city, the Dutch 'layered planning', (see chapter 19 by Jeroen van Schaick) regional design and area-based development planning have no grasp of the new relational significance of place. ICT has not abolished the city; the urban living environment, and neither has it strengthened it, in the traditional geographical sense, although it *has* in a relational sense.

Thirdly, many studies of the spatial Internet landscape – which show a progressive fragmentation and splintering of urban structures (Graham & Marvin, 2001), or the emergence of a new network of world cities with new hierarchies and urban relations (Sassen, 1992; Taylor, 2004) – are too traditional to reveal the possible contribution that spatial planning can make in specific circumstances. The relation between the centre and the periphery has become much too complex to be described in this dialectical way. Compare, for example, the 'shrinking cities' in eastern Germany and the northern United States with the booming areas in previous peripheral regions, such as Ireland, Finland and India. The physical availability of land, landscape quality or access (physically or virtually) is not so critical here. Computer and telecommunications technologies have made other factors, dictated by personal, business management or market considerations, more important.

These three perspectives on the spatial impact of Internet and telecommunications mainly fall foul of their own one-dimensional interpretation. The key shift in thinking still has to be made. Instead of a straightforward spatial planning based on flows, a context and/or location factor, what we need is a more subtle, actor approach to the planning and design of space. Justifiably, technocratic approaches are often dismissed as being too one-sided. Nevertheless, the most important shift we still have to make is not to view these technological innovations as 'driving forces', but as forces that 'enable' actual development on the ground (Dicken, 2003). In other words, they give companies, households and individuals greater spatial choice, but do not dictate or shape those choices. This observation turns out to be crucial. Because if we take commercial, (national) economic and personal considerations into account, not only do we see evidence of 'deterritorialisation', but also signs of a robust 'reterritorialisation'; not only do we see segmentation and fragmentation, but also fundamental processes of reclustering. These reterritorialisation and reclustering phenomena appear to be even more important than deterritorialisation and segmentation/fragmentation. They occur where new urbanism materialises and takes on spatial-economic, social, cultural and political forms (Boelens, 2005).

## Reclustering and reterritorialisation

Under the influence of commercial processes such as out-, in- and co-sourcing, a remarkable 'global shift' is taking place in spatial economics: business activities are being partly outsourced to subsidiaries, associated companies and other companies considered more suitable for the task in hand; the remaining core business activities are being broken down into the smallest possible units for relocation to places which offer the best cost advantages, conditions of production, market access, image, etc. Conversely, to spread risks or cover huge research and development costs, variable co-operative arrangements are being made – sometimes even with (former) competitors – in the form of equity investments, joint ventures, mutual licensing, or, in specific market sectors, mergers (Atzema & Boelens, 2004). New physical and organisational network clusters of independent and affiliated companies are being formed, with 'leader companies' setting the tone. These companies are willing to invest in projects that have a positive effect on other companies in a production chain, network or cluster. They are capable of cross-cutting the coordination of activities, deriving benefits from different geographical conditions and, if necessary or desirable, relocating their own activities.

Although of a different order, comparable trends are underway in the social and cultural domain of today's network society. There is talk of a decline in the more traditional forms of community and the emergence of segregated collectivities (sometimes fed by anxiety or by economic and cultural arguments), but also of the appearance of new 'home-grown social networks' that bring individual autonomy and community involvement back into balance – although now in a changeable form, involving multiple networks (Brinkgreve, 2004:51). The difference between 'strong and weak links' is an interesting distinction that has recently emerged (Flap *et al.*, 2005). This has something in common with the ideas of the political geographer Peter Taylor, who argues that satellite television, mobile phones and Internet make it much easier for residents of foreign origin to keep in touch with the home front. Instead of the McDonaldisation or the inevitable jihad, that Barber (1995) once maintained we would suffer, we have gained instead (or as well) seemingly limitless 'imagined communities' with an almost global reach, that have been enabled by the new ICT devices (Loopmans & De Maesschalck, 2004).

The institutional world is also influenced by the processes described above. We see evidence of a progressive and continuous re-scaling and re-bordering of political responsibilities: upwards, downwards and sideways (Brenner, 2004). There are also signs that decision-making, ideas, and support for spatial development are moving towards formal and informal bodies outside the political sphere. A vast (and changeable) aggregation of administrative responsibilities and formal and informal collaborative structures is emerging, as well as an enormous range of international governmental and non-governmental organisations (IGOs and NGOs) and various sorts of public–private and private–private partnerships (Held, 1999:53–54). The outside world has thrust the concept of governance forward by re-positioning both formal and informal structures to create working relationships among key actors, such as government, the market, the population, and leading firms and associations, to bring about effective action (Salet, Kreukels & Thornley, 2003). This requires a radical re-orientation of spatial decision-making and planning around, not only what Brenner calls 'scalar multiplicity' and 'administrative customisation', but also a more marked demand and alliance-oriented actor approach (Brenner, 2003).

## New challenges

All these trends and spin-offs from the 'telematics revolution' pose new challenges for spatial planners and designers. These challenges do not lie so much in the need to understand, and, where necessary, re-design the physical structures of telematics, nor in re-positioning places often considered unique and exclusive from the global-network perspective. Rather, they lie in the need to gain a better understanding of the forces and motives that underly the spatial processes of fragmentation *and* reclustering, deterritorialisation *and* reterritorialisation, in order to make a constructive contribution to these processes. New research is needed on the formation of distinct clusters in business networks to allow us to identify specific spatial strategies so that we can embed such clusters more securely within a dynamic regional and local context. Various parties now recognise that the European spatial-economic agenda, such as the Dutch 'Peaks in the Delta' report and the aspirations of the city and provincial authorities in the Randstad to create a metropolis in the west of the Netherlands (see previous chapter by Herman Rosenboom), have been increasingly wide of the mark (Amin & Thrift, 2002). The current urban development strategies also show a growing tendency to start from the wrong premises. Rather than using quantitative, location-specific typologies for residential, employment and living environments (such as the city centre, inner city, urban green areas, the village and rural areas), it is better to devise more dynamic, flexible, network-oriented typologies of diverse urban environments 'at a distance', than to fit in with the so-called 'split household', network company or 'mutimodal travel' of today. Even local politicians have finally acknowledged the existence of the network society. Over the last 20 to 30 years, the proposals for gradually redrawing administrative boundaries, positions and interventions have not, therefore, been plucked from the air; there has been a shift from top-down to bottom-up, from regulative planning (zoning and development control) to procedural, advocacy-based, entrepreneurial and development-oriented planning (see Part IV of this book). But the key shift that still needs to be made is to move away from 'inside-out' area-based planning priorities – key projects or 'issues' thought up by government – towards a more 'outside-in' strategy of 'steering through networking' that forges alliances and looks for innovative and stimulating spatial product-market combinations (Boelens, 2004).

In short, urbanism, urban policy, the urban economy and, above all, urban culture must be reconsidered. In the first instance, this is not a question of hardware but of software and orgware. In other words, it is not about the physical aspects as such, but, in the first instance, it is mainly about issues regarding other and better uses of urbanism or of its organisation. We can expect this to generate other issues in relation to urban design, planning and management; a challenge that will require a new urban and regional planning model.

I do not want to throw the current practice of urban design and planning standards and regulations overboard. We should supplement what we already have with a new vision by: amending actor-oriented research, basing new strategies for spatial interventions on projects rather than areas; and developing a new set of spatial planning tools and innovative, instrumental, product-market combinations of spatial development and Intelligent Operating System (IOS) technologies. In striving for a more efficient and dynamic use of space, much more is possible than current practice suggests. In addition to the first experiments about the digital district (Frissen, 2005) and initial innovations related to intelligent and self-learning transport systems, route guidance and up-to-date passenger information facilities (Instituut voor Maatschappelijke Innovatie, 2005), this new software and orgware could include:

- More flexible and dynamic land use;
- Variable and/or temporary land use;
- Amending the uses to which the public and semi-public domains are put;
- Moving definitively in the direction of urban renewal;
- Facilitating intermediaries for the flexible use of space.

What sort and type of new product-market combinations are formed is unimportant (because, in principle, the list is endless, and considerable innovation is possible); the crucial factor is how these innovations can be implemented, and by whom. In future, success or failure will depend on breaking down traditional blinkered interests and on how these innovations, as enabling forces, can stimulate innovative land use and a better quality of life. In the long run, this is what legitimises the work of committed urban and regional designers and planners.

Drewe (1996) has always kept the challenges associated with the growing use of computers, Internet and telecommunications and the 'network city' project on the practical and theoretical planning agenda, and has raised our understanding of the spatial impact of new technologies. The challenge now is to discover the underlying forces, and, above all, the actors behind them, which are reshaping our network-oriented and boundless society. Spatial planning will have to be streamlined. Design must move from an exclusive focus on the physical environment to an approach informed by behavioural sciences.

## References


Amin, A. & N. Thrift, 2002, *Cities - Reimaging the Urban*, Polity Press, Cambridge (UK)

Atzema, O. & L. Boelens, 2004, *Holland 2020 - A space Odyssey naar de Randstad als een concurrerende regio en hoe dat te bereiken*, Ministerie van Volkshuisvesting, Ruimtelijke Ordening en Milieubeheer, Den Haag

Barber, B., 1995, *Jihad versus McWorld - How Globalism and Tribalism are Reshaping the World*, Times Books, New York

Boddy, M. & M. Parkinson, 2004, *City Matters - Competitiveness, Cohesion and Urban Governance*, International Specialized Book Services, Bristol (UK)

Boelens, L., 1994, Een voortreffelijk lichaam; De uitdagende schizofrenie van de stad, *De Architect*, 3, pp. 22-31

Boelens, L., 2004, Sturen door netwerken, *S&RO*, 2, pp. 48-53

Boelens, L., 2005, *Network spaces - from planologic towards fluviologic*, Inaugural Lecture, Utrecht University, Utrecht

Brenner, N., 2003, Standortpolitik - State Rescaling and the New Metropolitan Governance in Western Europe, in: *DISP* 152/1, Zürich

Brenner, N., 2004, *New State Spaces - Urban Governance and the Rescaling of Statehood*, Oxford University Press, New York

Brinkgreve, C., 2004, Van huis uit - Individualisering en Opvoeding; In: Schnabel, P. (ed.), *Individualisering en sociale integratie*, SCP, Den Haag

Cairncross, F., 1997, *The Death of Distance*, Harvard Business School Press, Harvard Mass.

Cairncross, F., 2003, The Revenge of Geography, *The Economist*, March

Castells, M., 2001, *The Internet Galaxy - Reflections on the Internet, Business and Society*, Oxford University Press, New York

De Wit, O., 2004, Space adjusting technology - De opkomst van het mobiele telefoonnetwerk, *S&RO*, 2, pp. 34-37

Dicken, P., 2003, *Global Shift - Reshaping the Global Economic Map in the 21st Century*, Sage Publications, London

Dreier, P. *et al.*, 2004, *Place Matters - Metropolitics for the Twenty-first Century*, University Press of Kansas, Kansas

Drewe, P., 1996, *De NETWERKSTAD-VROM. Bijdrage van informatietechnologieën aan nieuwe concepten van ruimtelijke planning*, Verkennend onderzoek uitgevoerd in opdracht van het Ministerie van VROM, Delft

Flap, H. *et al.*, 2005, Gemeenschap, informele controle en collectieve kwaden; In: Völker, B., 2005, *Gemeenschap der burgers - Sociaal kapitaal in buurt, school en verenigingen*, Amsterdam University Press, Amsterdam, pp. 25-52

Frissen, V.A.J., 2005, http://www.stb.tno.nl/index.php?pointer=1-2-20-540-561-641 (January 2005)

Gillespie, A. & R. Richardson, 2000, Teleworking and the City; Myths of workplace transcendence and travel reduction; In: Wheeler, J.O. *et al.* (eds), 2000, *Cities in the telecommunications Age. The Fracturing of Geographies*, Routledge, New York/London, pp. 228-245

Graham, S. & S,. Marvin, 1996, *Telecommunications and the City - Electronic Spaces, Urban Places*, Routledge, London/New York

Graham, S. & S. Marvin, 2001, *Splintering Urbanism*, Routledge, London/New York

Held, D. *et al.*, 1999, *Global Transformations - Politics, Economics, Culture*, Polity Press, Cambridge

Instituut voor Maatschappelijke Innovatie, 2005, *Project Reiziger van de toekomst*, http://www.iminet.org/innovatie (January 2005)

Katz, L.E. & M. Aakhus, 2002, *Perpetual Contact - Mobile Communication, Private Talk, Public Performance*, Cambridge University Press, Cambridge (UK)

Koolhaas, R., 1995, *SMLXL*, Uitgeverij 010, Rotterdam

Linzmayer, O.W., 1999, *Apple Confidential - The Real Story of Apple Computer Inc.*, William Pollock, Berkeley

Loopmans, M. & F. De Maesschalck, 2004, Naar een kosmopolitische agenda - Peter Taylor over globalisering en politiek in het wereldstedennetwerk, *Agora*, 5/2, pp. 8-11

Mau, B., 2004, *Massive Change*, Phaidon Press, London

McCartney, S., 1999, *ENIAC -The Triumph and Tragedies of the World's First Computer*, Walker Publishing, Ontario

Ministerie van Economische Zaken, 2004, *Netwerken in cijfers 2004*, Den Haag

Pieterson, M., 1981, *Het technisch labyrint - Een maatschappijgeschiedenis van drie industriële revoluties*, Boom, Meppel (Neth.)

Salet, W., A. Kreukels & A. Thornley, 2003, *Metropolitan Governance and Spatial Planning - Comparative Case Studies of European City-Regions*, Spon, London/New York

Sassen, S., 1992, *The Global City - New York, London, Tokyo*, Princeton University Press, Princeton

Taylor, P.J., 2004, *World City Network - A Global Urban Analysis*, Routledge, London/New York

Thrift, N. & J. May, 2001, 'Introduction'; In: *Timespace; Geographies of Temporality*, Routledge, London, pp. 1-46

**Website**

http://nl.wikipedia.org/wiki/Geschiedenis_van_het_Internet (January 2005)

# 5 The Impact of Spatial Logistics in the Cross-border Spatial Corridor of Flushing-Terneuzen-Gent

Georges R.G. Allaert

## Introduction

In spite of the many scientific papers on transport economy and land-use planning, there is still very little research on the interferences amongst spatial organisations, logistics and infrastructure. 'Space' is nowadays a top-priority factor for development. It is a basic factor for production and consumption, for prosperity and well-being, for added value and mobility. In this chapter, we will reflect on the changing spatial-economic dynamics across the Bene-border (Belgium-Netherlands). Special attention will be given to the cross-border urban region of Gent-Flushing (Gent-Vlissingen) and the spatial-economic model VLISTERGENT (Vlissingen-Terneuzen-Gent). We will show that the cross-border infrastructural axis created by a canal and a tunnel stimulates the spatial impact of cross-border economic and social processes. Many seminars and congresses were organised by the International Scheldt Faculty and the Summer University of Zeeland in the 1980s and 90s, during which the huge added value of the maritime-industrial city-region Gent-Flushing was discussed. Every year since then, this topic has been a hot item on the political agenda of the cross-border region. Now, 25 years or so later, the scientific model is beginning to move, although still very slowly, towards more cross-border regional policies. It can be seen from different papers that VLISTERGENT is progressing 'from utopia', one could say, ' to reality'.

## A region as an active space: reflections on the growing importance of logistics

Growing internationalisation and globalisation are causing the regional and local levels to become more and more important. We use the term 'glocalisation' when referring to this process. At the same time, nations are losing power to global competition. Even in Europe, the growing integration process is causing regions, urban-regions, and cities, to become more competitive. Key strategies in the overall discussion are networking, clustering and complementarity. The importance of networking strategies can be seen in many land-use planning documents. The Dutch *Nota Ruimte* [National Spatial Strategy] (Ministerie van Volkshuisvesting, Ruimtelijke Ordening en Milieubeheer, 2004) and the Spatial Structural Plan for Flanders from 1997 are examples, but these policy documents are extremely 'introvert' in that they only focus on the spatial dynamics 'inside the country'. No, or hardly any, attention is given to cross-border 'spatial' networking. In particular, in the Dutch National Spatial Strategy, no attention at all is given to the socio-economic power of the Gent-Flushing urban region. Illustrations 5.1, 5.2a and 5.2b show the impact and location of this region.

| Powerlssue | Impact |
|---|---|
| *Social:*<br>· The direct consumer market | <br>· 550,000 inhabitants in a North-Sea corridor of 50 km |
| *Economic:*<br>· The 3 MIDA's<br>· Direct added value<br>· Indirect added value<br>· Jobs (direct)<br>· Jobs (indirect)<br>· Maritime traffic | <br>· Flushing, Terneuzen, Gent<br>· 5 billion euro<br>· 10 billion euro<br>· 45,000<br>· > 100,000<br>· 50 million tons |
| *Intellectual:*<br>· High density (universities, high schools and R & D) | <br>· > 100,000 people involved in a real 'brain corridor' |
| *Spatial:*<br>· Maritime industrial spaces<br>· Open spaces<br>· Infrastructure | <br>· > 4,000 ha (>1,000 ha still free)<br>· Blue-green, east-west corridors<br>· Laboratories in multi-dimensional developments and platforms (sea-rail-road-pipelines) on different scales |
| *Cultural/touristic values:* | A diversity of heritage /coast / water tourism |

**Illustration 5.1: Overview of the power of Gent-Flushing (VLISTERGENT). (source: G. Allaert, based on different documents)**

From my research on the Flushing-Gent urban region, I believe that the territorial space (dimension) must be seen as 'a force-field space' (Van Geenhuizen & Ratti, 2001) when taking action and as a factor for development. In the glocalisation strategies, we need to distinguish two mainstreams of force-field spaces: production spaces and market spaces. Both mainstreams have a specific spatial structure. In the case of VLISTERGENT, examples of production spaces are the three MIDA's (Maritime Industrial Development Areas). These are economic force-field spaces within the industrialised port regions.

The conurbation areas Gent and Flushing-Middelburg-Goes are examples of local/regional market spaces. Had these been territorial, in the sense of concentration and clustering, we would then have called them a *metroplex*. Metroplexation is an interference in glocalisation strategies (for its synergies, see Ill. 5.3). Midaplexation is the interference between a metroplex and an economic/industrial complex (or cluster) (Allaert, 1994, 1996). The Gent-Terneuzen canal zone is a good example of a growing midaplexation area, while the discussion about European Distribution Centres (EDCs) is an ongoing glocalisation process.

In the VLISTERGENT corridor, especially in the vicinity of the Westerscheldt Tunnel, there are ongoing talks, initiated by the big round-the-world-service customer-operator group Hessenatie/Noordnatie/Ports of Singapore (PSA), about operationalising the Westerscheldt Container Terminal, a huge EDC project. It is a good example of PSA's glocalisation strategy in the Rhein-Scheldt-Delta Region, the result of which is that, in the mainport development strategy for Rotterdam-Antwerp, Flushing will become a forefront node in global/regional competition. What we witness here is a global player setting up its world strategies in such a way that the local/regional/national policy-makers are unable to intervene with an overall strategy to protect the welfare and well-being of Flushing and the Rotterdam-Antwerp mainport area. This is not an isolated example; developments of this nature are occurring worldwide at the present time. The active space of Vlistergent is also a framework for logistic and

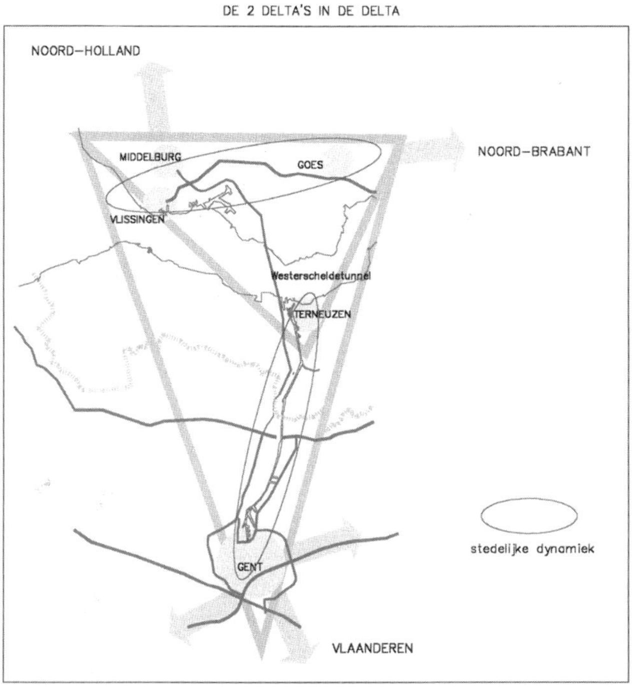


**Illustration 5.2a: Positioning of VLISTERGENT (NEI/Universiteit Gent, 2000)**

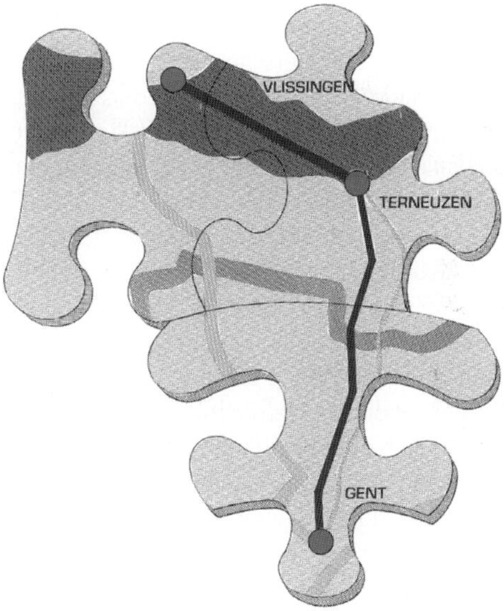


**Illustration 5.2b: Mental map of VLISTERGENT (Kamer van Koophandel & Fabrieken voor Midden- en Noord-Zeeland, 1992)**

transport fluxes: for goods, capital, and information inside and outside the region (see Ill. 5.3). In the cross-border region, however, logistic and transport fluxes operate on both global and local/regional levels. Logistic and transport strategies for the EPC/EDC-industrial clusters in the Flushing-Gent-corridor are based on the so-called 'networking' (of goods, capital, and information): chemistry, steel, automotive transport, agro/food. It is the multinationals who are still the driving forces in this region.

Looking at the general scheme of what logistic processing means (see Ill. 5.4), there are different 'stages' ranging from raw materials to components and assembly and onwards to wholesale marketing and distribution to the client (consumer) via 'transport' channels (port, road, rail, inland water, underground, air). Logistics can be seen as a systems process starting from the design of the product, followed by producing, to selling/retailing, within a space-time context. To an ever greater extent, transport is becoming the most important component of the supply/demand chain. In many cases, more than 50 % of the costs of logistics are transport costs... Although the context of physical space is very important in the systems process, there has been very little research so far on so-called spatial logistics or, better still, on the spatial control of the whole chain. Scientific knowledge and teaching on interference logistics and mobility (LoMo) is still undeveloped. In particular, there is very little methodology on LoMo integration and LoMo synergies at both micro and macro levels of scale.

The availability of space and the quality of that space are crucial topics when discussing logistics. To an ever greater extent, spatial logistics are at the forefront of development strategies concerning platforms, clustering, access, accessibility and intermobility. Logistics strategy is dependent on a whole range of external factors and constraints, beginning with the physical environment (infrastructure, mobility, the geo-location), and extending through the social environment (ecology, labour forces, stakeholders) to the political administrative environment (ranging from Europe to local authorities). To achieve sustainable efficiency, logistics, therefore, require more and more long-term thinking and strategic planning.

The huge investments required to build a multi-modal logistic platform is a good example of why it is necessary to couple a long-term vision with short-term actions (strategic planning). For instance, look at the strategy used for the Westerscheldt Container Terminal (WCT) in Flushing. In 2001, the company (Hessennatie/Noordnatie/Ports of Singapore) decided to build a 'world hub' multimodal platform for containers (150 ha, 1,5 million containers, 30 million ton traffic) in the Westerscheldt. The project will cost more than half a billion euros, and it will be more than 5 years before the first containers can be handled. The site will not be 'fully' operative before 2010 (almost 10 years after the company had made the decision) due to the *Milieu Effect Rapportage (MER)* [Environmental Impact Assessment] procedure and ongoing problems.

Looking at other company strategies in the region, the outsourcing and specialisation of the network economy have created other fields of work (in maritime industry and services), mostly carried out by SMEs- (small & medium-sized enterprises) outside the region rather than in the MIDA axes. We can call this 'spill-over and sub-harbourisation' and 'indirect added value'. To valorise hub-and-spoke developments (in ports, airports, European gates, etc. there is a need for both a trans-national spatial economic strategy based on the added value of transport and logistics, and a spatial design (see Ill. 5.5). I hope Belgium and the Netherlands will take this opportunity to design a cross-bordered hub-and-spoke network (from ports-airports and European gates to sub-harboured hinterlands) as a first mental map. A preliminary spatial drawing is given in Illustration 5.6.

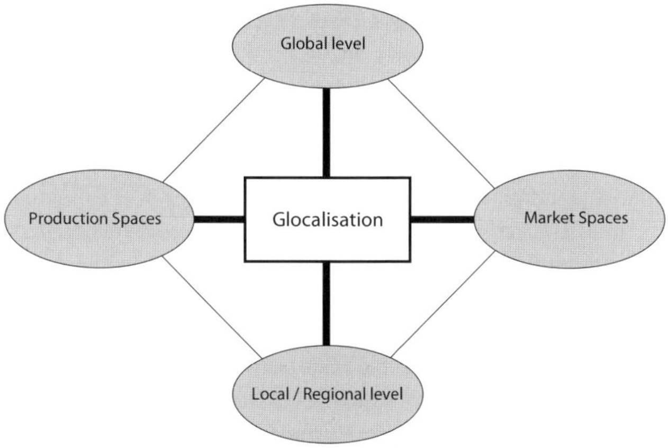


**Illustration 5.3: The diamond of the spatial economy as a frame for the knowledge on logistic and transport flux: goods, capital and information (Allaert, 2002)**

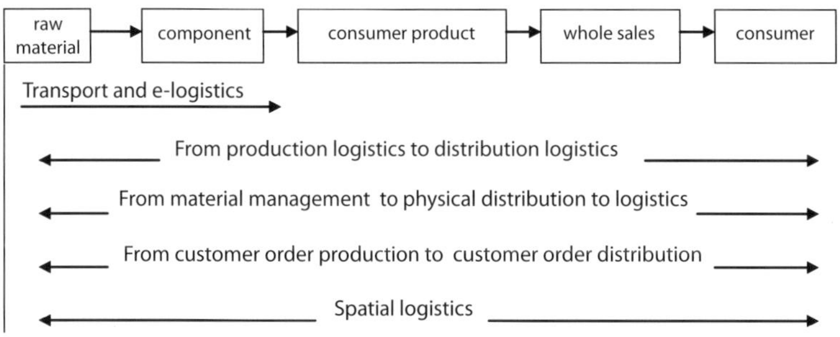


**Illustration 5.4: The chain of logistics on the 'company level'**

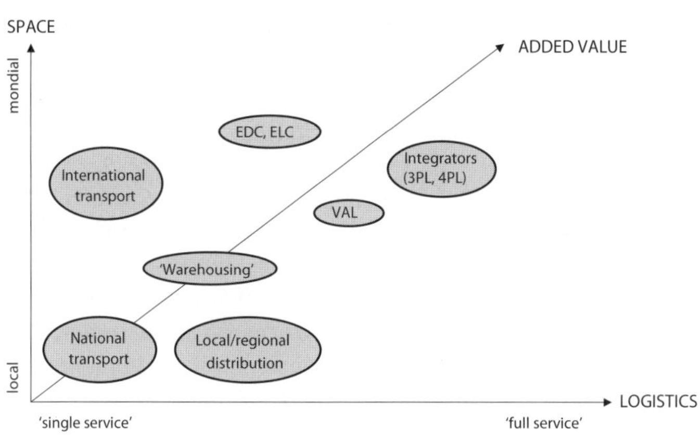


**Illustration 5.5: The added value of transport in a space-logistics diagram (Witlox, 2004)**

## Decision-making and governance in the region

Reality shows that there is still an enormous gap between companies' cross-border strategies (especially those of the big players; the multinationals) and governmental strategies. This is also evident in the cross-border region of VLISTERGENT. If VLISTERGENT is viewed as a regionally active space (RAS) in terms of territorial and support spaces, decision-making and governance are the final elements for success. Although there is a growing 'industrial' co-operation at company level, with respect to the 3 Cs (Concentration-Clustering-Complementarity), there is still an enormous gap at governance (political) level. The gap is evident on different levels of scale - national, regional and local. The national policy and administration in The Hague and the para-national administration (Flanders), with its opposite partner in Brussels, view the region as a national border and have very little interest in nationally oriented cross-border infrastructure. On the other hand, on the level of the Port Authority, Gent cannot set up a joint co-operation with Terneuzen without taking Flushing into account, because Zeeland Seaports is a bi-municipal Port Authority (Flushing-Terneuzen).

On the level of city-governance, there are friendly contacts between Gent and Terneuzen, but a joint cross-border EU-regional strategic plan is far from reality. Although governments are willing to take quick co-operative actions within the frontiers, we see that the business community (ports, chambers of commerce, etc.) has fewer problems with frontier actions. This is no surprise, when we realise that borders, for governments, have a completely different meaning from what they do for companies. For governments, the physical border demarcates the limits of their direct responsibility, but also of their competences, because the administrative organisation differs enormously across the borders. Companies have more in common in that they operate in a geo-economic network economy, which means that a border (whether the Belgian or the Dutch) often makes little, or no difference to them (Boekema & Allaert, 1999).

There are bottlenecks on three levels: financial, administrative and political. First of all, financial problems play an important role. Financially, the ports of Flushing/Terneuzen and Gent operate as separate units. By maintaining their own financial independence, an area of tension is created between serving their own interests and the common interests of the three ports. In the long run, the self-interest of the ports will always remain a matter of money, and it may take some time before co-operation becomes fruitful for all the partners involved. As long as the benefits remain uncertain, or too distant, the costs of co-operating will be viewed negatively. This leads to the conclusion that only when they have a common problem, such that all their interests, and the common interest, overlap, are these ports likely to co-operate.

Secondly, there are many administrative bottlenecks: differences in administrative culture, politics and in how public administration is oriented. All these aspects are so different on both sides of the frontier that co-operation is difficult. In the Netherlands, all interests must be taken into account, all competences are respected, everything is channelled into (precise) procedures, worked into plans. In Belgium, however, informal networks dominate. In Dutch spatial planning, the planning administrations formulate long-term visions and aspirations prior to drawing up plans. With this top-down approach, these plans are mostly accepted on lower levels. On the Belgian side, there is also a strongly centralised administrative style, whereby Brussels issues directives to the lower levels of government. In Belgium, the municipalities do not have the resources for implementing fixed, centralised plans. For constructing cross border and infra-structural works, this difference in financing can create problems. Because

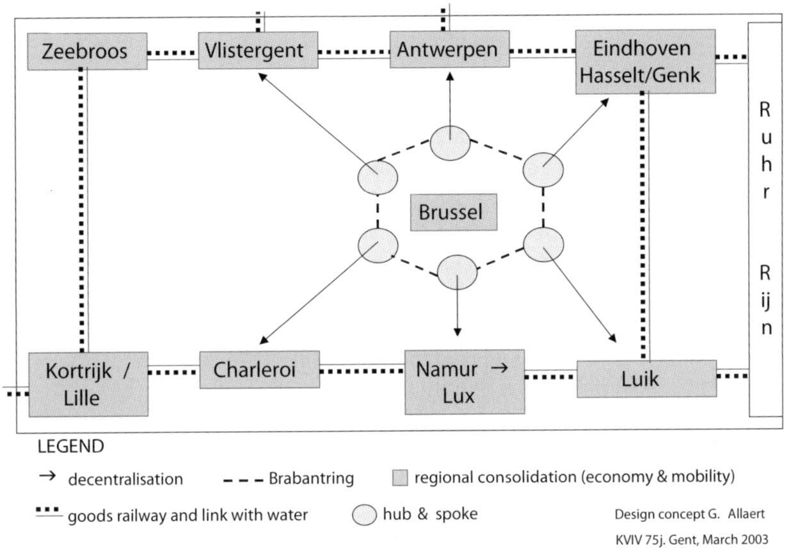


**Illustration 5.6: A mental map of a trans-national hub-and-spoke network (based on the outcomes of various stakeholder-group discussions, March 2003, published in Allaert, 2003)**

the two neighbouring countries are organised so differently on financial and administrative levels, it is difficult to establish a clear basis for cross-border co-operation.

Thirdly, there are bottlenecks on the level of strategic policies. Bottlenecks occur between the Belgian and Dutch actors because, in many cases, policy directives stipulated in Belgium are less strict than those in the Netherlands. Despite large differences in national legislation and business culture, there is a growing cross-border network of relationships between businesses in the Gent-Terneuzen canal zone. Companies seem to create cross-border clusters, which result in border synergies (of production and distribution) between companies, but because of the differences in policy directives, there is no cross-border policy of public governance, as yet.

## Towards an integrated planning strategy

In spite of the various bottlenecks which hinder the development of one single port-(city) structure stretching from Flushing, through Terneuzen, to Gent, as long ago as 1986-1990, physical planning specialists in port planning and management had already drewn up a cross-border planning figure in the VLISTERGENT model (Drewe, 1999).[1] In essence, this planning model is a framework for spatially controlling and managing the production structure and environment in a historic maritime industrial zone and network. The territorial optimalisation was based on four criteria: improving land-use qual-

1 See for the planning approach in this area also the next chapter by Jan Goedman.

ity, improving mobility, creating a sustainable economy (value added and employment places) and developing sustainable nature. With its economic-ecological strategy, the aim was, and still is, to wage a cross-border (bottom-up and top-down) offensive on the Scheldt region. In this corridor, which is known as a MIDA-PLEX (an abbreviation for three maritime industrial development areas in a North-South axis), different developments can be accommodated along side each other, ranging from agriculture to industry and services and even to many kinds of nature developments.

The VLISTERGENT concept should be seen as a pilot project (scientifically and visionary) based on the synergies envisaged in a city-network stretching from Flushing-Middelburg to Goes-Gent. Terneuzen (in the middle of this axis) is seen as the node for 'spill-over' maritime activities from both the Gent and Flushing sides. This new city-regional field will also generate extra added value with respect to new urban developments (urbanism) and new forms of tourism. However, the Westerscheldt tunnel, although in operation since mid March 2003, has not yet had a substantial (positive) impact in on industrial developments, urbanism and tourism. This means that the VLISTERGENT model has not yet completely left the planning board to enter the practical domain. The VLISTERGENT model will only become a reality if the policy-makers set up a cross-border strategy (Flanders/the Netherlands) focusing on optimising the following issues:

1. The road infrastructure on the left and right banks of the corridor, especially from Terneuzen to Gent;
2. The rail infrastructure (for both goods and passengers) on both sides of the Gent-Terneuzen canal zone;
3. Finding complementary strategies for bimodal and trimodal terminals in the axis, especially those based on water/rail and water/rail/road, taking the discussion about the Westerscheldt Container Terminal into account, and the political discussion, which is still open;
4. Maritime-industrial land use (a new MIDA-development). There is a potential 3000 ha, or more, that can be developed.

It should be mentioned that there are few places in Europe which have such potential. Initial calculations show that the four issues amount to investments of at least 5 to 6 billion euros, if a new lock at Terneuzen (estimated at 1 billion euros) is taken into account.

## Some research-based reflections

This section summarises the most important conclusions of our ongoing research. These conclusions constitute our reply to the question: what are the possibilities for cross-border co-operation?

### Reflection 1: Towards new ways of planning

The marketing and management of spatial (environmental) quality is an important issue in port planning nowadays. Every maritime industrial development area has specific territorial qualities, both physical and social. As a result of the new worldwide port-economic strategies, and the impact of public and private enterprising, the quality of the place (regional/local) has become more and more important in discussions on port strategies and management. Nowadays, port management is becoming more like advocacy management, and the port manager more like a marketing man for promoting its physical, environmental and social qualities. There are enough examples in the fields of land-use, and in improving the quality of the environment, communication and mobility.

On the one hand, there is the port manager, the person who facilitates negotiations; who is concerned with developing the port. He must have a basic knowledge of physical planning issues. On the other hand, a growing need is evident for long-term visions and commitments both in public and private sectors. This calls for a more open communication with the different 'players' in a port area or region, such as the local people, the owners of land, the environmentalists, the Chambers of Commerce, the individual entrepreneurs, and even the tourists who visit the port. This means a revolution in planning (Allaert, 2000). This new planning philosophy (as a communicative and co-operative action process) goes together with the new planning designs, the decision-making for which is centred on items such as landscape, symbols (ecological/sociological) green sites, public open spaces, and synergies of economies of scale and scope. To learn about a region, we now rely on bringing together stakeholders and shareholders in discussion groups and so-called force-field groups. In the light of these developments, the VLISTERGENT model must be seen as the mental map on top of the new map of collaborative management strategy for the region.

## Reflection 2: Towards complementary strategies

A structural, cross-border co-operation between the ports of Gent, Terneuzen and Flushing is logical from a scientific, spatial-economic viewpoint. Following the VLISTERGENT model, spatial and economic projects should complement port activities, economise on space and improve the competitive position of the area between Le Havre and Hamburg (Allaert, 2002). Because the ports in the Scheldt estuary complement each other well, they must be seen as a single region where each port helps to contribute towards the healthy competitive position of the whole region, rather than destroying the potential strength of the region by vying for each other's competitiveness. All the ports in the Scheldt estuary should formulate a common vision that interprets the maritime-industrial needs of the region more efficiently. This could be directed towards efficient land use, to opening up the area, and to environmental policy. The spatial-economic quality of the Scheldt estuary would then be improved. Scale enlargement would improve the competitive position of the region as a whole, making it one of the top five port regions in Europe. It would be undesirable, though, to develop the Scheldt estuary as a second Rijnmond (Rotterdam-Dordrecht). The complementary strategies of Flushing, Terneuzen and Gent should be seen as an industrial economy of scale and scope, aligned to such strategies as industrial networking, core business, co-manufacturing, and clustering. In this context, sustainable spatial economy becomes a much more feasible (scientific and policy-making) issue. Industry should also be linked with the agriculture potentials of the region.

## Reflection 3: Towards more research in science parks located within the port-city region

Some of the present policy documents contain directives that could act as a base for cross-border co-operation among the ports of the Scheldt estuary. However, the opportunities for putting these policy directives to good effect are being lost because of lack of commitment to the plans and the difficulties of aligning directives made by the Dutch government with those made for Flanders. An additional bottleneck is the lack of cross-border policies in certain sectors such as infrastructure and environmental policy. In their recent policy documents, the national governments, the Benelux and the European Union made very positive statements about cross-border co-operation, but yet they make no mention of this region. Organisations such as the RSD (Rhine-Scheldt-Delta) have tried to get more commitment from the EU, but progress is slow. Scientific influence in the whole discussion is still very weak, although academic studies on the strategies needed for the Westernscheldt are becoming more multidisciplinary (see the ongoing discussions in 2004 concerning the Process Reports). I am a great believer in the advantages of having one or more scientific parks in the MIDAPLEX. The reason is clear: Flushing-Gent is a brainport + greenport + seaport and it has 3 port-nodes. All of these generate R & D activities on which to base innovation and spin-off (Allaert & Drewe 1993; Allaert, De Klerck & Drewe, 2000).

## Reflection 4: Towards a more sustainable use of land in the port region

From a spatial planning point of view, co-operation would be a good thing for several reasons: the economics of the Scheldt ports is complementary, so co-operation would bring about a more efficient use of space and would increase the competitive (gateway) position of the region as a whole. However, the interests of the individual ports differ from the common interest and there is a lack of willingness to co-operate on many points. Improving the economic position of the region, based on a MIDAPLEX strategy is not the first priority of the current port management of Flushing-Terneuzen-Gent. Besides the economic possibilities that co-operation would offer, improvements could also be made in the sustainable use of space. In our present society, sustainability is not a subject to be neglected. When ports on both sides of the border develop areas separately from each other, it is very likely that land will be used inefficiently. If development is to be made sustainable, then clearly such a situation should be avoided. For that reason, a better tuning of spatial developments in the ports of Gent, Terneuzen and Flushing is an issue, which the authorities involved must take as their responsibility. Port communities, however, cannot be expected to put sustainability above economic issues. For port authorities this is different, because they have an obligation to handle issues of common interest with care. It is a responsibility that shouldn't be forgotten. Land is a rare commodity, and as such, demands good treatment. The expression 'Where there is a will, there is a way' comes to mind in this connection. It implies that where there is no interest in doing something, it will be difficult to find a way to do it. In a Europe where borders are fading away, here, the authorities on both sides of the border have the important task of devising a more efficient land-use planning for these border regions, and to do this better communication will be needed between governments on both sides of the border. The expression 'to understand is not the same as to comprehend' puts this clearly. It is also important for the authorities in the Netherlands and in Belgium, in Zeeland and in Flanders to put away their objections and try to understand each other, and not just to leave it to International Scheldt Faculty to act as the platform for thinking about the future of our ports and region in a more sustainable world.

## Reflection 5: The International Scheldt Faculty – from start to finish

In the summer of 1989, a foundation was created with the aim of establishing a network to transfer knowledge from Dutch and Belgian universities and polytechnics to governments, business and industries in the Scheldt region. Drewe and Allaert were at the frontier of this action (Drewe, 1998, 1999). From the very beginning, five groups worked on specific topics relevant to the Scheldt region. The activities of these groups were:

1. Environmental management, integrated preventative clean-process technology, integral sustainable growth;

2. Aquaculture – fishing, seafood technology, marine cultures;
3. Industrial control and maintenance technology – total quality management, maintenance and safety;
4. Management and administration – cross-boundary interregional management and administration;
5. Recreational and tourist product development.

The five study groups stimulated the spread of knowledge through regular meetings, seminars and long and short-term courses. There were representatives from research and education institutions, local and regional authorities and the business world in each of the five groups.

With prominent Flemish and Dutch members, the International Scheldt Faculty (ISF) could be described as a public/private partnership organisation. The universities of Gent and Antwerp (Belgium) and the universities of Rotterdam and Delft (Netherlands) and also the *Hogeschool Zeeland* were among the ISF's scientific members. The governors of the provinces of West and East Flanders and Zeeland were prominent members of the board, in addition to the Chambers of Commerce, and representatives from the business and industrial sectors. There were also two councils. The advisory council, consisting of prominent representatives of trade and industry in the Scheldt region, was the advisory body for the executive committee. The scientific council was made up of representatives from the teaching and research institutes, among them six universities. Both councils promoted the interests of the ISF at the trans-regional (Dutch-Flemish), national and European political levels. The ISF also stimulated policy activities in the EC Eu-region 'Scheldemond' and participated in different European programmes and organisations, such as COMETT, ESTURIALES, and INTERREG. After 15 years, the ISF Foundation was dissolved as an organisation. Since 2003, when the formal commitment was signed by two of the teaching organisations, some activities, such as certain research and teaching programmes, have been continued by the *Hogeschool Zeeland*, some in partnership with the University of Gent. It is amazing that although everybody is arguing in favour of more horizontal public/private actions and for more cross-border activities from organisations and foundations such as the ISF (which was the pilot organisation for this very region), the ISF as an organisation has not survived!

## References


Allaert, G., 1994, Gent: the Western node of the metropolitan network of Flanders; In: Kooij, P. & P. Pellenbarg (eds), *Regional Capitals: Past, Present, Prospects*, Van Gorcum, Assen (Neth.), pp. 19-35

Allaert, G., 1996, Sustainable Maritime Industrial Development in perspective, paper presented on the 11th International Harbour Congress, Antwerp, Proceedings, pp. 3-10

Allaert, G., 2000, Towards a new planning concept for maritime industrial development, paper presented on the 13th International Harbour Congress, Antwerp, pp. 22-31

Allaert, G., 2002, In the shadow of the Great VLISTERGENT, PIANC-Congress, Sydney, Australia, pp. 81-90

Allaert, G., 2003, *Wegwijs in de ruimtelijke economie*, Story-Scientia, Gent

Allaert, G. & P. Drewe, 1993, Gent Werkt, maar hoe innovatief is de Gentse economie, *Gent Werkt*, no. 94, pp. 31-44

Allaert, G., P. de Klerck & P. Drewe, 2000, Objective 2000 Gent, Towards an urban innovative environment; In: Crevoisier O. & Camagni R. (eds), *Les milieux urbains: innovation, systèmes de production et ancrage*, GREMI, Neuchatel, pp. 349-360

Boekema, F. & G. Allaert (eds), 1999, *Grensoverschrijdende activiteiten in beweging*, Van Gorcum, Assen

Drewe, P., 1998, The International Scheldt Faculty as an example of cross-border knowledge infrastructure; In: Brunn, G. & P. Schmitt-Egner (eds), *Grenzüberschreitende Zusammenarbeit in Europa*, Nomos, Baden-Baden, pp. 250-263

Drewe P., 1999, Grensoverschrijdende infrastructuur.De Internationale Scheldefaculteit en haar toekomst; In: Boekema F. & Allaert, G. (eds), *Grensoverschrijdende activiteiten in beweging*, Van Gorcum, Assen, pp. 141-150

Kamer van Koophandel & Fabrieken voor Midden- en Noord-Zeeland, 1992, *De betekenis en ontwikkelingsmogelijkheden van de zeehavens Vlissingen-Terneuzen-Gent in het Scheldemondgebied*, Vlissingen

Ministerie van Volkshuisvesting, Ruimtelijke Ordening en Milieubeheer, 2004, *Nota Ruimte: Ruimte voor ontwikkeling*, Ministerie van Volkshuisvesting, Ruimtelijke Ordening en Milieubeheer, Den Haag

NEI/Universiteit Gent, 2000, *Dynamisch centrum in de Delta,strategische planvisie voor de gemeente Terneuzen in het perspectief van de komst van de Westerscheldetunnel*, Rotterdam/Gent

Van Geenhuizen, M. & R. Ratti, 2001, *Gaining Advantage from Open Borders: an Active Space Approach to Regional Development*, Ashgate, Aldershot (UK)

Witlox, F., 2004, Invloed van de ruimtelijke ordening op de logistiek; In: *Proceedings 26th Flemish Scientific Economic Congress*, Commission (Chairman G. Allaert), Antwerpen, pp. 267-294

# 6 Mixed Scanning and Other Issues Related to Development Planning in the Netherlands

Jan Goedman

## Introduction

This chapter begins with a review of recent developments in Dutch spatial planning, and the proposed changes in planning procedures and methods. To understand the current planning debate in the Netherlands, we first look at its historical positioning. This goes back to discussions within the international planning community from the 1950s onwards. In the concluding section of the chapter, we will look at ways of developing new planning expertise. As sub-sections in this storyline, the following stages are discussed:

- Introducing mixed scanning: the comprehensive/incrementalist debate;
- Combining plans, parties and projects (mixed scanning);
- The growing importance of spatial discourses in spatial disciplines and doctrines;
- Combining design, research and development in planning practices;
- Facilitating the growth of tacit, explicit and scientific planning knowledge.

## Introducing mixed scanning: the comprehensive/ incrementalist debate

In a traditional comprehensive, rationalistic planning approach, projects are the ultimate outcome of a long procedure. This procedure became the subject of an important debate within the international planning community during the 1950s and 60s. Opponents of comprehensive planning favoured the more incrementalist approach of 'muddling through'. In 1967, as part of this debate, Amitai Etzioni proposed 'Mixed Scanning: a third approach to decision-making'. In decision-making, "vague commitments of a normative and political nature are translated into specific commitments to one or more specific courses of action" (Etzioni, 1967:217). In those days, there were two opposing conceptions of decision-making: the *rationalistic approach*, which assumed that the decision-maker had a high degree of control, and the *incrementalist approach*, which assumed much less command over decision-making. Mixed scanning, as proposed by Etzioni, combines components of both approaches, and by so doing, it

"is neither utopian in its assumptions as the first model nor as conservative as the second. ... In the exploration of mixed-scanning, it is essential to differentiate fundamental decisions from incremental ones. Fundamental decisions are made by exploring the main alternatives the actor sees in view of his conception of his goals, but – unlike what rationalism would indicate – details and specifications are omitted, so that the overview is feasible. Incremental decisions are made, but within the contexts set by fundamental decisions (and fundamental reviews). Thus, each of the two elements in mixed-scanning helps to reduce the effects of the particular shortcomings of the other; incrementalism reduces the unrealistic aspects of rationalism by limiting the details required in fundamental decisions, and contextuating rationalism helps to overcome the conservative slant of incrementalism by exploring longer run alternatives" (Etzioni, 1967:225).

Mixed scanning as a planning strategy was highly appreciated by planning theorists, such as Faludi: "A strategic planning agency employing planning strategies (the most sophisticated being mixed scanning) proceeds by setting a general framework within which other agencies operate" (Faludi, 1973:212).

Although, nowadays, how to combine plans, parties and projects in a more interactive way is an important planning issue, 'mixed scanning', as a planning method to facilitate this, is absent in most documents. Instead of returning to the comprehensive/incrementalist debate of the 60s, the development planning approach in the Netherlands is based on the influential *Ruimtelijke Ontwikkelingspolitiek* [Spatial Development Politics] published by the *Wetenschappelijke Raad voor het Regeringsbeleid (WRR)* (1998) [Scientific Advisory Council to the Government] as an aid for government policy-making.

## Combining plans, parties and projects

### The issue: Mixed scanning

Using the publication of the Scientific Advisory Council to the Government from 1998 as a frame of reference (Wetenschappelijke Raad voor het Regeringsbeleid, 1998), we analysed best practices in the Netherlands and abroad. This was used as a basis for the preferred Development Planning (DP) approach. This is presented in a nutshell, in the so-called DP 'turbo', as shown in Illustration 6.1. "This new method requires active and innovative co-operation in regional planning practice and the co-ordinated application of 'people, plans and money'. The model is process-driven; each plan is guided by its own external process managment" (Tetteroo, 2004:7). Van der Heiden (2003) adds two other links to this. Firstly: high spatial quality should be guaranteed by using the so-called layer approach (as proposed in the National Spatial Strategy) and secondly: using spatial design could be very helpful as a working method during the DP process.

In the Netherlands, the mixed scanning approach is seldom used explicitly. The relevance of this approach for development planning can be illustrated by a planning experience in Zeeland (Drewe, Goedman, Lockefeer & Roeleveld, 1991/1992 - see also the previous chapter by Georges Allaert), which can be evaluated in many ways as a 'DP approach *avant la lettre*', where development perspectives and strategic projects were elaborated interactively, in dialogue with the (public) parties involved.

### The case: Combining perspectives and projects in Zeeland (development planning *avant la lettre*!)

The research and design work for the urban region of Vlissingen (Flushing) and Middelburg in the province of Zeeland in the Netherlands was accomplished under the supervision of Paul Drewe (Drewe *et al.*, 1991/1992). The goal of this study was twofold: (1) to position Vlissingen/Middelburg regionally and (inter) nationally and (2) to identify strategic projects. Mixed scanning was chosen as the most appropriate approach to achieve this goal. The approach is presented in Illustration 6.2.

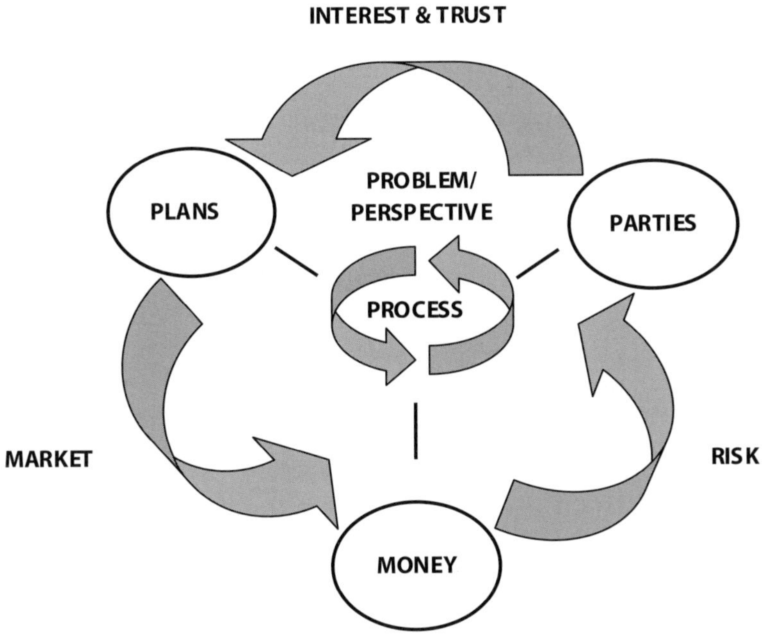


**Illustration 6.1: Development Planning in a nutshell (Twijnstra Gudde, 2003)**

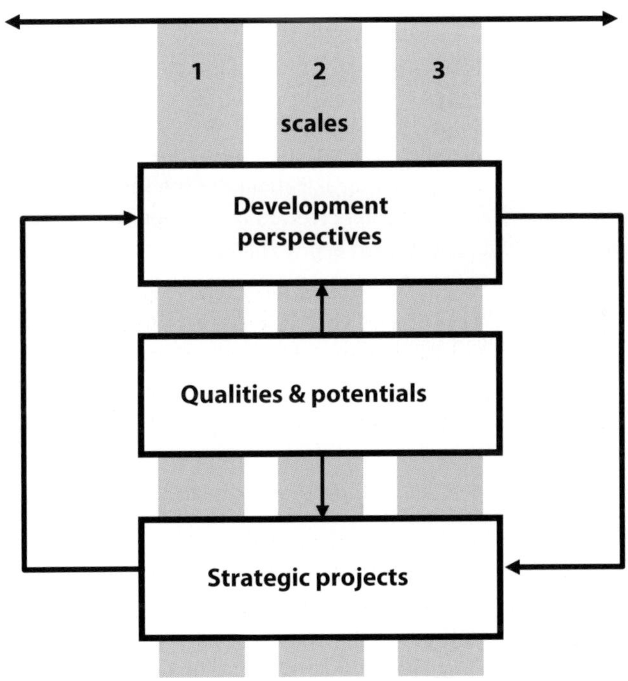


**Illustration 6.2: The Zeeland approach in a nutshell (Drewe *et al.*, 1991/1992)**

The work was organised in three steps:

1. Qualities and potentials: analyses of economic, ecological and social-cultural qualities were used as a base on which to gauge, development perspectives and strategic projects simultaneously. Alternative perspectives and projects were important considerations in this step.
2. Matching development perspectives and strategic projects: to prevent them from remaining too abstract, development perspectives are linked to concrete projects. Conversely: projects are only strategic if they are related to each other within the framework of a development.
3. Different levels of (spatial) scale: the working process ought to be structured via relevant levels of scale. Only in that way is the impact of developments on higher scale levels on Vlissingen and Middelburg conceivable, as well as their positioning.

The project was successful in supporting Vlissingen and Middelburg's goal, under the *Vierde Nota over de Ruimtelijke Ordening* [Fourth Memorandum on Spatial Planning], of combining their strengths to become a 'city region'. Many strategic projects included in the proposals were later implemented. We can mention here the tunnel connection under the Western Scheldt, the merging of the two harbour organisations to become Zeeland Seaports, and, last but not least, the start of the International Scheldt Faculty (ISF) as an EU Commett II projects. The strong advice to focus not only on industrial development and physical infrastructure, but also on an infrastructure for science and innovation, has been important in shaping a climate which facilitated the recent start of the English-speaking Roosevelt Academy. This was seen as a natural step in the development of the region (Goedman, 1990). Also relevant here is the way the portfolio of strategic projects was presented (see Ill. 6.3). Most projects were related to one level of scale, but, in many cases, projects had important implications for other scale levels as well. In Illustration 6.3, this is indicated by different dotted lines.

Some years after the publication of Drewe *et al.*'s Zeeland study, the *Rijks Planologische Dienst (RPD)* [National Spatial Planning Agency] started a study, to facilitate their own work, on the relevance of mixed scanning, especially in relation to gaining support for computer aided planning support devices. In this study, mixed scanning was perceived as a method for combining the strategic and operational foci of spatial planning. The so-called diagonal planning used in the regional responses to the Fourth Memorandum on Spatial Planning is perceived as an (implicit) mixed-scanning method (Van Biezen & Koedam, 1993). In the curriculum of the Chair of Spatial Planning of the Faculty of Architecture at Delft University of Technology, mixed scanning is still perceived as an important component of the planning cycle (Hulsbergen & Kriens, 2000) (Illustration 6.4). A characteristic feature is the systematic SWOT analysis. The interactive relation between development perspectives and strategic projects is operationalised by specifying a 4-part mini cycle: development vision, development aims, strategic projects and a feasibility study. This is mixed scanning in a more strict sense than that conducted in the Zeeland study.

## The implications

Mixed scanning, which focuses on (re)combining perspectives (spatial vision) and projects, can be applied in many ways to recent developments in planning, especially to linking Structure Schemes [*structuurschema's*], projects and land-use plans. The new Spatial Planning Act [*Wet op de Ruimtelijke Ordening, WRO*] – to be implemented in 2007 – contains new rules for spatial planning. These rules focus on combining goals for developing efficient and transparent policy with those for simplifying the legislation for maintaining strict law and order. It is relevant here to mention that on each of the Dutch administrative levels (national, provincial and local), there are legislative frameworks for three

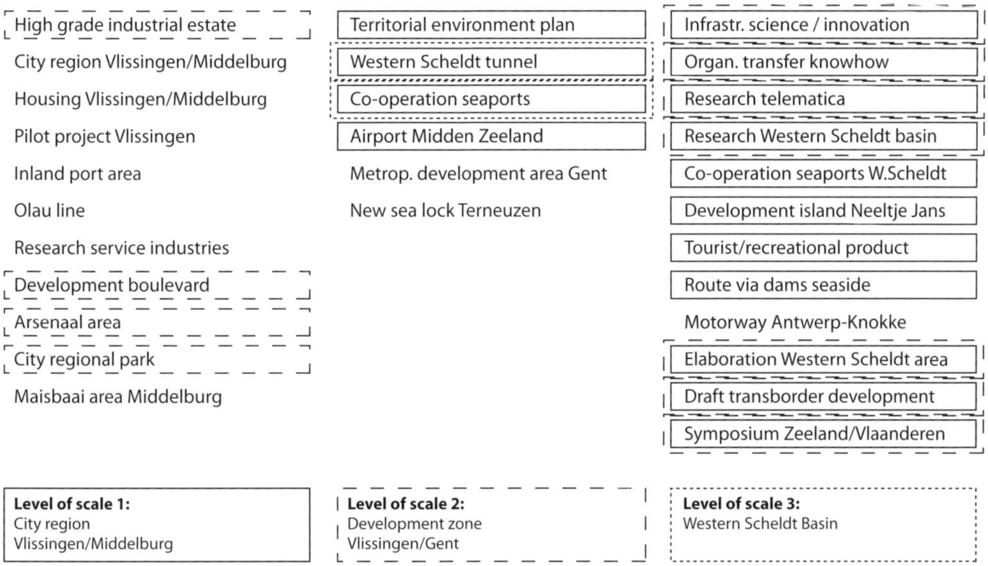


**Illustration 6.3: Portfolio of projects grouped under the three levels of scale; different dotted lines indicate different levels of scale (Goedman, 1990)**

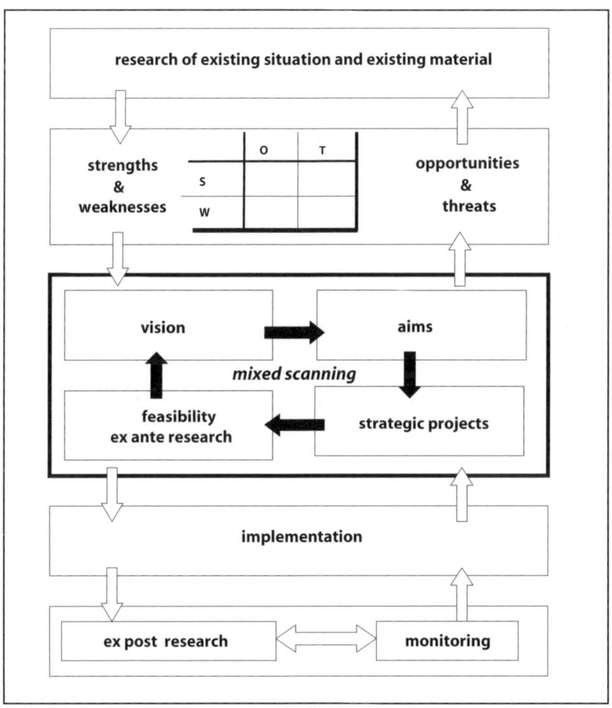


**Illustration 6.4: Mixed scanning as a component of the planning cycle (Hulsbergen & Kriens, 2000)**

components of spatial planning: Structure Schemes, Project Decisions, and (digital) Land-use Plans. Structure Schemes are the strategic policy documents that will replace the present formal national, regional and local plans. It is hoped that (digital) Land-use Plans will facilitate both Spatial Developments and legal maintenance, simultaneously. A decision-making procedure for a project can be combined with a decision to revise a land-use plan (this is possible on all three levels of scale). On the national level, this provision will replace the current National Projects Procedure.

Related to this is a recent evaluation, undertaken by a Dutch Parliamentary Commission, of defects in decision-making procedures for national projects, in ventures such as the high-speed passenger train, and the Betuwe Track for goods trains. A major conclusion from this evaluation was that national projects of this kind should be evaluated *ex ante* within a vision of the structure that allows different alternatives (e.g. a mixed scanning approach). This new thinking will clearly facilitate a more efficient and interactive (re)combination of development perspectives and strategic projects on the three levels of scale, than is possible under existing regulations, even though this does not extend, as yet, to project planning (Ministerie van Volkshuisvesting, Ruimtelijke Ordening en Milieubeheer, 2003).

## The growing importance of spatial discourses in spatial disciplines and doctrines

### The issue: The break-through of discourses as frames of reference

The importance of discourse was first stressed by Hajer, as early as 1989 (Hajer, 1989), but it was not until Hidding, Needham and Wisserhof's (1998) review of discourses on the city and the countryside that discourse came to be considered as a potential frame of reference in spatial planning. These new frames of reference were first applied in publication of the Ministry of Housing, Spatial Planning and Environment, to sketch the development of the European Spatial Development Perspective (ESDP). This was viewed as the integration of three discourses – cohesiveness, sustainability and competitiveness – at EU level (cf. Goedman, 2003; Ministerie van Volkshuisvesting, Ruimtelijke Ordening en Milieubeheer, 1999; Faludi, 2004).

Discourse analysis, as a tool for studying spatial perceptions and proposed interventions, was utilised by Zwanikken in his study of these discourses, published under the title *Ruimte als voorraad? Ruimte voor variëteit!* [Space as stock? Space for variety!] (Zwanikken, 2001). Among the many references that Zwanikken includes in this study is one of Hajer: "a specific ensemble of ideas, concepts and categories that are produced and transformed in a particular set of practices and through which meaning is given to physical and social realities". This definition corresponds very closely with Hidding and Wisserhof 's perception that discourses are social constructions produced by the interaction of researchers, planners, designers, policy-makers politicians and stakeholders. In the early stages of constructing a discourse, they call this a 'proto-discourse' (Zwanikken, 2001:44).

In Denmark, Vagnby and Jensen have presented an interesting review of Danish policy development for housing improvement and urban renewal using an analytical framework that deals with doctrines and discourses. In their opinion, a discourse is a social process embedded in space and time. It is transformed into social reality by the action of social agents in institutional settings. Their approach is interesting, although they tend to see discourses as an alternative to the doctrines defined by Faludi and Korthals Altes, rather than the more complementary frames of reference that we talk about in this chapter (Korthals Altes, 1995; Vagnby & Jensen, 2002). Zwanikken defines discourses as "frames of reference with which actors interpret and construct physical and social phenomena". Although I agree with this short definition, it should be emphasised that this interpretation includes (possible) intervention strategies. I also agree with the clear division that Zwanikken makes between the different frames of reference of disciplines (i.e. paradigms) discourses and doctrines (Zwanikken, 2001:45).

## The case: 'Towards a new geography' and 'spatial perspectives'

The *Wetenschappelijke Raad voor het Regeringsbeleid* [Scientific Advisory Council to the Government] has not only published the report *Ruimtelijke Ontwikkelingspolitiek* [Spatial Development Politics], but also a review on the outcomes of three disciplines in connection with important transitions in spatial structure at different levels of scale. Based on recently published results produced by the economics, sociology and geography scientific communities, the authors presented three main trends towards what they call 'a new geography' (Wetenschappelijke Raad voor het Regeringsbeleid, 2002):

- Ambiguous, changing spatial structures, brought about by the concentration of certain social and economic activities and the dispersal of others (with ICT as the main driver for both);
- The changing perception of space: ideas about the characteristics of a place are becoming ever more diverse, and less directly defined by the residents of that place;
- Important effects on the physical 'design' of places: e.g., historic inner cities, outlet centres, gated housing environments, etc..

They conclude that, although these developments are well addressed in the analyses of the *Vijfde Nota over de Ruimtelijke Ordening* [Fifth Memorandum on Spatial Planning], this is not so in its dominant discourse. In their opinion, the town/country dichotomy – so prominent in Dutch spatial planning tradition – is outdated.

## 'Spatial perspectives' revisited

Officially, the Spatial Perspective 'City Land (plus)' was, and still is, the starting point of the *Nota Ruimte: Ruimte voor Ontwikkeling* [National Spatial Strategy: Space for Development] (Ministerie van Volkshuisvesting, Ruimtelijke Ordening en Milieubeheer, 2004). The starting for 'City Land (plus)' is to make a clear distinction between town and country, to highlight the differences between the built environment and open space. This, of course, is interesting in connection with the critical comments of the Scientific Advisory Council presented above.

In the 'Netherlands 2030' project, three other 'perspectives' were discussed:

- 'Palette', in which citizens and companies have a large freedom of choice in where they locate;
- 'Park Landscape', in which the characteristics of the Dutch landscape are the starting point for the mix of town and countryside;
- 'Flow country', in which the future spatial structure of the country is based on water and traffic flows.

Many stakeholders discussed and commented on these perspectives, focusing mainly on the following aspects: economics, sustainability, (social) liveability and culture. In the end, most participants favoured the 'City Land' perspective: "Sparing land use and the conservation and strengthening of nature and open areas are widely accepted policy targets. Although this perspective meets some resistance because it is presumed that it will bring about lower housing quality, 'City Land' is the perspective which most people support" (Ministerie van Volkshuisvesting, Ruimtelijke Ordening en Milieubeheer, 1998). Thus it was obvious that 'City Land' would be chosen as the starting point for developing new policy.

How should this be understood? In the first place, the perspectives of the 'Netherlands 2030' project should not be perceived as alternative proposals for the spatial structure of The Netherlands, but as four spatial discourses, each having substantial social support, which together, combine different ways of understanding and intervening. Thus, it is more a matter of thinking in terms of a coalition of discourses, than of choosing between alternatives (Hajer, 1993). In fact, this is exactly how the new policy documents were formulated. More implicitly, the urban network concept, including some components of 'Park Landscape' and 'Flow Country' can be traced in the layer approach (for the layer approach, see also chapter 19 by Jeroen van Schaick).

However, other perceptions are also possible. Drewe's proposal was to link three of the perspectives to different ministries. 'City Land' to the Ministry of Housing, Spatial Planning and Environment; 'Park Landscape' to the Ministry of Agriculture, Nature and Food Quality and 'Flow Country' to the Ministry of Transport, Public Works and Water Management. This is questionable, because by so doing, he implicitly qualifies them as doctrines (Drewe, 2002). Healey suggests that the power struggle between the ministries, and the much weakened position of the Ministry of Housing, Spatial Planning and Environment, has "reinforced a retreat in policy content in the Fifth Memorandum on Spatial Planning, in which the innovative spatial vocabulary of the analyses (in terms of the layer approach and urban networks) has been captured, reinterpreted and positioned back into the established spatial planning policy discourse (which makes a clear division between 'urban' and 'rural' areas)" (Healey, 2004:54).

## The implications

Healey, in her description of the role of strategic spatial planning, also stresses the possibilities of coalitions on lower levels of scale:

> "Articulating a strategic orientation with a spatial dimension may have direct material benefits in capturing resources from a higher government level. It may also help the formation of active coalitions among an array of small municipalities, or mobilize active stakeholder groups important to an area's development who can move perceptions (and hence actions) from just 'being in a area' to a recognition of an area as having an identity (a city or region 'in itself'), and beyond this, to having the capacity to act 'for itself'." (Healey, 2004:45).

There is no doubt, that discourses, as frames of reference, are important tools for analysing the effectiveness of combining development, research and design. This argument is expanded further in the next section.

What we want to underline here is that the different frames of reference of disciplines, discourses and doctrines can be very helpful in gaining a clear picture of the division of labour and the extent of co-operation that might be possible between three major Dutch spatial planning institutions: the *Ruimtelijk Planbureau (RPB)* [Institute for Spatial Research], the *VROM-raad* [Council to the Ministry of Housing, Spatial Planning and Environment] and the *Directoraat-Generaal Ruimte (DGR)* [Directorate General for Spatial Policy]. We can use the terms 'understanding' and 'intervening', to illustrate the

| | understanding | |
|---|---|---|
| **Disciplines** | | Start: RPB |
| **Discourses** | | Start:VROM-Council |
| **Doctrines** | | Start: DGR |
| | intervening | |

**Illustration 6.5: Disciplines, discourses and doctrines**

specific mix of understanding and intervening within the frames of reference, in the following way (see Ill. 6.5). The starting point for the work of these three institutions is, or should be: doctrines, in the Directorate General for Spatial Policy; discourses, in the VROM Council; and disciplines in Institute for Spatial Research. The way to accomplish this was clearly illustrated in a meeting held in October 2004, under the title: 'Creativity a production factor for urban regions?', during which complementary scopes were presented (Ruimtelijk Planbureau, 2004b; Ministerie van Volkshuisvesting, Ruimtelijke Ordening en Milieubeheer, 2004; Ministerie van Volkshuisvesting, Ruimtelijke Ordening en Milieubeheer *et al.*, 2004).

A second implication of the new focus on spatial discourses is connected to planning and policy concepts, such as 'mainports' and 'urban networks'. This can be illustrated by Van Duinen's 'Planning Imagery' (Van Duinen, 2004). In this study, she analyses the emergence and development of new planning concepts in Dutch national spatial policy. Her analyses of the meaning and images of, and support for, the 'mainport' clarify that, in the course of its development, the mainport concept has meant different things to different people:

"depending on the problem definition, the development of mainports was subsequently approached as a *transport*, *economic* or *spatial* issue, ranging from a traffic-generating node, a traffic-economic generating node, or a traffic-economic generating node that is spatially structuring for the working and living climate. (...) What is interesting, however, is that these problem definitions did not mutually conflict, but reinforced each other (...) The underpinning mutually reinforcing frames (...) led to three coalitions accumulating in one super coalition that all united under one and the same banner of the mainport from different motives, backgrounds, and arguments." (Van Duinen, 2004:105-109)

Van Duinen uses a triangular figure to illustrate the approach of the mainport concept (see Ill. 6.6a). Inspired by Van Duinen's *ex post* analysis, we can apply her approach to the urban network concept, in a more *ex ante* manner. A quote from the English summary to the National Spatial Strategy can be used to give a short motivation for introducing urban networks:

"The Netherlands is developing into a network society and a network economy. On the one hand, individualisation continues to progress; on the other hand, all those individuals are increasingly closely interconnected in numerous networks. This development also has major consequences for spatial planning. There is more and more coherence between the various cities and urban areas. The government applauds this development towards urban networks. Partnerships between such networks expand the support base of public facilities and services and open up opportunities for the optimal use of scarce space. To respond to this trend, the national government has designated six national urban networks." (Ministerie van Volkshuisvesting, Ruimtelijke Ordening en Milieubeheer, 2004:14)

The core, label and problem definitions can be reduced to the components shown in Illustration 6.6b. An important question for the future of the Urban Network Concept is: whether and how a development similar to that of the mainport coalition(s) can bring about a coalition of discourses between the three problem definitions, the related institutions and actors, and the solutions they have in mind. Van Duinen has analysed other policy concepts, such as the 'western wing of the Randstad' and the so-called corridor concept, both of which were unsuccessful in bringing about such a coalition, due to contradictions between the underpinning discourses and their supporters.

# Combining design, research and development in planning practices

## The issue: Combining research and design

In another study by the Scientific Advisory Council to the Government, *'Zijlicht op toekomstonderzoek'* [Side-light on future research], (unfortunately less well known than *'Ruimtelijke Ontwikkelingspolitiek'* [Spatial Development Politics]), different ways of combining and dividing 'knowing', 'wanting to' and 'being able to' were analysed (Wetenschappelijke Raad voor het Regeringsbeleid, 2001). Two of the approaches used in that study are relevant here:

- An analysis of discourses and institutions in the South Wing of the Randstad carried out by Hajer and Salet (*ibid.*:15-111);
- An analysis of combining research with design (researching while designing) that was focused on analysing and (re) designing the water system of the so-called Deltametropolis (another name for the Randstad), carried out by Blom, Kuypers and Lemmers (*ibid.*:225-322).

The first research project analysed the process of wrestling to form a coalition of discourses to enhance the range of living environments, bearing in mind the position of that region with regard to international (economic) competition. With so many actors and institutions involved, this was clearly quite an undertaking! In the second study, the practice of surveying while planning is analysed within the context of an assignment to redesign the water system of the Deltametropolis and, in particular, the Zuidplaspolder. Both studies have been important as a reference and input for new planning practices in this area.

## The RZG Zuidplas

In projects such as the Zuidplas, some of the innovations suggested in this chapter are already being implemented. This is a project aimed at restructuring the area between the cities of Rotterdam, Zoetermeer and Gouda (RZG area) and, in particular, the Zuidplas polder. It is one of two areas, targeted for transformation, which were 'taken out' of the greenbelt of the Ranstad conurbation, the so called 'Green Heart', and 'transferred' to the South Wing project.

Moreover, in the RZG Zuidplas project, both of the approaches listed above – trying to form a coalition of discourses and combining research and design – are being used. This is why the first planning results of the RZG Zuidplas project are so important (www.driehoekrzg.nl). In this project, an inven-

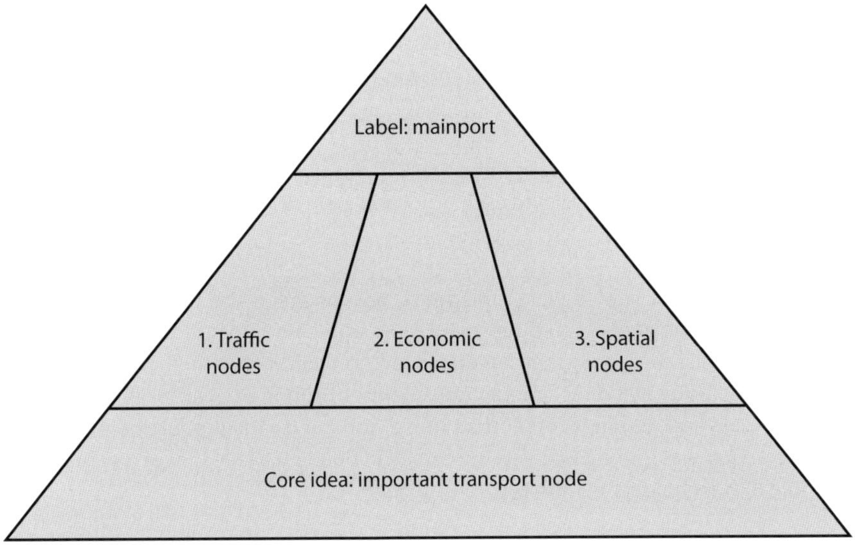


**Illustration 6.6a: The discourses that underpin the mainport concept (Van Duinen, 2004)**

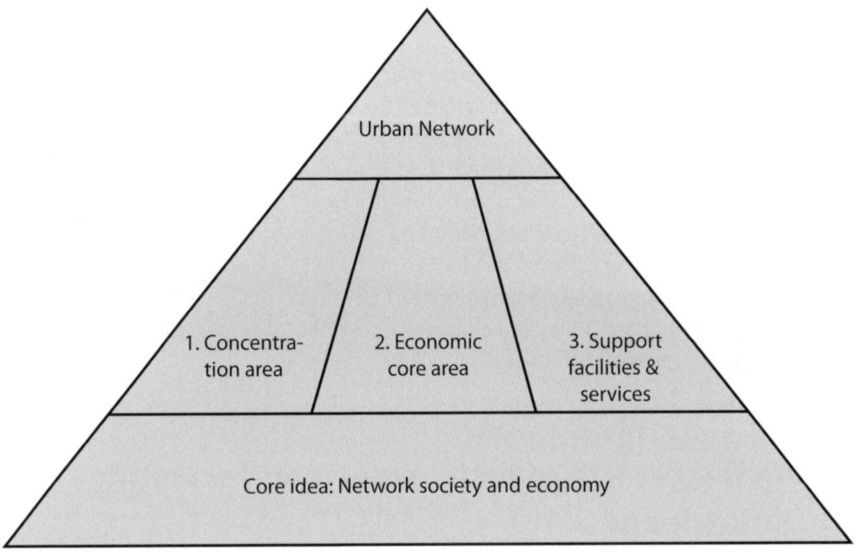


**Illustration 6.6b: The discourses that underpin the urban network concept**

tory was first made of the qualities and potentials of the area. This was achieved by surveying and designing (potential solutions) using the layer approach, and by using the results of earlier work on the Deltametropolis networks (see Ill. 6.7). The outcome was the so-called RZG Zuidplas Atlas (Stuurgroep Driehoek RZG Zuidplas, 2003).

The planning process was organised along three 'tracks '– Plans, Governance and Finances – from the very beginning. These tracks ran parallel to each other and were very similar to those proposed in the reviews of development planning.

An interesting feature is the way in which the interaction between the tracks was organised. The idea was that the different tracks would represent language communities, who would work apart, but yet together. 'Translators' would then be employed, to coordinate the interaction between the tracks. Only on one occasion was direct interaction organised, during a so called 'visualisation day'. On this occasion, planners and designers other than those working on the production of the Atlas were invited. This interaction was found to be very successful in building a coalition of discourses between the twenty-three (!) stakeholders/institutions.

In the next phase of the Structural Plan with these three municipalities (within the context of existing legislation), the intention is to put the above ideas into practice by interrelating development perspectives with the various part plans of the RZG Zuidplas project as a whole.

### The implications

The RZG area project is an important part of a larger plan, the so-called *Zuidvleugelprogramma* [South Wing Programme]. This is a portfolio of twenty projects for the South Wing of the Randstad. This South Wing Programme is one of four portfolios in the National Spatial Strategy. (The other portfolios are: the North Wing, the Green Heart of the Randstad, and the south-eastern part of the Netherlands). This portfolio of projects was used as the point of departure for developing a frame of reference. The first step was to make a quick scan of related documents and available research, so that any planning problems could be identified. Then, to collect input for the frame of reference, different planning sessions were organised. Because a frame of reference compiled in this way consists of a number of component parts, the aim was to make it possible to manage the portfolio of projects in many different ways (output). This approach is presented in a nutshell in Illustration 6.8.

## Facilitating the growth of tacit, explicit and scientific planning knowledge

In periods of instability, the frames of reference need to be partly re-developed. To identify 'best practices', the proposals of the Scentific Council to the Government have proved to be important, and, because they place emphasis on the necessity of system innovations, the Institute for Spatial Research's development planning analyses also use different planning practices as their starting point (Ruimtelijk Planbureau, 2004a). The work of Habiforum, a Dutch organisation working with 'communities of practice' (CoP) oriented towards innovation in multiple and intensive land use, is relevant in this respect. This strategy makes a distinction between tacit, explicit, and scientific knowledge. The core element of

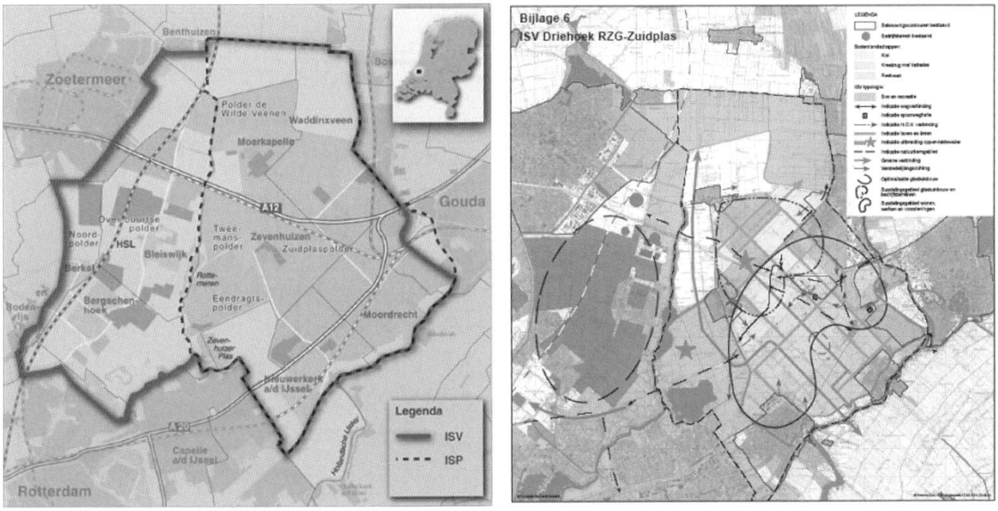


**Illustration 6.7: A review of RZG Zuidplas (Stuurgroep Driehoek RZG Zuidplas, 2003:4,64)**

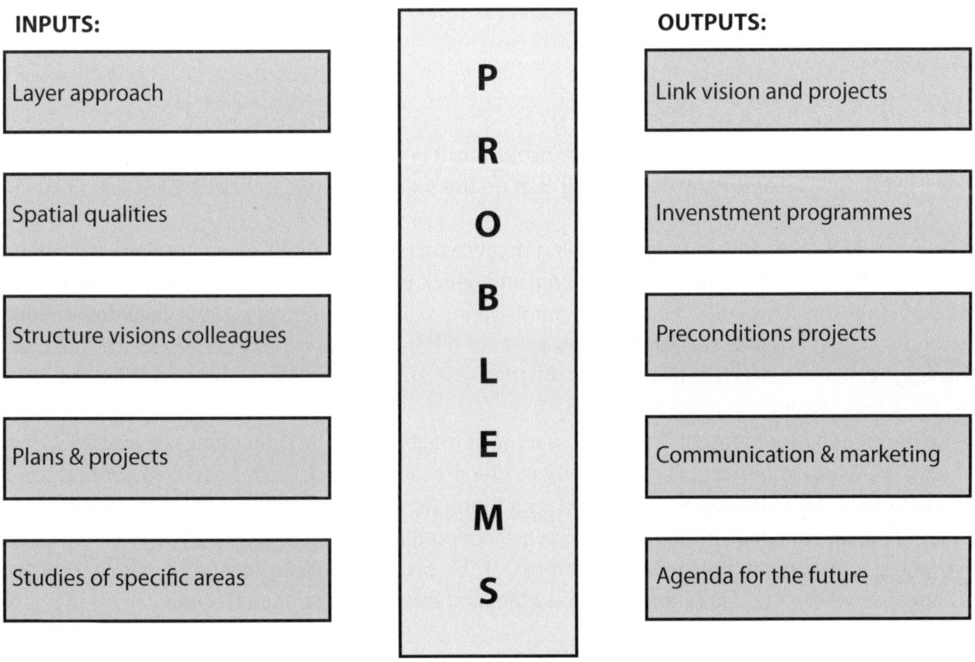


**Illustration 6.8: The planning approach in the South Wing Programme**

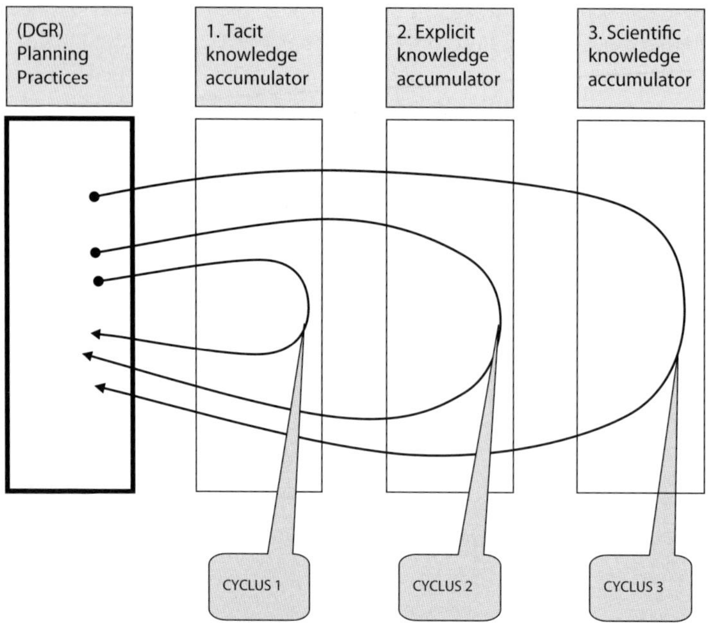


**Illustration 6.9: DGR Planning Practices: developing and storing planning knowledge**

the CoP strategy is the predominance and priority of tacit knowledge as the starting point for efficient ways of learning – 'second-hand knowledge' is of no use for actors working in their own practices.

On the basis of its own development planning practices, the Directorate General for Spatial Planning supports new combinations of design, research and development. This is seen as a way of supporting spatial planning, which is still viewed as the 'undisciplined discipline' (Pinson, 2004). By differentiating between tacit, explicit and scientific knowledge, the different cycles and accumulators of planning knowledge can be located (see Ill. 6.9).

The exchange and development of tacit knowledge as practiced by The Directorate General for Spatial Planning is organised into a '3 O Community of Practice', (from the Dutch: *Onderzoek, Ontwerp and Ontwikkeling*, meaning Research, Design and Development). Below, we reflect on our experiences with different planning practices in this type of practice community in:

- working with the strategic planning context of the project portfolio for the South Wing of the Randstad, where the role of the Directorate General for Spatial Planning is strategic (see previous section);
- working with an informal planning context related to the development of the Green Heart, where the role of the Directorate General is advisory;
- working with a selection of so-called example plans selected by provincial authorities for developing a Development Planning approach, where the role of the Directorate General is facilitatory.

In recent years, planning support devices have been developed to support a better integration of environmental and spatial planning processes. One of these is *'MIRUP'* (an acronym for *Milieu in Ruimtelijke*

*Plannen*, meaning Environment and Spatial Planning), the structure and content of which was developed by the urban region of Haaglanden (the region containing The Hague and surrounding municipalities), with the support of the Ministry of Housing, Spatial Planning and Environment (Stadsgewest Haaglanden, 2003). This product has been transformed into a digital support system consisting of five portals: planning process, layer approach, typology of areas, qualities, and measures. The information behind these portals can be used by different planning groups. What is important for this chapter is that the development of this product shows that combining different components of knowledge behind the different portals and their related tools has come to be recognised as a necessity in every planning practice. Explicit knowledge can then be filtered from the tacit knowledge resulting from these practices.

Finally, in one of the projects of the Directorate General carried out in 2004, discourse analysis was applied as a scientific tool for gaining a better understanding of the effectiveness (politically) of combining research and design to bring about coalitions of the relevant parties in a project. This initiative was based on the planning practice of Sijmons and H+N+S landscape architects (Sijmons, 2003; Gomart & Hajer, 2002; Hajer *et al.*, 2004). The conclusions, so far, stress the importance of interweaving content and process-management development, mixing research, participation and plan development, and mobilising different representation media. It is also important to realise that the first product is a story (line), not a spatial plan; that the second product is a discourse coalition; and that there can be no talk of operational interventions (projects and plans) until the third product has been formulated.

## Concluding remarks

Inspired by Etzioni's mixed scanning approach, the task ahead is to try to find new combinations of plans, parties and projects (the 3 'P's, discussed in the third section), new interactions among disciplines, discourses and doctrines (the 3 'D's, discussed in the fourth section) and, last but not least, new ways of integrating spatial research, design and development (the 3 'O's, discussed in the fifth section). Only a few of the (27) possible connections are presented in this chapter. In particular, we have indicated that the most innovative linkages are those among the three 'middle' components – parties, discourses and design.

So there is no need to go all the way back to De Casseres, who, in the 1920s, invented the *urbanistiek* [urbanism] and the *planologie* [spatial planning] in opposition to the *stedebouwers* [urban designers] approach, which he considered as being too unscientific (De Casseres, 1926). With the more recent experiences in the spatial planning discipline, it should be possible to set aside the old dialectics and find new ways of reconciling spatial research, design and development.

### References

De Casseres, J.M., 1926, *Stedebouw*, S.L. van Looy, Amsterdam

Drewe, P., 2002, Recente ontwikkelingen in de Franse ruimtelijke ordening als inspiratiebron voor een nieuwe aanpak in Nederland; In: *Academische Reflecties*, Ministerie van Volkshuisvesting, Ruimtelijke Ordening en Milieubeheer, Den Haag, pp. 115-125

Drewe, P., J. Goedman, W. Lockefeer & G. Roeleveld, 1991/1992, *Vlissingen & Middelburg in ruimer perspectief, Portfolio van strategische projecten*, Bureau Micropolis, Middelburg (Neth.)

Etzioni, A., 1967, Mixed Scanning. A third approach to decision-making; In: Faludi, A., 1973, *A Reader in Planning Theory*, Pergamon Press, Oxford, pp. 217-229

Faludi, A., 1973, *Planning Theory*, Pergamon Press, Oxford

Faludi, A., 2004, Europese planning in historisch perspectief, Supplement to *S&RO 2004/5*

Goedman, J. C., 1990, De Internationale Schelde Faculteit in Europa; In *Liber Amicorum Prof. M. Anselin*, Gent

Goedman, J.C., 2003, Urban environments, regions and networks in transition. Layers and dimensions of sustainable urban development in the Netherlands, *Die Zukunft der Stadte, Rhein-Mainische Forschungen Heft 124*, Frankfurt am Main

Gomart, E. & M.A. Hajer, 2002, Is that politics? An inquiry into forms in contemporary politics; In: Bernward, J., Y. Ezrahi & H. Nowotny, *Looking Back, Ahead – The 2002 Yearbook of Sociology of the Sciences*, Kluwer Publishers, Dordrecht

Hajer, M., 1989, *City Politics: Hegemonic Projects and Discourse*, Avebury/Gower Publishing Ltd, Aldershot (UK)

Hajer, M., 1993, *The Politics of Environmental Discourse: A Study of the Acid Rain Controversy in Great Britain and the Netherlands*, PhD Thesis, Trinity College, Dublin

Hajer, M.A. *et al.*, 2004, *Naar een Plan dat Werkt!*, Final report on the 'Planning in praktijken' project commissioned by the Ministerie van Volkshuisvesting, Ruimtelijke Ordening en Milieubeheer, Den Haag

Healy, P., 2004, The Treatment of Space and Place in the New Strategic Spatial Planning in Europe, *International Journal of Urban and Regional Research*, 18/1, pp. 45-67

Hidding, M, B. Needham & J. Wisserhof, 1998, *Stad en Land, een programma voor fundamenteel-strategisch onderzoek*, NRLO-rapport 98/17, Den Haag

Hulsbergen, E.D. & I. Kriens, 2000, *Plancyclus*, Chair of Spatial Planning, Faculty of Architecture, Delft University of Technology, Delft

Korthals Altes, W., 1995, *De Nederlandse planningdoctrine in het fin de siècle*, Van Gorcum, Assen

Ministerie van Volkshuisvesting, Ruimtelijke Ordening en Milieubeheer, 1998, *Nederland 2030 - Debat*, Den Haag

Ministerie van Volkshuisvesting, Ruimtelijke Ordening en Milieubeheer, 1999, *Spatial Perspectives in Europe*, Spatial Reconnaissances 1999, Ministerie van Volkshuisvesting, Ruimtelijke Ordening en Milieubeheer, Den Haag

Ministerie van Volkshuisvesting, Ruimtelijke Ordening en Milieubeheer, 2004, *Nota Ruimte: Ruimte voor Ontwikkeling*, Den Haag

Pinson, D., 2004, Urban planning: an undisciplined discipline?, *Elsevier Futures*, 36, pp. 503-513

Sijmons, D., 2003, *Voldongen fictie, De potenties van het regionaal ontwerp*, Ontwerpatelier Deltametropool, Ministerie van Volkshuisvesting, Ruimtelijke Ordening en Milieubeheer, Den Haag

Stadsgewest Haaglanden, 2003, *MIRUP, Handreiking voor milieu in ruimtelijke plannen*, Den Haag

Stuurgroep Driehoek RZG Zuidplas, 2003, *Interregionale Structuurvisie*, Projectbureau Driehoek RZG Zuidplas, Gouda (Neth.)

Tetteroo, N.H.M., 2004, Concerted development planning policy and implementation, *Nova Terra*, 4/2, June, pp. 6-10

Twijnstra Gudde, 2003, *Best practices ontwikkelingsplanologie*, Twijnstra Gudde, Den Haag

Ruimtelijk Planbureau, 2004a, *Ontwikkelingsplanologie*, NAi Uitgevers, Rotterdam

Ruimtelijk Planbureau, 2004b, *Kennis op de kaart, Ruimtelijke patronen in de kenniseconomie*, NAi Uitgevers, Rotterdam

Van der Heiden, N., 2003, Best practices van Ontwikkelingsplanologie, Er kan al veel!, *Rom-special September 2003*

Vagnby, B. & O.B. Jensen, and co-referents, 2002, From slum clearance to urban Policy: discourses and doctrines in Danish inner city redevelopment, *Housing, Theory and Society*, 19, pp. 3-13

Van Biezen, M.P.W. & G.R. Koedam, 1993, *Mixed scanning als besluitvormingstechniek, Ruimtelijke Planvorming met behulp van GIS*, Studierapport Rijks Planologische Dienst, Den Haag

Van Duinen, L.B.J., 2004, *Planning Imagery*, PhD Thesis, Universiteit van Amsterdam, Amsterdam

VROM-raad, 2004, *Nederlandse steden in internationaal perspectief: profileren en verbinden*, VROM-raad Advies 043, Den Haag

Wetenschappelijk Raad voor het Regeringsbeleid, 1998, *Ruimtelijke Ontwikkelingspolitiek*, Sdu Uitgevers, Den Haag

Wetenschappelijk Raad voor het Regeringsbeleid, 2001, *Zijlicht op Toekomstonderzoek*, Vier casus, 2nd. Working Document, Stuurgroep Toekomstonderzoek en Strategisch Omgevingsbeleid, Wetenschappelijk Raad voor het Regeringsbeleid, Den Haag

Wetenschappelijk Raad voor het Regeringsbeleid, 2002, *Stad en land in een nieuwe geografie, Maatschappelijke veranderingen en ruimtelijke dynamiek*, Voorstudie van Asbeek Brusse, Van Dalen en Wissink, Sdu Uitgevers, Den Haag

Zwanikken, T.H.C., 2001, *Ruimte als voorraad?, Ruimte voor variëteit, De consequenties van discoursen rondom 'ruimte als voorraad' voor het rijks ruimtelijk beleid*, PhD Thesis, Radbout Universiteit, Nijmegen

**Websites**

Ruimte X Milieu, http://www.vrom.nl/pagina.html?id=17819

RZG Zuidplas, http://driehoekrzg.nl

# cummings plans qualities

Sierksma

Let me commence with a declaration of honesty. This little essay intends to fool the reader into believing that it is all about planning. Of course, it is not. It is an essay on poetry and on the inimitable Paul Drewe. If by chance the reader is clever enough not to be fooled, I beg of him what aesthetics so gently terms as 'a wilful suspension of disbelief'.

What else is poetry but the routing of words by way of punctuation. The bard is a master planner negotiating meaning, balancing each word on his razor's edge. No blue prints, mind you! The result, a poem, is a many-splendoured thing. To be read in many ways – planned nevertheless.

The poet uses an array of traffic signs, allowing his reader to navigate his verse. Then again, he is free not to use them at all. Take for instance the title of this piece, cummings plans qualities. It indicates precisely what the poet e. e. cummings does. You will note the lack of apostrophes. I might also have written cummings' plan's qualities, which signifies something different. In that case cummings is said to have a plan – a plan with qualities.

This absence of apostrophes makes my title ambivalent, suggesting both meanings. It is a poetic device, so to say, this time used by an essayist. Then again, an essay in itself is a little ambivalent, its writer trying out an idea on the reader in an endeavour to make him see his perspective – of course one of many. Unlike poetry, but very much like a scientific article, an essay must not be ambivalent in using commas, apostrophes and dashes. To indicate punctuation's importance, I relate the following little episode.

Some years ago I published a book with pieces on English Gardens in Germany. Being the son of Friesian father, who between '43 and '45 fought in the Resistance, you will understand that I was pleased with its title. Tropical North – a Hot Reception in the Country of our Eastern Neighbours. To a friend, a translator by profession, I sent a copy.

Within a week the book was returned. It had changed colour. Added in red, on each page many a comma was scratched - according to my critic, they were superfluous or even irritating. Some absent commas were added, again in red. It took me a day to reach that indispensable meta-level from which a writer is able to read his own text through the eyes of a critical outsider. The man had been so right – too bloody many nauseating little flies, buzzing amongst my words, making many a sentence illegible.

When somewhat later, together with a colleague of mine, I was writing another book, I decided to become radical in matters linguistic. I left out all but a few sparse commas. I sent the co-author my part of the manuscript, which was returned within a day. On it, once more in red, were stamped the words 'Unreadable – where are your bloody commas?!' And so right this reader was!

From then on I became a comma nut. In whatever text I read, whether a novel, an essay or a poem, I scratched out comma after comma, or I added one. Reading became a slow, tortuous activity. Then, one day, I opened a little book with verse by Emily Dickinson. Its introduction relates how, in earlier editions, the poetess' punctuation had suffered from the editor's hand, which had replaced all her dashes with… commas. The new edition presented the poetry in it original punctuation, rerouting as it were her lines. An example:

'Speech' – is a prank of Parliament –
'Tears' – a trick of the nerve –
But the Heart with the heaviest freight on –
Doesn't – always – move –

Indeed – what would be left of the poem without these desolate dashes! Just contemplate the one at its very end. They also suggest a solution to my own problem – why not simply alter all my commas into dashes…

Poetry, like urbanism, is planned mystery. Author and planner both want to get something across, to communicate the meaning bestowed on a word or a street. However, the choice of poetical verse or a street as the meaning's vessel tends to obscure the author's intention. It can never be as clear as analytical language can be. Thus, both poet and planner can never be all that sure of themselves. They 'rage against the dying of the light'.

A good poem is as full of sense as it is senseless – it mirrors man's life, which as we know is full of sound and fury - not signifying all too easily. Just like our cities. The critical reader must be sympathetic with the poem he reads. However, where a poem may fully succeed, even the best urban plan fails both town and citizen - not all too badly, though. Thus, making sense of an urban plan demands irony, as well as an indispensable sense of disappointment.

The urban critic, who researches such plans, must be kind and critical – he cannot be ruthless, though he must. To survive this double bind, the urban critic himself must be punctuated, like good verse. To know when you can be severe - and to know when not.

Paul Drewe is such an amiable man – a grandmaster of irony, both distanced and involved. In the unpoetical environment of our university the two of us once met, immensely pleased to recognize in one another a lover of e. e. cummings' poetry. That poet often does not punctuate his lines at all – or does so in strange ways that may teach the urban critic and the essayist reserve.

progress is a comfortable disease:
your victim (death and life safely beyond)

plays with the bigness of his littleness
- electrons deify one razorblade
into a mountain range: lenses extend

unwish through curving wherewhen till unwish
returns on its unself.

# Part II

## Spatial Developments and their Societal Effects

# Introduction to Part II

Part II: *Spatial Developments and their Societal Effects*, takes the opposite point of departure from that in Part I. Urban and regional changes and developments can deeply influence the way space is used and therefore people's short and long-term opportunities. Spatial developments can take many forms and have a diversity of functions. To gain a good view of what happens, the spatial scale of reasoning about developments and effects must be unambiguous. The selection of experiences and theories from different countries and cultures covered in this second section of the book shows that in order to understand their effects and to make generalisations that can be applied in the future, or in other places, a thorough knowledge of the spatial changes is needed.

A major spatial trend is the change in urban environments. Peter Roberts focuses on *Urban and Regional Regeneration: Principles, Practice and Lessons from Experience in the United Kingdom*. Indeed, the evolution, scope and content of regeneration policies and initiatives in the UK are important sources of information and experience. The spatial scale of the regeneration has increased, and is nowadays often city-wide or even regional in extent. The increased complexity calls for comprehensive and sophisticated governance and accountability arrangements, local or regional partnerships, and community-based engagement and management. To link spatial developments to societal needs, comparative national and international research and the education and training of regeneration professionals are requirements that appear to be needed more than ever before.

In *Networks and Urban Planning: the Evolution of a Two-way Relationship*, Gabriel Dupuy looks back at the development of running water, sewerage, power, transport and communication networks from the mid-19th century onwards, noting that they had little influence on the work of urban planners and designers at that time. Nowadays, confronted with the new information and communication technologies (ICTs), history seems to be repeating itself. Urban planners and policy-makers, who in the meantime have learned to work with the mid-19th century networks, are now being slow to respond to the opportunities offered by the ICTs. The difference, however, is that the pace of the current technological change is extremely fast, and it is developing in a large, changeable competitive, free-market environment.

Approaches for dealing with spatial and social processes are discussed in *Does Space Matter? Spatial and Socio-economic Segregation in the Oude Noorden District of Rotterdam* by Paul Stouten. Because architectural constructions can be duplicated in different environments, this leads to the idea that physical space is irrelevant. At the same time, however, the increasing urban spatial and social complexity cannot be denied. The analyses of the housing situation in the Oude Noorden district, and urban policies in Rotterdam, clearly show that recent housing and spatial approaches in that city are causing counterproductive controversies. Even though not all problems can be solved on a neighbourhood scale, investments for social programmes need to be linked to public-space improvements.

Roberto Rocco's *Reason, Capital and Urban Space: The New Role for Urban Planning in Triggering Societal Change* deals with the perception of urban space in the last decades of the 20th century. The present 'aesthetisation' of the public spaces and their reduction to a 'mathematical' construction excludes and alienates underprivileged population groups from city life, city administration and urban planning itself. The participatory Porto Alegre model in Brazil, and comparable approaches elsewhere, inspired by this project, show that there are alternatives that can be given priority and developed further to strengthen the public sphere, without excluding urban residents and users.

Finally, Joost Schrijnen in *Spatial Developments in the Netherlands, 1975-2005 – Scale Increase, More Actors, More Disciplines* analyses Dutch spatial policies and practices since 1970. The context for urban planning and design has changed drastically during this period, as testified by the series of national Memorandums on the spatial future of the Netherlands, the debate on urbanisation strategy and the form and function of the delta region and peripheral parts of the country. All participants at all levels of the planning process are subject to redefinition. To maintain its practice, the discipline of urban planners and designers needs to take a firm position within this changing context.

# 7 Urban and Regional Regeneration: Principles, Practice and Lessons from Experience in the United Kingdom

Peter Roberts

## Introduction

Regeneration is now a well-established and widely-accepted area of policy activity, which attempts to deal with a range of physical, environmental, social and economic problems encountered in urban and rural settlements. Although chiefly associated with urban improvements, regeneration initiatives have also been extended to rural areas and, as a consequence, it has increasingly become an important method of spatial intervention at various spatial scales. This extension of the regeneration mode of policy design and delivery from the local to the regional level is indicative of the difficulties encountered in attempting to co-ordinate policy interventions and initiatives, be they urban or rural, and of the need to ensure the most effective overall use of resources.

In the United Kingdom (UK), regeneration initiatives are now considered to be essential tools for the planning and delivery of comprehensive solutions to problems associated with the multi-causal decline of local and regional areas. Put simply, regeneration seeks to deal in a co-ordinated way with a range of issues and is aimed at the establishment of lasting solutions. This latter point reflects an essential characteristic of regeneration: it is not about providing a short-term, temporary or 'quick fix' solution, rather, it is intended to provide the foundations for a permanent resolution of the problems encountered in an area and for the long-term progress of the place to which the regeneration action has been applied. This essential distinction between 'quick fixes' and long-term, lasting solutions has also been recognised as challenging many prevailing attitudes towards political and financial policy. Regeneration, at least in an ideal sense, cuts across many established constraints and demands the introduction of a much more comprehensive and certain approach than is normally encountered. Furthermore, regeneration solutions, if they are to offer a lasting approach, should be based upon an established partnership or collaboration between the public, private and voluntary sectors, and this characteristic can be used to enable the more efficient and effective design and delivery of a whole range of policy initiatives.

Three key topics are explored in this chapter. Each of these topics is worthy of a chapter in its own right and, as consequence, all that can be offered herein is an introduction to the subject. Following this introduction, the next section will provide a brief summary of some of the formative factors and forces that have led to the establishment of regeneration as a discrete area of policy activity; this section will consider the evolution, scope and content of regeneration policy and will identify a number of the distinguishing features of regeneration as against other approaches to urban, rural or regional intervention. The following section will review the key principles, characteristics and structure of regeneration policy and practice; an important consideration here is the way in which regeneration can act to deliver a wide range of policy objectives in a given area. A fourth section of the paper will assess regeneration practice and will offer some lessons from practice that can be used to guide the future formulation and delivery of initiatives. The final section will present conclusions and briefly examine the possible future evolution of regeneration.

## Regeneration policy: Origins and evolution

The origins of modern regeneration activity in the UK can be traced back to the late 1980s. However, although it is generally agreed that a form of intervention in the (mainly) urban policy system, which carried the label 'regeneration', is some 15 to 20 years old, a more generic concern with the planned and managed adjustment of urban areas can be seen to have a longer history. Whilst some authors have pointed to urban (and sometimes rural) interventions which occurred many centuries ago, this earlier form of intervention was generally sporadic, partial and limited in scope when compared to the more comprehensive attempts at urban, rural and, increasingly, regional restructuring which have taken place since the middle of the 19th century.

In one sense the roots of modern regeneration can be identified in the late 19th century civic renewal movement. This broad civic renewal approach was typified by the emergency of major programmes of urban improvement which were evident at various levels of scale – some very local, others city-wide or sub-regional. Such programmes were seen both as a necessary intervention in order to reduce the problems associated with rapid industrialisation and urbanisation, such as overcrowding, unsanitary conditions and poor public health, on the one hand, and as a means of expression of civic pride and distinctiveness, on the other hand. Practical improvements in order to reduce the problems evident in the Victorian slum 'city of dreadful night' (Hall, 1988:14) were introduced alongside grand civic demonstrations of prosperity (outward and actual) and urban distinctiveness. The present author has every reason to be aware of this Victorian inheritance of regeneration vision and skills, which were honed to a high level of excellence in Birmingham under the guidance of the legendary Joseph Chamberlain. Chamberlain's 1870s programme of urban improvement was promoted through a 'civic gospel' (Browne, 1974:7) aimed at eradicating living conditions which, in Chamberlain's view, had created a situation whereby "it is no more the fault of these people that they are vicious and intemperate than it is their fault that they are stunted, deformed, debilitated and diseased" (Browne, 1974:30). This heritage of extensive civic urban intervention – it extended to the creation of municipal housing, open space, water, gas and electricity services, and even a bank – reappeared in the 1980s with the desire of the City of Birmingham to develop a comprehensive regeneration strategy, an activity with which the present author was involved.

What this brief excursion into the history of regeneration demonstrates, is that there is little that is really new. Many of the basic principles and areas of activity associated with modern urban regeneration were developed and practiced during the late 19th century; the real problem is that for much of the intervening century these basic collective understandings and skills appear to have been either lost or abandoned in the race for personal and institutional profit and advantage. It is only relatively recently that the 'lost' art and science of regeneration has emerged from the 'hidden' forgotten experience of the past. The real test now is to translate this recognition of the importance of comprehensive regeneration into permanent practice.

A number of important themes can be identified from the varied history of regeneration – both urban and regional – and these themes can be seen as the essential building blocks of much of current theory and practice. The first theme reflects the continuing presence of a concern to plan and manage the physical conditions of towns, cities and regions, especially in relation to changing socio-political and economic values. At the very heart of regeneration is the question of changing spatial function: what activities cluster in particular places and what factors drive the transformation of those functions over time? As Fainstein has argued there are significant differences of perception regarding the role and function of places: some consider places as goods or commodities, other regard places as cultural artefacts. This range of views can be summarised as the difference which is reflected in the evolution of the tension between urban areas as places for human activity and as assets, or as the 'difference between use and exchange values' (Fainstein, 1994:1). These tensions and representations of the varying interpretations of the place versus asset debate can be seen in attempts to resolve the challenges of regeneration through physical improvement – better civic spaces, improved housing, enhanced economic activity facilities, etc – and the more comprehensive management of spaces and places. In short, the physical response is one which is about creating an improved collective sense of wellbeing; a response to the changing demands made upon places by a continually evolving (post) industrial society (Roberts, 2000).

A second theme from the past, which has survived the present, reflects the concern to ensue the public health of places and the provision of satisfactory living conditions. Many of the early urban renewal schemes in the UK during the late 19th century were concerned with the replacement of unfit housing and the promotion of satisfactory dwellings. The 'model village' and 'garden city' movements represented the high points of these initiatives, considered by Ebenezer Howard (1902) to be a solution to the problems associated with both urban and rural living, in short, a garden city represented the best of town and country. Many Victorian housing developments took place in situ; an element of regeneration that has reappeared as a policy preference in recent years, and these developments were accompanied by public health enhancements. The overall consequence of a concern with housing and health was a rise in life expectancy and a major improvement in the quality of life for urban residents. Quite correctly and to be expected, this theme continues to provide a challenge to would-be regeneration initiatives in both urban and rural areas, with the added complication that housing has to both be in the correct location and affordable.

The third overarching theme is related to the latter factors – housing supply and location – and also reflects the growing concern during the 20th century to ensure that urban areas do not expand in an uncontrolled and inefficient fashion. The suburbanisation of towns and cities, especially when combined with problems of social segregation as a consequence of economic imbalance and misguided housing policies, generated massive urban growth in the period between 1918 and 1939. Hall notes (1974) that urban areas grew as a consequence of the improvement of transport and the introduction of new building techniques, with the built-up area of London trebling in size between 1918 and 1939, even though the population only increased by 2 million from 6.5 million to 8.5 million. Quite clearly this type of urban growth was both unacceptable in terms of the inefficient use of land and, perhaps more importantly, because it led to major difficulties in ensuring the provision of infrastructure and other services. As a consequence, planning controls were imposed under legislation introduced in 1947. Whilst controls over suburbanisation through the initial designation of green belts – a formal category of environmental protection for defined areas of land around towns and cities – were effective in checking undesirable urban expansion in areas of particular concern, the overall conception of green belt has been diluted over time and has become a much more extensive form of restriction than was originally envisaged. During the 1950s, 60s and 70s, green belt designations increased in scale and number, but with often perverse consequential effects: there had been containment of the urban areas, but suburbanisation had increased and land and property values had witnessed massive inflation (Hall, Gracey, Drewett & Thomas, 1973). The real problem was that the first consequence was the most cherished result, whilst the second and third effects were largely ignored. By the late 20th century the consequences of green belt policy had led to the effective fossilisation of large areas of land, often land with low ecological values. Conversely, much urban regeneration activity was on abandoned sites with high bio-diversity. A wider debate on the merits of rural land protection versus urban 'brownfield' regeneration continues, but this is outwith the scope of this chapter (Roberts, Joy & Jones, 2002). However, it is important to recognise the tensions that the containment debate have generated and the influence of such tensions on regeneration policy and practice.

At one level, the third theme is a product of the fourth theme – the search for social welfare and economic progress – although the effects of attempts at the physical control of urban areas have often been a response to factors other than those associated with socio-economic matters. Nevertheless, the various attempts made to promote social welfare and economic progress have been notable by their influence on urban (and rural) form, including the use of containment policies in order to attempt to separate urban areas and the efforts made to link homes to work through land allocations, layout and the provision of transport infrastructure. Again, in a similar manner to the three preceding themes, many elements of regeneration policy have emphasised the need to pursue the objectives of social inclusion and economic progress, sometimes at the cost of a degraded environment. It is with the latter point in mind that the final theme discussed here – sustainable development – has emerged in recent years.

Before moving to a consideration of sustainable development, it is helpful to consider a fifth theme : increasing policy intervention in order to promote regeneration. Although regeneration policies were established during the 1920s and 30s, the real impetus for major intervention was the need to enable post war reconstruction. From the 1940s to the present day the scale and scope of regeneration policy have widened to cover a growing number of topics or subject areas, on the one hand, and a range of spatial initiatives, on the other hand. Activities as diverse as rural diversification, urban health campaigns and small town cultural festivals are now seen as falling within the compass of regeneration activity. A policy 'rich' environment has ensued, but this carries with it the danger that the plethora of policies will fail to bring about effective and efficient regeneration. Scale and scope are of little account in the absence of quality strategic visions, and this danger is frequently evident.

One way of attempting to reconcile the conflicts which are inherent in many aspects of regeneration policy is to emphasise the importance of place as a central focus of concern. Rather than considering the needs, requirements and characteristics of individual subjects or topics at the outset – housing, transport, economic development, social welfare services, recreation etc – the best practice of regeneration suggests that the initial establishment of a regeneration strategic vision and plan is a pre-requisite for effective action (Burwood & Roberts, 2002). As a consequence of these concerns it is possible to identify a sixth theme – strategic vision – which has become an increasingly important motif of good regeneration theory and practice.

Finally, but most important, all regeneration efforts should be designed to help to deliver sustainable development. In one sense the demands of sustainable development dwarf the delivery capacity of any individual regeneration initiative, but seen from another perspective, the contribution made by millions of small regeneration initiatives represents the only satisfactory way of really ensuring the delivery of lasting sustainable development. Irrespective of the scale of action – from the individual neighbourhood to international – the tasks of sustainable development remain constant: the promotion of effective environmental management, the delivery of social justice, and the enablement of responsible economic progress. Even without the presence of an agreed global agenda for sustainable development, these tasks reflect the central purposes of comprehensive regeneration. Three issues flow from this discussion which are of particular relevance to regeneration. First, there is the question of how best to ensure the creation and ownership of a regeneration strategy which is designed to deliver sustainable development: the broad consensus here is that this is best achieved through community engagement, wide and robust partnership and collaborative working, and the establishment of a system of monitoring and review at the outset in order to ensure accountability (Carter, 2000). Second, it is also essential to ensure that the various barriers to social engagement are recognised and overcome at an early stage: chief among such barriers are educational under attainment, exclusion from employment and income, and what is increasingly described as spatial exclusion (Roberts, 2003). Third, it is essential to develop individual regeneration programmes as part of a wider mosaic of initiatives: no village, town, city, region or country is an island. Most territorial systems are permeable, not closed, and this should be seen as an opportunity to promote regeneration that is of benefit to other places and people – this is true intergenerational equity.

Having briefly examined the evolution of regeneration theory and policy, and having considered by implication the major characteristics of current regeneration activity from an historical perspective, the final element to be considered in this section examines the distinguishing features of regeneration as a discrete activity system. Although the details of regeneration are covered more fully in subsequent sections, it is important to consider regeneration within the broader context of spatial intervention activity. A key distinguishing characteristic of regeneration is a concern with the comprehensive adjustment of the conditions evident in a particular area or place; this suggests that regeneration is by nature interventionist and seeks to bring together a wide range of policy fields and individual actions in order to achieve a lasting transformation of a place. Furthermore, it is also evident that regeneration is concerned with a variety of time horizons, from short-term immediate rectification to long-term transformation. This question of timescale is crucial to effective regeneration, because many initiatives require considerable financial, human and resource investment, many elements of which take a considerable time to assemble, allocate and utilise. Finally, it is also essential to recognise that regeneration is a multi-sector activity. Even if the public sector wishes to take the leading responsibility in the design and delivery of a regeneration scheme, it is apparent that many factors will hinder or inhibit such an approach. An absence of resources, ownership or competence may be the most visible impediments, but such factors represent the tip of the iceberg of impossibility. Without the active involvement of the private, community and voluntary sectors it is unlikely that a regeneration programme will be implemented or survive. As a consequence, regeneration has to be planned and practised as a cross-sector activity that both mobilises the resources of many partners and delivers results that are of benefit to the full range of stakeholders.

## Principles, Characteristics and Structure of Regeneration

This section of the chapter offers a brief summary of some of the key organising and operating principles of regeneration policy in the UK. Through a discussion of the distinguishing futures and characteristics of regeneration programmes, this section provides a basis for the assessment of the performance of individual regeneration schemes in the next section of the chapter. A final topic in this section considers the broad structure of regeneration activity in the UK.

Although it is difficult to provide a brief summary of the distinguishing characteristics of regeneration policy in the UK, or anywhere else for that matter, it is possible to isolate and assess a number of common characteristics that define the field of policy activity. The cause of this problem in identifying the characteristics of regeneration is the very varied nature of the individual places for which policy is designed. Each village, city or region possesses a number of unique features, opportunities and problems and, as a consequence, the requirements of each place demand that a particular blend of policy actions is developed and applied in order to introduce and maintain a regeneration programme. As a footnote to the preceding point, it is also important to note that regeneration implies constant attention to the management of change; as noted earlier, to engage in regeneration is to go beyond the inherent limitations of a 'quick fix' individual solution to a single element of the fabric or functioning of an area.

The common characteristics of regeneration, and more importantly regeneration in areas experiencing significant restructuring challenges, can be identified from a number of sources. These sources represent attempts to identify the best practice of regeneration which has stimulated and delivered the successful response of areas to a variety of restructuring challenges. The summary presented below draws upon work undertaken in the UK by a number of different researchers, including studies at different levels of scale and in different locations (Burwood & Roberts, 2002; Jeffrey & Roberts, 2005; Maginn, 2004; Office of the Deputy Prime Minister, 2004; Regional Development Agencies Secretariat, 2002; Roberts, 2000). In brief, the key characteristics of successful regeneration practice include:

- Most importantly and essential if regeneration is to be comprehensive, lasting and 'owned' by all stakeholders, the need for any regeneration action beyond a small isolated action to be guided by a clear strategic vision and plan – this plan or strategy should be based on extensive research, consultation and other evidence, and should be truly visionary and offer a foundation for the preparation of detailed sectoral or sub-area implementation programmes;

- The requirement that action on the economy should be undertaken from the outset in order to ensure the long-term strength and diversification of the economic base, including the creation of an appropriate range of employment opportunities and the provision of an attractive and valued range of services – the functionality and prosperity of a place is still determined by the provision of services and jobs, more than any other factor;

- The establishment of ways of linking economic activities and jobs to the resident population – too many regeneration initiatives in the past have become 'cathedrals in the desert' providing jobs for the wealthy and advantaged who live outside the regeneration area, rather than for those who live within it; this requirement implies the provision of education, training and access to work facilities;

- The need to ensure that local residents and communities as a whole are engaged in the regeneration process and that they benefit from the activities undertaken in any programme – this requirement implies the creation of community capacity and competences, the provision of a range of social and welfare services and the establishment of appropriate social infrastructure that will ensure the long-term delivery of social justice and inclusion;

- The requirement to promote plurality and quality of opportunity – this has been described as the combination of a "pluralistic turn … complemented by a turn to community" (Maginn, 2004:171) and can be seen as an essential characteristic of regeneration policy which is aimed at reducing discrimination and the marginalisation of particular ethnic, social or income groups;

- The need to promote a range of cultural, leisure, tourist and heritage features and attractions which are linked to the needs of the community and which allow for the creation of jobs and income for local people – the key here is to provide such attractions and facilities in a way that reflects the heritage of the individual town or region and through the establishment of a distinctive form of provision which can act as a symbol for a place;

- The desirability of encouraging the effective management of all land and property resources, including the need to re-use areas of brownfield land, disused buildings and other infrastructure or facilities that have fallen out of use – the effective management of real estate is no longer solely a matter for the private individual or organisation, it is now an issue that involves all sections of society – as noted earlier, places and property are not simply assets to be traded, they are the location of human activities and cultural artefacts;

- The need to ensure the effective and responsible management of the environment and the most efficient possible use of all environmental resources – this characteristic refers to urban and rural areas; and to the full range of activities, including all elements of the built environment, transport, energy use, waste management, the provision of open space, countryside and landscape, water resources and the promotion of bio-diversity;

- The desirability of encouraging the repopulation of central urban areas in order to re-establish a resident community and to promote a sense of neighbourhood – this is an especially important matter in English towns and cities where the migration of population to the suburbs and the countryside has been the dominant trend in recent years;

- In order to deliver and maintain a number of the above characteristics, the need to provide a range of basic facilities – quality and choice in housing provision, excellent and efficient transport facilities (both local and long distance) and an appropriate range of economic and social infrastructure;

- The provision of engagement, collaboration and partnership structures that enable the effective planning, implementation and management of regeneration actions – this is more than consultation and communication, although these are important methods, it is also about the effective and democratic governance of places;

- The establishment of adequate facilities to monitor, evaluate and review the operation of a regeneration programme – this is a frequently neglected aspect of regeneration policy, but it is an essential feature of any attempt at regeneration which has as its aim the long-term restructuring and management of the transformation of a previously failing place.

The above-noted characteristics reflect what the Egan Review, of the skills necessary for the development of sustainable communities, has described as the components of sustainable communities (Office of the Deputy Prime Minister, 2004). Equally, assessments of best practice in regeneration – from the neighbourhood to the regional level – have emphasised the need for all regeneration programmes to take account of, and display, these characteristics. This is not to suggest that all places should construct a regeneration response which is identical in terms of either appearance or the facilities made available, rather it offers guidance with regard to the choice of desirable processes and outcomes. Far from promoting a blend or homogeneous approach to regeneration, the various attempts to review UK regeneration experience which have been referred to in this section, all point to the overarching importance of establishing place-specific regeneration visions and approaches and introducing comprehensive portfolios of action.

A final topic in this section of the chapter is concerned with the overall structure of regeneration activity in the UK. As in most European Union nations, regeneration policy is a matter of concern and action at various levels of scale, from the local to national, with the added complication that the European Union itself also contributes to regeneration, especially at regional level, through the operation of the Structural Funds policy portfolio. However, whilst strict legal powers and resources are evident at various levels of scale, the specific delivery of regeneration programmes in the UK is generally undertaken at local, sub-regional and, occasionally, regional levels. The vast majority of regeneration programmes are local in scale and content, but they call upon financial and other resources from higher levels of government. A further complication exists in any discussion of regeneration activities in the UK due to the presence of different government and governance arrangements in the various nations and regions. Devolved governments have been established in recent years in the three Celtic nations – Wales, Scotland and Northern Ireland – and in one of the nine English regions – Greater London. In these territories local government is, therefore, closer to the formulation of policy than is the case elsewhere in England. In the eight remaining English regions non-elected, consultative regional assemblies have been established, but these assemblies lack the power of either policy formulation at the ability to bend budget allocation in order to give priority to particular regeneration objectives. For a fuller discussion of regeneration and associated policy activities in a devolved nation, the reader is referred to Newlands, Danson and McCarthy (2004), whilst a summary of regeneration action in the English regions outwith Greater London is provided by Maginn (2004) and the Core Cities Group (2001). Although the structure and organisation of regeneration activity is complex and somewhat fragmented, thereby sometimes exacerbating the problems encountered in the delivery of regeneration, the reality is that the partnership or collaborative model of operation, which draws resources from all levels of government and from the private and voluntary sectors, is now dominant.

The partnership model is not new in the regeneration arena, indeed it can be argued that partnership has been the norm for many years. What is new in the UK, is that the range of stakeholders and partners has been extended to include ordinary citizens. At local level, Local Strategic Partnership (LSPs) have been established throughout England; these organisations often involve more than one local authority and always include representatives from the private, voluntary and community sectors. Other enhancements of the partnership model have been introduced into a variety of areas of public policy, including land use planning, health and social welfare arrangements, local education provision and certain aspects of economic development. In some towns, cities and rural areas partnership-based models of regeneration management are now well-established with extensive competence across all areas of regeneration activity. Two examples of such arrangements are provided here. In the case of the City of Leeds, with the recent publication of a partnership regeneration strategy which seeks to build-on the previous two decades of collaborative working, the Leeds Initiative has marked its coming of age as a comprehensive partnership. *The Vision for Leeds II* (Leeds Initiative, 2004) is a well-argued and realistic strategy for the second largest city in England. Another excellent example of extensive partnership working can be seen in the case of South Tyneside. Here, in the North East of England, a once less prosperous and somewhat marginalised metropolitan area, has sought to introduce a step change in both the style and structure of its approach to regeneration. Building on the visionary transformation strategy commissioned by a strong LSP, South Tyneside has witnessed a radical change in its potential and prospects through the actions brought about under the highly innovative 1,000 day transformation programme (South Tyneside, 2004). All of these examples demonstrate the power of partnership working and provide evidence that effective area governance can be established.

What can be seen from the evidence provided in this section, is that regeneration as a discrete activity system is now less dependent on the central state for delivery than it was in the past. Although the 'new localism' is still somewhat constrained by rules and procedures imposed by central government, effective local partnership arrangements are the real key to successful regeneration. Equally, whilst central government urges that LSPs and other collaborative arrangements at local level should be designed to operate in a comprehensive and integrated manner, it is still the case that individual central government departments continue to design and implement separate area-based initiatives. Indeed, in one recent study it was claimed that some central government area-based initiatives duplicate other actions or even cancel-out other initiatives (Regional Co-Ordination Unit, 2002). It is the case, therefore, that, at least in most of England, local, bottom-up regeneration partnerships act as a crucial design and delivery mechanisms, preferably displaying the key characteristics outlined in the first part of this section.

## Lessons from Experience

This penultimate section of the chapter provides a brief summary of some of the more important findings from recent research on the performance of regeneration initiatives. Particular emphasis is placed here on regeneration activities in the North West England region, a region with a varied inheritance of socio-economic, environmental and physical conditions. The assessments presented here draw on three research studies: a national evaluation of some 49 cases (Burwood & Roberts, 2002) which attempted to identify lessons from the experience of successful regeneration schemes and projects, an assessment of selected cases of regeneration supported by Regional Development Agencies (Regional Development Agencies Secretariat, 2002) and an in-depth assessment of regeneration schemes and projects identified as potential 'role models' under the Exemplar Programme promoted by the North West England Centre of Excellence for Regeneration, RENEW (Jeffrey & Roberts, 2005). Although these research studies make use of different classification frameworks and evaluative criteria some helpful lessons emerge from the varied practice which they assess.

The first, and arguably most important, finding from the three studies in that in general they confirm the accuracy of the wider assessment of the characteristics of regeneration outlined in the preceding section. The twelve characteristics outlined in this section are found in many of the individual cases examined in the three research studies. Furthermore, the case studies also confirm a number of the critical observations recorded elsewhere in this chapter, including the importance of ensuring that regeneration actions are tailored to fit the circumstances present in an individual locality or region, and the importance of promoting the local 'ownership' and control of a regeneration initiative.

Particular features of the regeneration scheme and projects examined in the three studies that are of value for a wider audience, include elements that can best be classified under the five headings of: people, planet, prosperity, politics and place. As will be seen in the following paragraphs these headings broadly reflect the essential building blocks of sustainable development. They also all incorporate one overarching consideration in their approach to regeneration, this is the need to ensure that quality is prioritised and encapsulated in all aspects of activity – this is a variant on the Total Quality Management (TQM) approach to business operations, which can be considered as TQM for places or Territorial Quality Management. This is an essential finding, because it reinforces the view that the design and delivery of regeneration strategies is at least as important to the success of places, and to the businesses which are located there, as is the use of TQM in the management of business. Furthermore, this finding reflects the well-established conclusion that failing places are both the consequence of previous economic and business failure and likely to be the cause of future economic or business failure. Although the distinction between 'rustbelt' and 'sunbelt' cities is less relevant in the UK than it is in the USA, there is considerable historic and contemporary research evidence of the operation of either 'downward' or 'upward' patterns of spatial development; most notably in the extensive literature on the 'cycle of deprivation' from the 1960s onwards (Shelter Neighbourhood Action Project, 1972; Mossberger & Stoker, 1997).

Developing the first of the five headings – people – it is evident that the majority of successful regeneration initiatives place considerable emphasis on the empowerment of individuals and communities. Social welfare and the promotion of social justice are essential components of regeneration, irrespective of the spatial scale of the initiatives or the location of the individual place. Two cases illustrate the significance attached to people factors. The REEL Salford project, based in the City of Salford, has attempted to engage young people and adults in failing communities in regeneration activities through the use of artistic and creative media. Although this is an unconventional approach to community engagement, the consequences have been extensive and lasting; films have been produced which involve the community and which have enabled regeneration problems to be identified, options to be discussed, and the implementation of solutions assisted through the visual communication of the selected option. Associated benefits of the initiative include the provision of training and employment in the visual and media sector which would otherwise not have existed. A second and very contrasting case is that of the Oxford Place Methodist Centre in Leeds. In this case, an important under used religious building in Leeds City Centre was converted into a multi-activity centre providing for a variety of needs, including office and activity spaces for 16 charities and care organisations, amenities for homeless people, a children's centre, meeting rooms, worship space and a café. Oxford Place has provided essential facilities to organisations that need to be located in the city centre, but it is unlikely that they could afford to rent adequate floor space had the initiative not been developed.

The second heading – planet – is a reference to the needs of the environment as an essential component of regeneration. Irrespective of the location of a regeneration activity, it is now regarded as important that the regeneration strategy and the processes of implementation safeguard and enhance environmental conditions. Whilst these environmental considerations vary considerably from place to place, it is now accepted that factors such as promoting energy efficiency, the greater use of public transport, the effective management of waste and water, the quality of the built environment and the protection of bio-diversity are central elements. The overall aim here is to minimise the ecological 'footprint' of the place where regeneration is occurring. Two examples offer an illustration of the ways in which this priority has been addressed. The Callanbridge Residents Association regeneration programme has provided much-needed improvements to this housing estate located on the edge of Armagh City, Northern Ireland. An important element of the programme has been the direct involvement of local people in improving and managing their built and natural environment. Equally, the Northwich Community Woodlands scheme has emphasised the importance of providing for the long-term management of habitats and eco-systems through community involvement.

Prosperity, the third heading, refers to the need for successful regeneration to make a major contribution to the economy of an area. Most regeneration programmes place emphasis upon the creation, retention or expansion of economic activities and jobs, but the regeneration challenge goes beyond the simple creation of economic opportunities. As noted earlier, it is equally important to link local people to existing or new jobs. Schemes such as the Lancashire Foundation Apprenticeship project aim to link young disadvantaged people to employment through training and the provision of jobs in local firms. Other initiatives emphasise the provision of education, training and mentoring in order to help people to establish their own businesses. Above and beyond the traditional provision of support for the creation or maintenance of conventional firms, many regeneration programmes, such as the Pennywell Community Business in Sunderland, place particular emphasis on the establishment of alternative economic structures, including community businesses and co-operatives.

The fourth element, politics, refers to the need for appropriate governance arrangements to be put in place to ensure that a regeneration scheme is adequately governed. In the past many regeneration initiatives were imposed upon communities; such initiatives were frequently dictated by central or local government with only token consultation and little or no real community engagement. Such an approach is no longer considered acceptable and the majority of regeneration schemes now emphasise the need for full community engagement at the outset. Equally, most initiatives have now incorporated arrangements for the monitoring and review of performance, thus ensuring full accountability. Local regeneration schemes are normally managed through either a LSP or a bespoke local governance arrangement. At regional level, recent reforms have introduced non-elected regional assemblies in the eight regions outwith London. In the London region an elected body – the Greater London Assembly – is responsible for ensuring democratic accountability. The engagement and accountability arrangements for the Oldham Beyond initiative demonstrates the importance of involving the communities and all stakeholders from the start of the process of regeneration. This initiative is now subject to independent scrutiny and is also attempting to provide enhanced governance in all of the areas with which it is involved.

The final element, place, is the enduring keystone of successful regeneration. Without a real sense of place, regeneration can become a process that lacks identity and heart. Most successful regeneration programmes are distinct and distinctive (Roberts, 2000); they build on the opportunities evident in a place and they emphasise the establishment of solutions that are specific to the area. Such approaches are also deeply rooted in local and regional communities; they are comprehensive and lasting. Good places are sustainable and they are able to face new challenges with confidence. For these and other reasons, the best practice of regeneration goes beyond the bland and mundane replication of solutions from elsewhere; rather than simply replicating the success of other places, the objective is to transform an area and to ensure that what is established is a mechanism for regeneration that will develop alongside the competences of the community. This is the approach that has been adopted by South Tyneside (this example was discussed in the previous section of the chapter) which decided three years ago to embark on the radical transformation of the locality (South Tyneside, 2004). The Transformation Plan was introduced in order to promote a step change in attitudes and actions; the end result is a comprehensive and innovative 1000 day programme of regeneration which is distinctive and deeply embedded in the various local communities. Another excellent example of the transformation of a place is the Eldonian Village in Liverpool. This initiative, which dates back to the early 1980s, is about reinventing place within a major city, and has transformed the environment and life chances of a previously failing community. One acknowledgment of the success of the Eldonian Village has been the winning of a World Habitat Award from the Social Housing Federation of the United Nations; a rare achievement for a once marginalised area and an important model of how to establish and express the quality of place (Coxon, 2005).

## Conclusion and Future Prospects

In one sense the lessons from the experience of regeneration provide the conclusions to this chapter. However, there are a number of other issues that are of importance with regard to both the theory and practice of regeneration in the UK.

The first set of conclusions relates to the scale and scope of regeneration. Clearly the spatial scale of regeneration has increased over the past twenty years: from the situation in the early 1980s, when much regeneration was site or building-specific and mainly urban, it is evident that many current regeneration initiatives are now often city-wide or even regional in scale. Equally, the scope of regeneration has widened: in the early 1980s the physical dimension was emphasised, whilst most regeneration initiatives nowadays are aimed at delivering sustainable development.

A second set of conclusions represents the response to the increasing spatial scale and wider scope of regeneration. As scale and complexity have increased, it has become important to introduce more comprehensive and sophisticated governance and accountability arrangements. In the devolved nations and regions – Wales, Scotland, Northern Ireland and London – this is a somewhat less contentious matter than it is in the other eight English regions. In the absence of elected government, governance and accountability are provided through, indirectly elected regional assemblies and elected local government, supplemented when necessary by formal partnership arrangements through mechanisms such as LSPs. The question of complexity is directly related to the governance issue because, in the absence of devolved government (and the presence of powers to enact legislation and issue policy guidance) national regeneration policy in England is often too general to fit local and regional circumstances and depends upon co-ordinating the actions of many central government departments. Again this means that co-ordination often has to be retrofitted at local or regional level and depends upon the ability of a local or regional partnership to enable this.

The third set of conclusions reflects the importance of learning from experience, both within the UK and more widely. All regeneration theorists and practitioners benefit from comparing experience and engaging in attempts to evaluate the comparative performance of both the 'processes' and 'products' of regeneration. However, insufficient time and resources are available to support such activities and, as a consequence, too much effort is wasted on 'reinventing the wheel'. Much more national and international comparative evaluation research is required.

Finally, there are a number of conclusions related to the likely future trajectory of regeneration theory and practice. In this category it is possible first to point to the greater emphasis which has been placed on community-based engagement and the management of regeneration in recent years; this trend will continue and it is likely that greater powers of decision and operational co-ordination will be given to communities to allow them to dictate the style and form of regeneration. Secondly, it is likely that greater functional and administrative devolution will occur in the eight English regions outwith London; this will enable the introduction of more place-specific policies for regeneration. Thirdly, it is also possible to point to the increased emphasis which is now placed on the role of spatial strategy as an overall 'corporate plan' for regeneration at local and regional levels; this new emphasis on spatial guidance and co-ordination will continue. Fourthly, the adoption of the goals of sustainable development to provide overall guidance for regeneration has already produced positive results and this approach will become a permanent feature of UK regeneration policy.

Considerable progress has been made in recent years, and communities at local and regional levels have benefited from the introduction and operation of more comprehensive and better-resourced regeneration policies than existed in the past. There are still gaps in provision, selective shortages of resources and, most importantly, some regeneration schemes still suffer from the absence of long-term funding. A final task to be undertaken is the need to engage more fully than in the past in the education and training of regeneration professionals; this is an urgent need and is now the subject of considerable debate and action, including the introduction of a new national academy to help to reinforce the positive recent history of designing and delivering regeneration. This new initiative is a task to which colleagues elsewhere in Europe are well qualified to contribute.

## References


Browne, H., 1974, *Joseph Chamberlain: Radical and Imperialist*, Longman, London

Burwood, S. & P. Roberts, 2002, *Learning from Experience*, Office of the Deputy Prime Minister, London

Carter, A., 2000, Strategy and Partnership in Urban Regeneration; In: Roberts, P. & H. Sykes (eds), *Urban Regeneration*, Sage, London

Core Cities Group, 2001, *The Challenge for Cities in the 21st Century*, Core Cities Secretariat, Birmingham

Coxon, A., 2005, Razing Questions, *Housing Today*, 425, 5

Fainstein, S., 1994, *The City Builders*, Basil Blackwell, Oxford

Hall, P., 1974, *Urban and Regional Planning*, Pelican, Harmondsworth (UK)

Hall, P., 1988, *Cities of Tomorrow*, Basil Blackwell, Oxford

Hall, P., H. Gracey, R. Drewett & R. Thomas, 1973, *The Containment of Urban England*, Allen & Unwin, London

Howard, E., 1902, *Garden Cities of Tomorrow*, Swan Sonnenschein, London

Jeffrey, P. & P. Roberts, 2005, *Renew Exemplar Programme*, RENEW, Liverpool

Leeds Initiative, 2004, *Vision for Leeds II*, Leeds Initiative, Leeds

Maginn, P., 2004, *Urban Regeneration, Community Power and the (In) Significance of Race*, Ashgate, Aldershot (UK)

Mossberger, K. & G. Stoker, 1997, Inner-city Policy in Britain: Why it Will Not Go Away, *Urban Affairs Review*, 32, pp. 378-402

Newlands, D., M. Danson & J. McCarthy, 2004, *Divided Scotland: The Nature, Causes and Consequences of Economic Disparities Within Scotland*, Ashgate, Aldershot (UK)

Office of the Deputy Prime Minister, 2004, *The Egan Review: Skills for Sustainable Communities*, Stationary Office, London

Regional Co-Ordination Unit, 2002, *Review of Area-based Initiatives*, Office of the Deputy Prime Minister, London

Regional Development Agencies Secretariat, 2002, *Driving Urban Regeneration in the English Regions*, Regional Development Agencies Secretariat, Birmingham

Roberts, P., 2000, The Evolution, Definition and Purpose of Urban Regeneration; In: Roberts, P. & H. Sykes (eds), *Urban Regeneration, A Handbook*, Sage Publications, London, pp. 9-36

Roberts, P., 2003, Sustainable Development and Social Justice, *Sociological Inquiry*, 73, pp. 228-244

Roberts, P., V. Joy & G. Jones, 2002, Brownfield Sites: Problems of Definition, Identification and the Evaluation of Potential; In: Rydin, Y. & A. Thornley (eds), *Planning in the UK*, Ashgate, Aldershot (UK)

Shelter Neighbourhood Action Project (SNAP), 1972, Another Chance for Cities: *SNAP*, 69, 72, Shelter, London

South Tyneside, 2004, *Transforming Together: South Tyneside's Regeneration Strategy*, Transformation, South Shields (UK)

# 8 Networks and Urban Planning: the Evolution of a Two-way Relationship

Gabriel Dupuy

## The early networks: discreet beginnings

The first of a series of technical systems designed to supply urban populations with the services, that are now regarded as the basic necessities of city life emerged in the mid 19th century. These systems included running water, sanitation, power, transport, and communications. In view of their spatial organisation – meshed or branching cables, conduits and so on – they came to be known as 'networks'.

The new networks emerging in the mid-1800s often stemmed from technological breakthroughs of the Industrial Revolution, for example, using compressed air technology to create a supply of motive power, or gas for public lighting (*'gaz de ville'*, or 'city gas'). Subsequent breakthroughs have continued to spawn a succession of networks, of which, perhaps the most outstanding example in recent times is the Internet, which came from the telephone. This spawning process that has been gradual, and at times extremely slow, in extending across the entire urban area. For instance, it was more than a century before every building in Paris was connected to the main sewers.

Many of the early networks appeared in the days when urban planning was beginning to establish itself as a doctrine and a professional practice, and for quite some time, urban planners failed to recognise the importance of their role. Indeed, all but a handful of visionary pioneers (e.g. Ildefonso Cerda) consigned those networks to a lower order of logistics. The technicians who worked on them were considered to be there strictly to ensure that the networks were properly maintained, and not to meddle in the more noble business of planning or policy-making. To be fair to the mainstream urban planners of the day, it must be said that those early networks were really rather discreet. Their construction caused barely a ripple in the urban fabric. Conduits and cables were slipped discreetly into its unoccupied chinks and crevices – on, or often beneath, the public highway – and, equally discreetly, fed into buildings and homes, without significantly altering the architecture or urban shape.

Take the garage, for instance: with the advent of motorised vehicles in the early 20th century, this became a terminal of the road system. Previously, people usually went out and returned home on foot, or sometimes by horse or horse-drawn carriage. Having to park a car before entering a building marked a dramatic departure from earlier practices. Le Corbusier was one of the first architects to grapple with the ramifications of the automobile revolution. His Villa Savoye, built in Poissy near Paris in the 1930s, was designed with the motor car in mind: it had a driveway so that passengers could be dropped off at their front door, before driving the vehicle into a built-in garage. By and large, however, the main entrances of the vast majority of buildings continued for a long time to be designed strictly with pedestrians in mind. Architects seemed to want to banish the car to covert garages concealed at the bottom of a back garden or in a basement. The façades of Eugène Hénard's modern Parisian apartment buildings, designed at the turn of the 20th century, scarcely give a hint of a garage, even though the architect-planners, bent on reform, were very aware of the parking-space factor. Mathieu Flonneau, in his thesis on the history of the motor car in Paris, observes the many ways in which public spaces had to be adapted to cater for the ever-growing volume of automotive traffic (Flonneau, 2002). He also draws attention to the deference paid to the urban shape, which continued through to the 1960s and the presidency of (the much maligned) Georges Pompidou.

While there may have been a great demand for efficient urban networks, in the modern sense, architect-planners were 'put in their place' and made to feel that they could not encroach upon the built environment and/or significantly alter the aesthetics, structure or shape of the city. At the time, these considerations were regarded as being far more important than being able to move around town at speed, or even having a home equipped with modern conveniences. For quite some time, then, urban planners could sit back and watch the new networks spread through the city without feeling any need to revise their approaches or ways of thinking.

## New network urbanism

From the 1970s and 1980s onwards, the work of historians began to create an awareness of the true role of these new networks in the organisation and functioning of cities (Tarr & Dupuy, 1988). Engineers started to describe how their techniques had developed over time, and how they were related to urban space. In fact, a host of historians addressed the network issue: Thomas Hughes and Alain Beltran with respect to electricity; Letty Anderson and Jean-Pierre Goubert to water; Joel Tarr and André Guillerme to sanitation; Clay McShane and Dominique Larroque to transport; and Seymour Mandelbaum and Pascal Griset to communications, to name but a few American and French examples. All of them reveal the importance of these networks in the making of the modern city. Most of these writers contrast the unobtrusive way in which networks were physically incorporated into the built environment, with their considerable impact on urban life. This relatively recent observation has certainly prompted contemporary urban-planning practices to acknowledge the significance of networks and take their role into account. Examples of the ways in which the old networks have benefited from progress can be seen in the new tram lines or parking restriction policies in some cities, and the enhancement of public lighting or running water systems in others.

Despite this, the old networks are now being replaced or completely transformed by the new. Partial or total automation is improving efficiency, reducing operating costs and radically changing the relationship between operators and users, as, for example, in underground transport systems. The newly enriched range of Information and Communication Technologies (ICTs) is also providing urbanites with untold means of harnessing the potential of the city's physical space and of becoming integrated into its social space.

In the 1960s, the idea that progress in the realm of ICTs would do away with the need to travel to work every day, leaving people free to choose where to live or locate a business, gave rise to the concept of the 'electronic cottage'. Then, after the advent of the Internet, all the talk was about the 'cybercity'. Nowadays, attention has turned to the 'digital city' and the links between metropolisation and the major nodes of global communication networks.

It is unrealistic, however, to say that the immediate and tangible effect of ICTs on a city will bring about its rapid transformation. True, the shopping malls of American cities may be losing some 2,000 travel agencies a year due to direct bookings over the Internet; but urban commerce as a whole continues to stand firm in the face of e-commerce. The numbers of teleworkers remain relatively small, and the structure of the 'digital city' is still one of bricks and concrete.

Let there be no mistake, though, ICTs are making a dramatic difference to the ways in which urban space is used. So we really do need to reappraise the relationship between cities and networks; root and branch. The changes are enormous.

## A change of scale

The networks serving urban populations used to be called '*urban* networks' or '*urban* technical networks'. Those names meant something in the past, in the early days of DC electricity networks, pressurised water supply, sewerage systems, *gaz de ville*, trams, metros and so on. Not only were their users urbanites but also, as they were technically self-contained, they were managed by the municipal authority. Their scale was resolutely urban. Since then, major changes have taken place in most sectors, but nowhere more markedly than in that of the ICTs. Take the Internet, for example. The growth of the Internet hinged on the capillarity of the telephone network, which was first deployed in cities. But the Internet is a global, not an urban, network. Its main infrastructure, like many of its operators, is not urban but international. Internet operators do not decide to move into a particular region without considering the picture at the global level: they operate in other countries too; they have to comply with international technical standards; they are competing with other international operators; and they need access to global financial markets. Their capacity to supply broadband Internet services at a reasonable price, at local level, depends on global Internet-related trends. Events at the local level cannot deviate too far from those global trends without giving rise to a digital divide. This means that urban centres now have to keep up with global trends. At first, this requirement was confined to large global metropolitan areas, but, subsequently, this requirement has spread to smaller well-placed cities that now act as waypoints or gateways.

Geographical scale is not a recent problem. It was already an issue in the 19th Century, with the railways, especially with regard to the siting of stations. The railways were a large-scale network and urban planners and policy-makers learnt to work with it. In France, they produced solutions such as the Paris metro, which was implemented in 1900, and then, some 60 years later, the RER high-speed city–suburban link. Nowadays, some interesting new solutions are being found for the problem of how to combine the urban with the national (or international) scale. The Karlsruhe TramTrain, for example, Milan's *Passante* or their equivalents in Brussels and Barcelona have effectively turned the metropolis from a terminal into a waypoint.

## Technological progress

With the ICTs, even more than with transportation or drinking water supply technologies, the pace of technological change is fast. In the not so distant past, people were amazed at the speed of Internet penetration. Less than 10 years later, with the arrival of ADSL and broadband connections via an ordinary telephone line, the situation has changed again. Before long, it looks as though UMTS will make fixed-line Internet access obsolete.

The effect of these rapid technological changes has been great. In the past, ongoing network expansion, with the same technologies and standards for all, maintained the principle that, in the course of time, every citizen would gain access to a universal service. Right of access to the networks went hand in hand with the right of access to the city. Such considerations now seem almost meaningless. In France, the deregulation of the telecommunications sector led to the introduction of a universal, strictly fixed-line phone service (with the continuing provision of call boxes and directories) that just happened to coincide with the cell phone and Internet boom. The service instantly lost appeal. Everyone today expects broadband. Tomorrow, it will be UMTS.

This issue of the timing of technological developments has led to marked differences in service provision to different urban areas or population groups. Graham and Marvin (2001), albeit with some arguable generalisations, expose the undesirable effects on urban cohesion of premium networks – water, electricity, telecommunications, transport – that offer better services to those who can afford to pay. Some may persist in imagining or expecting cities eventually to enjoy a full and homogenous range of services, but the fact is, as far as networks are concerned, new technological developments and the socio-economic demand are giving rise to the 'two-track' city. Such differentiation is not a passing phase. Each successive new development opens up a new rift, often before the last can be bridged. The ICTs come first; then the general trend.

## Deregulation versus planning

More importantly, ICTs are deployed in an amazingly competitive, free-market environment. The world of networks has changed greatly since the early 1980s; since the days when Judge Harold Greene ordered the dismantling of the monopoly of the American Telephone and Telegraph Company (AT&T). Rutherford (2004) shows how, in the case of Paris and London at least, the operators, old and new, and the political authorities are playing a difficult game in an ever-changing context where, for want of perfect competition, mistrust is the rule. Notwithstanding the operators' penchant for discretion, there is no hiding how they act: hunting for niches, claiming territories, putting up passive resistance when the regulators step in. As for the political actors, they use their powers to change laws frequently, to advance such legitimate issues as social cohesion and national and regional development, to run pilot projects and so on. In a period of transition, however, one cannot always be sure that the self-declared actors are actually up to the part that they claim they are meant to be playing. In many cases, the institutional environment has changed, not only due to the deregulation of the telecommunications sector, but also for other reasons (such as decentralisation in France, reunification in Germany, new administrative regime regulations in London, etc.). This affects the distribution of roles. To take another French example, the Institute for Urban Planning and Development for the Ile-de-France Region (IAURIF) was an active promoter of the *Francilian* teleport project of the early 1990s. One would therefore have been inclined to take the agency seriously, at the time, when its public statements hailed the teleport as a balanced means of coordinating regional telecommunications infrastructures. In fact, the project hinged on an agreement between the (then monopolistic) operator and the public authorities, and it was very soon blown out of the water by deregulation. So nobody talks about the *Francilian* teleport any more, and it is clear that IAURIF is not up to the job of regional telecommunications development planning today.

There is nothing historically new about deregulation. The first railway and electricity companies were privately owned and in competition with each other. Nevertheless, the public authorities managed to work with that situation. In Paris, in the early days of electricity, planners mapped out geographical areas for which concessions – or partial monopolies – were awarded. Each operator was allotted a share of urban territory known as a 'sector'. Each sector's boundaries were determined by the public authority in such a way that they encompassed zones that contained a mixture of dense, animated and wealthy areas, and less dense or poorer ones. The hope was that it would be in the operator's interests to deploy its network across the entire sector and thereby ensure a degree of spatial equalisation. But it later became common in Europe, and especially in France, to see the authorities assuming responsibility for both urban planning and the administration of networks run by public-sector monopolies. For a while, they managed to dovetail the two with relative ease. It was the golden age of network urbanism (Dupuy, 1991; Dupuy, 2000). Now, when the networks have only just come to occupy a more recognised place in urban planning, everything is changing again: geographical scales are dilating to the edge of infinity; technology is developing at ever faster speeds; and, with the liberalisation of the network utilities, monopolies are giving way to almost uncontrollable competition. Urban planning may be able to afford to overhaul its practices and thinking enough to adapt to the first two (changes in scale and technology), but the third (network liberalisation) has sent its paradigms reeling and destabilised its action. Can a new network-oriented form of urban planning take shape in this new context? History offers a wealth of encouraging examples. But first of all, European planners will need to abandon their nostalgia for the post-Second World War golden age, for urban planning that embraces public networks as a basic requirement. Tomorrow's network urbanism is bound to be very different from that of yesteryear.

## References


Dupuy, G., 1991, *L'urbanisme des réseaux*, Armand Colin, Paris

Dupuy, G., 2000, A revised history of network urbanism, *OASE*, 53, Uitgeverij SUN, Nijmegen

Flonneau, M., 2002, *L'automobile à la conquête de Paris, 1910-1977. Formes urbaines, champs politiques et representation*, Thesis, Université Paris I, Paris

Graham, S. & S. Marvin, 2001, *Splintering Urbanism: Networked Infrastructures, Technological Mobilities and the Urban Condition*, Routledge, London and New York

Rutherford, J., 2004, *A Tale of Two Global Cities, Comparing the Territorialities of Telecommunications Developments in Paris and London*, Ashgate Publishing, Aldershot (UK)

Tarr, J. & G. Dupuy (eds), 1988, *Technology and the Rise of the Networked City in Europe and America*, Temple University Press, Philadelphia

# 9 Does Space Matter?
## Spatial and Socio-economic Segregation in the Oude Noorden District of Rotterdam

Mr. Ibrahim (not an Arabic grocer in Paris, but Turkish) and Momo (a Jewish boy, perhaps, whom Ibrahim has adopted as a son) are on the road from Paris to Istanbul: "If you want to know whether you are in a rich or poor neighbourhood, you have to look at the trash-bins. If you don't see any trash and you don't see trash-bins either, then the people are filthy rich. If you see trash-bins but no trash, then they are ordinarily rich. If you see trash next to the trash-bins, then the neighbourhood is neither rich nor poor, but touristy. If you see trash but no trashbins, then it's a poor neighbourhood. And if you see people living between the trash, then they're poverty-stricken."
"Oh well, here they are rich."
"Yes, naturally, this is Switzerland !"
(E.-E. Schmitt, 2003, *Mijnheer Ibrahim en de bloemen van de Koran*, Atlas, Amsterdam, p. 76)

Paul Stouten

## Introduction

There are different approaches to the possible relations between spatial and social processes in a city. This was an important aspect of the research carried out on the living conditions of households in the Oude Noorden district of Rotterdam between 1975-1993 and after 1993 (Stouten, 2004). Today, social life moves through nodes in different networks, through points of power, convergence or transition. It is not necessarily anchored to any one place or space (Gieryn, 2000). Does that mean, then, or could it be, that space just does not matter? Architectural constructions look more and more like duplicates of other constructions – buildings, shopping malls, office complexes, housing estates, etc. – elsewhere. This may fuel the idea that space does not matter anymore. Nevertheless, the increasing complexity of urban development is undeniable, as are dynamic systems overloaded with images driven by social, environmental and economic changes. Moreover, they themselves are prime generators of many such changes that impact on the local situation. No town or city is immune from either the external forces that dictate the need to adapt, or the internal pressures that are present within urban districts (Roberts, 2000).

Cities, in general, are confronted with growing problems associated with social exclusion; problems that are enlarged spatially by segregation and transition processes. Despite the structural character of those problems, the spatial concentrations of disadvantaged groups is an undeniable fact, although the concentrations themselves – the unemployed, the young, the unskilled, immigrants and ethnic minority groups – may differ from city to city and within individual cities (Drewe, 2000). The analyses in this article are directed towards answering the question of why, especially in urban renewal districts, space *does* matter. First we analyse the residential situation in the Oude Noorden district. After that, questions about the city and urban space are related to policies on exclusion that are mainly based on ethnic characteristics. Such policies are aimed at impacting the demand side of the housing market, but they also impact the supply side too, to give residential differentiation. Both issues are at the centre of spatial politics. In Rotterdam, recently, measures have been proposed by the local authorities aimed at decreasing the number of ethnic minority groups within the city. This has grown into a big issue, which has caused much controversy.

## Problem definition

In Dutch literature, the concept 'segregation' is usually linked with specific household characteristics. In politics and the media, concentrations of higher incomes and highly-educated people are seldom seen as problematic. Recent local authority arguments have linked concentrations of low-income groups, people with a low education level, and ethnic minority groups, with fears of safety, criminality and poor quality residential environments. The fear that American-style ghettos and segregated neighbourhoods may, indeed, emerge on Dutch soil haunts politicians and journalists. It is generally assumed that ghettos are social prisons, where generations of inhabitants are excluded from the wider society and upward mobility. However, Dutch researchers have found hardly any evidence of ghetto formation in the Netherlands (Droogleever Fortuijn, Musterd & Ostendorf, 1998). In combination with residential segregation, it is the sharpening attention given to ethnicity, and increasing spatial segregation, that will lead to the further separation of different social and ethnic classes, and population categories. Whether ghetto-like developments arise will depend on how this attention is tuned, but if they arise, then the end result could lead to the disintegration of urban society. However, segregation is not a one-dimensional problem. For instance, in Rotterdam in recent years, (B&W Rotterdam, 2003), as in the 1980s (Drewe, 1986), the problem has been defined on the basis of just one or two characteristics of the population; mainly on the percentage of immigrants in a district.

In Metropolitan policies carried out since 1995, spatial segregation has been assumed to be the cause of serious problems in certain urban districts, such as a worsening quality of life. At the same time, these definitions have obvious limitations. The spatial approach, for instance, is limited because not all problems,such as unemployment, low education, and health care, are due to spatial matters. The concept of residential segregation is found to be closely related to the development of industrial society. Whereas in the USA, anxiety and concern arises when an underclass starts to develop in particular districts, in the Netherlands, a variety of measures, such as grants and social housing schemes, have been implemented to reduce these sharp contrasts, but since the mid 1990s, these measures have been the centre of heated discussions in central government.

## Urban regeneration in Oude Noorden district

### A dynamic phenomenon

Urban regeneration is not a static phenomenon. In 1975, building and modernisation were started in the Oude Noorden district, one of the twelve assigned urban renewal districts. This 'building for the neighbourhood' strategy was aimed at solving the housing problems of the present residents. This strategy was replaced in the 90s by several separate, mainly economic and social, programmes. By fragmenting the renewal programme in this way, the action that was taken did not cover all aspects of the problems that had arisen. So, at the end of the 90s, in this, and five other urban renewal districts, a new concept of strategic planning gained priority. This new concept included the physical, economic, social and environmental dimensions of urban regeneration and the implementation and management of the regeneration process. About 65% of the housing stock was modernised or re-built as a result of the 1975-1993 urban-renewal plan. As in other Rotterdam urban renewal districts, the housing stock consists of 76% social housing, mainly in a good condition. This high share was achieved due to the local government's vigorous buying campaign from private landlords in the 1970s.

The most fundamental change that took place during the urban renewal period was the removal of the indigenous population and the settlement of new residents, most of whom had Moroccan, Turkish or Surinam backgrounds. At the same time as the change in population structure took place, there was a sharp rise in unemployment. This was because the economic restructuring, which had started in the 1960s, mainly led, from the beginning of the 1980s onwards, to heavy unemployment in the Port of Rotterdam, and in shipbuilding and accompanied industries. Many of both the indigenous and new residents of the district had been employed in these sectors. In the second half of the 1990s during a period of economic growth the unemployment level was still very high.

## Urban renewal and the production of space

The implementation of the 1975-1993 urban renewal programme was based – and that was crucial – on the production of spatially and temporally defined entities, such as the first ring, second ring, etc. These entities were called 'communities', and the way in which spatio-temporality was produced became in itself a vital component of the social process (see also Harvey, 1996). The idea was to solve urban problems by using a community approach that was mainly based on the spatial aspect specific to each urban renewal district. The spatial and social frontiers of districts were seen as being equally important in attacking the problems faced by each district and in implementing the programme to solve it. A disadvantage was that the 'communities', well-founded in this way, then had the tendency to organise themselves differently from, and to the exclusion of, others (Stouten, 2004). In the beginning, as they had no access to social housing, it was particularly starters on the housing market and immigrants who were excluded. An advantage of this approach, however, was that there was no displacement of the indigenous population, and many people have benefited from improved housing.

Then, a couple of years into the urban renewal process, the neighbourhood started to become less homogenous than planners had anticipated, and the aspiration of 'building for the neighbourhood' had to be replaced by building for the present neighbourhood, and later on, by building for the city. The aim of the 'building for the neighbourhood' strategy had been to improve the housing conditions of the population living there at that time by building good and affordable housing in their own neighbourhood. However, there were a number of contradictions in those aims, mainly due to compromises between local government and tenant organisations. Both tenant organisations and some local government departments wanted to stimulate the number of families in the neighbourhood, and reduce the numbers of immigrants and single people, these being groups characterised by high mobility. Other local government departments wanted to strengthen the economic position of the city by stimulating the return of middle and higher income groups. Then a stop was placed on attempts to achieve a balanced population structure because the poorer households were finding it more and more difficult to pay their housing expenditures.

Another growing problem was how to plan demolition and the production of new housing. A specific problem was how to reallocate homes to households belonging to one of the ethnic minority groups.

So it was the urban renewal process itself that came to dominate the list of priorities, and with that the shift from building for the indigenous population to building for a population of newcomers. Since the 1990s, there has been a further shift towards making it easier for households living outside the neighbourhood to move to the Oude Noorden district. The district is neither a ghetto, nor do the inhabitants experience it as being one (Stouten, 2004). Although there is no sign as yet of ghetto formation, there are nonetheless small concentrations (not larger than housing blocks and streets) of socially disadvantaged residents, despite government efforts and measures to prevent this. These concentrations occur in those parts (mainly in the northern section) of the district that have a relatively higher population density, smaller streets and not enough public space. The problems related to this are due not so much to ethnic concentrations as to household composition. These groups mainly have large families, and this, combined with the dense population, puts great pressure on public space and playing grounds, leading to their degeneration. If the district were to remain heterogeneous, with different groups living next to each other, there would be fewer tensions on the use of public space. Polarisation and duality would also be inadequate to describe this situation. The district offers a large number of positive elements for low-income groups. Due to urban renewal, much of the housing stock is of good quality. There is a large and varied range of shops, and specific facilities and shops for ethnic groups. Moreover, the district is very near to the city centre so it is not isolated spatially.

## Heterogeneity: the meaning and value

The most striking characteristic of the social composition of the district is the over-representation of unemployed and low-income groups compared with the national level, and, to a lesser extent, the city level. And within these groups, there are relatively more people of Turkish or Moroccan origin. However, not all residents of the newly built and modernised houses have a low income; there are Turkish and Moroccan middle-class households too. It should be noted that these concentrations of low-income groups are mainly the result of external developments, and have little to do with the outcomes of urban renewal policies. Not all households were in a position to profit from economic growth during the 1990s. In particular, older job-seekers with lower education levels profited less from economic prosperity. Taken as a whole, the education level in the district has risen in recent decades to a level that almost equates with that in the city. This is mainly due to an increase in the number of residents with a middle and higher level of education.

As almost all the groups targeted by social policies are strongly present in the population of this neighbourhood, a large number of the low-income households – one-parent families, single people, families with children and the elderly – have profited from the newly built and modernised social housing that was part of the urban renewal programme. However, due to the increase of running costs for utilities and rent, combined with austerity policies on housing subsidies, they now form a risk group. The housing associations are faced with large and structural arrears in rents.

Viewed from the angle of social aspects such as 'exclusion' and 'keeping abreast of advances in information technology', this district does not differ very much from the rest of the Netherlands. Although households older than 65 years lag behind in these respects, the situation for the unemployed and for people unable to work is on a par with the national level.

An interesting finding was that although most of the households with the highest incomes had only lived at their present address for just a short time, prior to that, they had been living in the neighbourhood for more than 10 years (Stouten, 2004). This is proof that higher income households are happy to continue their housing career in the neighbourhood itself. This demonstrates a level of social and material satisfaction with the district, and particularly with their present newly built and modernised social housing. Most of the respondents in this group do not want to move, and about 12% said that they would like to buy their present apartment. Only a very small number of households has concrete removal plans, mostly because the present dwelling is to considered to be 'too small' or of 'insufficient quality'. About half of those planning to move would like to stay in the district. Most of the inhabitants have been satisfied with the social cohesion, but there are nuances; half of the respondents to the survey rated their contacts with other people as pleasant, while the other half complained that people hardly know each other. Despite structural problems with drugs, criminality and intimidation, more than half of the respondents thought that the inhabitants of the neighbourhood 'live there peacefully'.

## City: divided spaces

Castells defined space within an urban context as follows:

> "Space is a material product, in relation with other material elements – among others, men, who themselves enter into particular social relations, which give space to (and to the other elements of the combination) a form, a function, a social signification. It is not, therefore, a mere occasion for the deployment of the social structure, but a concrete expression of each historical ensemble in which a society is specified." (Castells, 1972:152)

Changes in production, employment structure, distribution of goods and services, social conditions and relationships also cause changes in urban structure and fabric. These developments have led to the networks to and from a few areas within the city becoming more and more internationalised. This affects the whole city, however. Important in theories on urban development is that these economic restructuring processes are thought to have led to a shift in industrial and post-industrial labour markets. In relation to the social dimension, two different approaches can be distinguished (Burgers & Musterd, 2002): the first of these, represented by the work of Castells and Sassen, and particularly the former, considers polarisation to be the central force, and the second approach, represented by Wilson, focuses on the spatial mismatch between supply and demand on the labour market.

Though the concept of the dual city is inadequate to represent the Dutch situation, especially in old city districts, it allows some scope for discussing the analysis of large post-industrial cities in the Netherlands (Stouten, 2000). In Dutch research on spatio-social urban inequalities and divisions, the notion of segregation is often used, particularly with respect to spatial aspects. In classical human ecology, according to Saunders (1981), this concept has been used neutrally to reveal the internal spatial differentiation in cities along lines such as social class, life cycle and ethnicity. However, during recent decades, the focus of this idea on the ethnic dimension has given it negative connotations. Because of differences in payment and quality of work, and the status of being employed or not, the labour market is one of the main sources of inequality (Burgers & Kloosterman, 1996). In the last few decades, marked changes have occurred in the labour market because of the fundamental restructuring of urban economies due to globalisation, de-industrialisation, a concentration of economic management, and the related growth in service industries. But there is no evidence for claims, such as those made by Sassen (1991), that the economic restructuring processes are leading to a growing polarisation of the occupational and income structures (Stouten, 2000). Important factors in this debate are changes in the occupational, educational and income structures compared with the positions of the unemployed and immigrants. Below, we concentrate on the last two aspects. There are several problems in analysing the socio-spatial structure:

- Choosing the most appropriate level of scale – the building block or the neighbourhood, the core municipality or the total region, or the Randstad;
- Judging to what extent neighbourhoods, such as Oude Noorden, with its large share of non-Dutch groups, are isolated spaces, spatially, economically and socially (also with respect to advances in information and communication technology).

## Unemployment

During the last 20 years, there has been a different pattern of employment structure in the four major cities of the Randstad than in other parts of the Netherlands. The relatively high share of unemployment there has been caused by strong post-industrial changes which have brought about new employment (and thus educational) requirements. The presence, in the major cities, of groups of workers whose position on the labour market is weak, coupled with strong competition from commuters, has reinforced the mismatch. Wilson's mismatch scheme appears to explain the concentration of unemployment in the large Dutch cities. It is largely the poorly educated who are unemployed, many of them belonging to one of the minority groups. These are people who had difficulties in finding jobs, even in the expanding urban labour markets of the 1990s.

Long-term unemployment remains a major aspect of social division, and the division between the employed and the unemployed is indeed a sharp one, but not more so than elsewhere in the EU. The high rate of long-term unemployment, especially among the low-educated and non-Dutch population, is linked with the transition from traditional manufacturing industry to economies based on advanced service industries. As long as the supply of women returning to the employment market and school-leavers is sufficient to fill the increasing number, of more highly qualified, part-time and flexible jobs, especially in the service sector, there will be no chance for the long-term unemployed, and they will not be able to benefit at all from that growth in jobs (SCP, 1998). Since 1999, this situation has worsened, not only for those already unemployed, but also for the newly unemployed – their numbers grew from 11% in 1999 to 14.2% in 2004, which approached the figure for 1996. On top of that, between 2003 and 2004, a particularly problematic development took place in that unemployment also increased – by almost 20% – among the under-23s (De Stem, 2004).

## Immigration

An important element in the discussion about unemployment is the participation rate of non-Dutch workers. The proportion of non-Dutch in the working population has increased over the past few years, but the working population among the non-Dutch of working age in the Netherlands is still about 9% lower than among the Dutch of working age. As in other rich countries of Europe, current ethnic segregation has its roots in international migration, especially during the 1990s. Before the 1990s immigrants from (former) colonies such as Surinam and labour migrants came to the Netherlands from poorer countries such as Turkey, Morocco, and, during the 1990s, Eastern Europe; refugees came from politically unstable parts of the world. During the 1980s, the number of non-Dutch registered as unemployed continued to rise, while unemployment among the Dutch population fell.

In 1997, 40% of the non-Dutch population was concentrated in the four large cities. Some of this group aimed at getting jobs in the primary labour market, which offers high salaries and provides career prospects through hierarchical business structures. A second group aimed at small enterprises with poor working conditions and low payment, while a third group, the so-called ethnic entrepreneurs – like Mr. Ibrahim – established their businesses mainly in districts of urban renewal, and still do. Combating unemployment is clearly one of their main motives for doing this.

The informal economy fits in with, and in fact meets, the demand for a flexible organisation of production within the formal economy. Flexibility is important as it allows the employment of casual workers, and the use of flexible contracts. Illegal workers also come within this employment structure (Sassen, 1991). Because the division and inequalities are more structural, the positions on the labour market are more difficult to correct for those in the lower employment scales. But at least the urban renewal programme in Oude Noorden has given the unemployed, and at long last minority groups too, improved housing conditions.

## Globalisation and space; centres and margins

The dominancy of information industries, strongly linked with the growth of a global economy, has contributed to a new geography of centrality and marginality (Sassen, 1999). In part, this new geography reproduces existing inequalities. Rather than making space unimportant, the globalisation of economic activity has given rise to new sorts of places such as the 'global city'. Space and place entrepreneurs act like 'growth machines' by pursuing even more intensive land use to increase exchange values even more (Gieryn, 2000). In the 1960s, community organisers and tenant groups in Oude Noorden, fearing that the quality of life and the environment would be harmed, resisted this kind of activity to protect the value of the district for the users.

Nevertheless, the investments in social urban renewal here appear to have been worthwhile (Stouten, 2004). The quality reached by enormous investments in social housing, facilities, and later, also private-housing-sector subsidies for modernisation and the refurbishment of the public space in general, were judged positively by the users, especially the floor plans of the housing. However, due to the changing politics regarding housing, and moves towards privatisation and market development, the exchange values are sharpening the differences between locations more and more. New policies of the central government related to market development are forcing housing associations to increase their rents even more – made easier once tenants have moved out. But the higher rents have led to arrears in rent payments becoming structural, and this is threatening the present social quality of the neighbourhood (Stouten, 2004).

The focus in this section has been on two issues that are central to spatio-social politics. In addition to these, the municipal authorities of Rotterdam believe that, as well as the recent differentiation in housing tenures, living conditions and integration can also be improved if minority groups are forced to disperse.

## Residential differentiation

Globalisation leads to more fragmentation and localisation of housing and consumption. According to Harloe, Marcuse and Smith (1994) there is a trend towards developing new residential enclaves that are defensively cordoned off from the environments in which they are embedded. Up until the mid-1990s, the out-migration of the original population decreased due to urban renewal schemes, and areas have been made more attractive. The measures that have been directed specifically towards these areas have improved the quality of dwellings and the residential environment. Despite this, the concentration of ethnic minorities and unemployment has shown a further increase, but this has not been accompanied by an increase in spatial and social isolation (Roelandt, 1994; Burgers & Kloosterman, 1996; Stouten, 2004).

These concentrations can be found in both old urban renewal districts and in high-rise, post-war housing districts. Far from segregation increasing, in urban renewal districts there is insufficient room to accommodate the continuing growth of concentrations. Although sharp contrasts in the Netherlands have mainly been removed by measures to provide housing, a person's social position is increasingly determined by their position in the labour market. Because changes in the Rotterdam economy have been less favourable than elsewhere, coupled with the fact that there are more lower-educated job-seekers there than, for example, in Amsterdam, Rotterdam is likely to be one of the cities that will experience social exclusion. Of the large cities in the Randstad, Rotterdam is also the one with the highest number of problematic neighbourhoods and the greatest number of social problems in the Netherlands (Burgers & Kloosterman, 1996). One of these is social exclusion. This is a problem that is often defined as being due to the formation of an underclass, but this too general a definition. Those excluded from the formal labour market are anything but a homogeneous group of residents; there are many differences among them. Until the mid-1990s, because of strong state intervention, districts such as Oude Noorden were not isolated, either socially or spatially. The most striking feature of city districts in the Netherlands at the present time is heterogeneity, socially and ethnically, combined with a certain level of fragmentation (Stouten, 2000). And the most striking division within these districts is the predominance of unemployment and low incomes. This does not automatically imply that there is a large ethnic population in the district, because there are also large numbers of unemployed among the Dutch. The differences in income and higher education occur mainly within the indigenous population. However, in Oude Noorden, research has found that there is also some differentiation among Turkish and Moroccan residents, some of whom belong to the middle class (Stouten, 2004).

Current municipal strategic planning tends to overlook changes in residential differentiation, not recognising them as one of the key elements. Rotterdam has largely copied the aim of central government policies to increase the owner-occupied sector. At the end of the 90s, in districts such as Oude Noorden, small pilot projects with low and middle priced owner-occupied houses showed that there were households in the district that were able and interested to switch from the rental to the owner-occupied sector. Households in the ethnic groups were also found to be among those able to pay for low-priced housing, but for the more expensive units, the buyer profiles were young Dutch singles and Dutch couples where both partners earned above average incomes. Due to growing demand, low interest rates on mortgages and tax relief, housing prices have increased enormously, so for most of the present tenants, only low-priced housing would be accessible.

## Social engineering and space

An extreme form of selectivity has recently been introduced by local authorities aimed at stopping migrants with poor economic and social perspectives from settling in districts where there are disproportionate numbers of minority residents. This strong belief in engineering communities, spatially and socially, is based on the hypothesis that one can prevent problems such as the degeneration of housing and residential environments, poor social coherence and integration, and low education, by excluding and blaming specific groups. These possibilities ignore, for instance, that not all households belonging to one of the ethnic groups have a low income and make no progress. As long ago as 1972 and 1980, legal and social obstacles appeared that made it impossible to influence the composition of the population. Nevertheless, present right-wing parties continue to call for extreme measures to disperse low income, mainly ethnic, groups. However, even the most extreme proposals cannot stop concentration. It will hardly have any impact on the composition of a population in which the present share of indigenous people is as high as 56%, decreasing to an expected 50% ten years later. Using housing assignment as an instrument for settling immigrants in the so-called ethnic and poor concentration districts should only be carried out to a limited extent, and even that is discriminatory. Paradoxically, at the same time as attempts are made to realise this policy, the main trend in housing provision is towards privatisation.

The solution for these urban problems will not be found in dispersing the immigrant population, but more in adopting approaches to increase safety (and reduce criminality), education, and investments to refurbish public space and modernise the building stock. As proved by measures taken in the past, such investments are very worthwhile (Stouten, 2004).

# Conclusions

At first, spatial and social entities were seen as being equally important in combating problems and carrying out programmes in urban renewal districts. Later on, though, it became clear that problems such as unemployment cannot be solved on the district level, and even housing problems and the degeneration of public space need a wider approach. What is needed is to integrate the physical, social and economic aspects. In the light of these lessons learnt from the past, the recent local-government proposal to decrease the proportion of low-income, migrant residents in these urban districts is all the more astonishing. Again, as in the 1970s, for all intents and purposes, local government policies (B&W Rotterdam, 2003) are focused on ethnicity, which, in particular, discriminate against low-income immigrants (Mik, Drewe, Hulsbergen, Van der Knaap, Rodgers, Rosing & Sleegers, 1979; Priemus, 2004). These policy proposals are mainly based on fear. In the 1970s and 1980s, it was the fear of riots and tension between immigrant and Dutch groups. What happened at that time was that organisations of both groups of tenants took measures to combat ghetto formation, attack discrimination and the degeneration of housing and facilities. In the 1980s, as now, there was also a period (more severe, even) of economic decline. The current policy measure, introduced in 2004, is based on middle-class fears about the worsening social climate, and about losing economic security and voters. Under this policy, access to cheap housing is not refused to people with too high an income, but to those whose income is too low. However, this is in conflict with current legislation such as the *Besluit Beheer Sociale Huursector (BBSH)* [Management of the Social Sector Decree]. Consequently, a situation has arisen in which not only are the problems inadequately defined, but existing policy measures are in conflict with each other. According to Mik *et al.* (1979), rent arrears, housing degeneration and poor success on the labour market within ethnic minority groups are caused by socio-economic limitations such as income and education, the inadequate instrumental integration of these ethnic groups and indirect institutional discrimination on district and local government levels. With regard to the first issue, the classification low-income/low education group and that people in this group tend to live in the same districts, is a general classification, whether people belong to a minority group or to the indigenous inhabitants (Drewe, 1986).

It is the social and ethnic heterogeneity, combined with a certain level of fragmentation that is the most striking characteristic of these districts. The different groups here compete for social facilities, housing and (public) space. This means that to help prevent ghetto-like formations arising, what is needed are investments for social programmes and for modernising and refurbishing public space. Viewed from this angle, therefore, space *does* matter, even though not all problems can be solved on the scale of a block of buildings, a street or a neighbourhood. By building low-cost, owner-occupied housing, residential differentiation will be a positive force in holding districts together as a heterogeneous entity in which both potential movers and buyers seek to continue their housing career within the district because they feel comfortable there, socially, materially and environmentally.

## References


B&W Rotterdam, 2003, *Rotterdam zet door. Op weg naar een stad in balans*, Gemeente Rotterdam

Burgers, J. & R. Kloosterman, 1996, *Dutch Comfort: Post-industrial Transition and Social Exclusion in Spangen*, Rotterdam, Area, 28/4, pp. 433-445

Burgers, J. & S. Musterd, 2002, Understanding Urban Inequality: A Model Based on Existing Theories. An Empirical Illustration, *International Journal of Urban and Regional Research*, 26/2, pp. 403-413

Castells, M., 1972, *La question urbaine*, Maspero, Paris

De Stem, 2004, Steeds meer jeugd werkloos, *BN De Stem*, October 7

Drewe, P., 1986, *Minderhedenbeleid – een bijzondere toetssteen van goed bestuur. Verkenningen van het binnenlands bestuur*, No. 22, Ministerie van Binnenlandse Zaken, Den Haag

Drewe, P., 2000, European Experiences; In: Roberts, P. & H. Sykes (eds), *Urban Regeneration, A Handbook*, Sage Publications, London, pp. 9-37

Droogleever Fortuijn, J., S. Musterd & W. Ostendorf, 1998, International Migration and Ethnic Segregation: Impacts on Urban Areas - Introduction; In: *Urban Studies*, 35/3, pp. 367-370

Gieryn, T. F., 2000, A Space for Place in Sociology, *Annual Reviews Sociology*, 26, pp. 463-496

Harloe, M., P. Marcuse & N. Smith, 1994, Housing for people, housing for profits; In: Fainstein, S., I. Gordon & M. Harloe (eds), *Divided cities: New York & London in the contemporary world*, Blackwell, Oxford, pp. 175-203

Harvey, D., 1996, *Justice, Nature and the Geography of Difference*, Blackwell, Oxford

Mik, G. (rapp.), P. Drewe, E.D. Hulsbergen, G.A. Van der Knaap, H.M. Rodgers, K.E. Rosing & W.F. Sleegers, 1979, *Segregatie in Rotterdam, feiten en beleid. Samenvatting van het onderzoek*, Erasmus Universiteit, Rotterdam

Priemus, H., 2004, Stedelijk beleid uit balans, *Tijdschrift voor de Volkshuisvesting*, 3, pp. 6-12

Roberts, P., 2000, The Evolution, Definition and Purpose of Urban Regeneration; In: Roberts, P. & H. Sykes (eds), *Urban Regeneration, A Handbook*, Sage Publications, London, pp. 9-37

Roelandt, Th., 1994, *Verscheidenheid in ongelijkheid*, PhD Thesis, Thesis Publishers, Amsterdam

Sassen, S., 1991, *The Global City; New York, London, Tokyo*, Princeton University Press, Princeton

Sassen, S., 1999, The City: Strategic Site/New frontier; In: INURA, *Possible Urban Worlds; Urban Strategies at the End of the 20th Century*, Birkhäuser Verlag, Basel/Boston/Berlin, pp. 192-200

Saunders, P., 1981, *Social Theory and the Urban Question*, Unwin Hyman, London

SCP, 1998, *Sociaal en Cultureel Rapport 1998: 25 jaar sociale verandering*, SCP, Rijswijk

Stouten, P., 2000, Strategic Planning, Sustainability and Urban Regeneration: In: Carmona, M. & J. Rosemann (eds), *Globalisation, Urban Form & Governance*, Delft University Press Science, Delft, pp. 29-45

Stouten, P., 2004, *Duurzaamheid van de stadsvernieuwing*, PhD Thesis, Delft University Press, Delft

# 10 Reason, Capital and Urban Space: The New Role for Urban Planning in Triggering Societal Change [1]

Roberto Rocco

## Introduction

The emergence of a renewed 'market-oriented reason' has dramatically changed the way public urban policies are discussed and carried out. This has had equally dramatic consequences for how urban space is planned and designed. It is not only urban 'form' that has changed, but also how architects and urban planners perceive the role and function of urban space. This new perception is the product of ideological constructions that conceal the conflicts and contradictions that permeate and conform with urban space. These ideological constructions are based on disjointed discourses.

Following the new Marxist thinkers, especially Chauí (1981), ideology is not merely a collection of false ideas, but a societal representation of reality that, in the end, reinstates reality itself. Ideology is therefore always an incomplete or partial explanation of reality; it is an explanation full of 'necessary' gaps. Were these gaps to be filled, then ideological discourse would crumble and actual conflicts would be exposed. In relation to urban space, one of these 'gaps' has to do with the 'objectivity' of space.

Space can be measured, quantified and submitted to all kinds of scientific estimation and quantification, without which it would be impossible for urban planners and architects to design and build. Nonetheless, the multi-layered nature of urban space cannot be derived merely from its immediate physical aspects. The French historian E. Leroy-Ladurie said that *"le présent ne s'explique pas par le présent"* (the present [time] cannot be understood by the present [time itself]). Similarly, urban space cannot be understood *just* by measuring or quantifying it. It is necessary to look at the constituents of space, which do not necessarily reside in physical space itself, e.g. its history, economy, and even ideological constructions. Therefore, urban space should be both physically measurable and socially intelligible.

The ideological discourses that aim at presenting urban space just as a measurable and quantifiable unit, neutralise the conflicts that take place within it and contribute towards turning it into a commodity. The commoditisation of urban space is at the root of the recent shift of public policies in a more pragmatic and market-oriented direction. In order to legitimise this process, a new discourse of public management has come into being.

---

1 A preliminary version of this text has appeared as a short article in the Indonesian architectural journal *Jurnal Dimensi*, 32/1, July 2004.

This chapter deals with the perception of urban space within the framework of globalisation, neo-liberalism and the reorganisation of the public sphere during the closing decades of the 20th century. We carry out our analyses based on a critique of the renewed instrumental reasoning that underpins recent trends in urban planning. This applies especially to planning in Developing Countries, where trends such as strategic planning can assume an entirely different meaning, relying on the reification of urban space by turning it into a commodity.

Finally, we present some alternatives to the present trends, as well as some remarks on the role of urban planning as a tool for public participation and for constituting a new public arena in the debate on the future of cities.

## The emergence of a 'new reason'

During the last decade, urban problems have been approached and studied using a 'new reason' that originates from the repositioning of market values (neo-liberalism) that came about in the wake of the retreat of the public sphere (Sennett, 1992), the birth of 'post-modern thinking' (Harvey, 1990) and the general acceleration of socio-economic phenomena, generally branded as globalisation.

The reasons for this are many and complex. It would be impossible to understand the phenomenon without trying to comprehend the making of a new kind of public sphere and the nature of the State in a world dominated by new flexible modes of production, the growing power of Trans-National Corporations (TNCs), and the persistent cross-border movements of speculative capital, all of which are supported by the increasing sophistication of business services and the techno-information revolution.

We can trace the origin of some contemporary urban and spatial theories to the resurgence of a 'new reasoning', which we choose to call 'new instrumental reasoning'. Put differently, much of what has been said and written about urban space, urban renewal and urban management in the last decade is based on a reasoning that is instrumental to the necessities and the logics of late capitalist accumulation. This trend is apparent in issues such as the 'Minimum State' and 'Poverty Alleviation'. It is also present in apparently innocent and valuable concepts such as 'Sustainability' (e.g. when used because necessary compensation and subsidisation policies have been abandoned), and 'Strategic Planning' (e.g. when used by people who forget to enquire who the true beneficiaries are of the strategies adopted).

## The reification of space

Partnerships between public and private sectors can be justified in the context of the social contract proposed by Hobbes (1651, rep. 1962) and Rousseau (1762, rep. 1968). However, they are not so easily explained under the critique of the social contract made by Hegel and carried out by Marx and others. Marx considered the modern State to be the institution that, above all others, has the task of ensuring and preserving exploitation and class domination (Bottomore, 1985). For orthodox Marxist thinkers, the state is the institution of organised violence used by the ruling class of a country to maintain the conditions of its rule. Despite Marx's formulations and the subsequent elaboration of his ideas by an endless string of theorists (e.g. Marx & Engels, 1970), there is still a heated debate about the true role of the State in modern society. Most thinkers along Marxist lines would agree, however, that the capitalist State is largely based on ideological assumptions that operate in order to ensure the domination of the ruling classes over the rest of society. Of course, those who see the State as the mediator of class struggles and the guardian of law and individual rights under constitutional governments vehemently deny this view. Despite the fact that constitutional governments in advanced Western societies were relatively successful in solving social struggles and promoting general welfare after World War II, there is now a new phenomenon at hand: a general political disengagement and the internal corrosion of the egalitarian social utopias that have dominated modern Western thought since before the French Revolution (Harvey, 1990). This has had immediate impacts on various fields of knowledge, including those related to urban management and design.

Put briefly, this process has taken place due to the internal corrosion of the core idea of equality and social justice, due to the decay of egalitarian utopias, and the departure from ideas centred around the Welfare State, especially after the fall of the Berlin Wall in 1989, when the blatant failures of 'real socialism' became evident to everyone. At present, the various modes of structuralism (Marxism and Keynesianism) are being denied. Concomitantly, the fundamentals of the Welfare State are being demolished, to be replaced by a new version of 19th century *laissez-faire* (Singer, 1999).

Within this framework, where, hand in hand with a firm belief in market infallibility and technical development, the economy is being liberalised, there is a striking reoccurrence of irrationality in politics and planning, regarding certain aspects of new instrumental reasoning.

Contemporary capitalism relies increasingly less on labour and the products of labour, but more on the rapid and unforeseeable movements of speculative funds and production that is fragmented and dispersed around the world (a product can be designed, assembled and sold in different countries, and made from parts that come from different corners of the globe). Therefore, the point of reference is no longer labour (i.e., the labour force itself and labour materialised in the form of products), but rather competitiveness, efficiency (particularly of logistics) and the ability to play the games of late capitalism. This trend has been brilliantly explored by Sennett (2000, 2003).

The change of emphasis from labour to competitiveness has led to a move from humanistic values to pragmatic standards related to market-oriented effectiveness. Consequently, there is an increasing rejection of egalitarian values (in Western Europe, expressed in terms of the Welfare State) as valid parameters for decision-making. During the 1990s, in forming public policies for the city, technical considerations associated with competitiveness and efficiency began to gain precedence. Increasingly, the city came to be viewed as a commodity that had to be 'marketed'; to use Ashworth's (1998) term, the city was becoming the 'market city'.

The blind belief in science that dominated the imaginary and ideological constructions justified violence perpetrated in the name of Western Civilization and its self-proclaimed superiority. Today, the world of appearances, imagery and representation, so important to the marketing of products, has almost replaced the products themselves. At the same time, this trend has displaced core thinking and judgment processes. Urban projects have become contaminated with 'commodity fetishism'; they have been given appearance without substance (pure marketing). We refer here to large urban intervention projects carried out in various cities all over the world, where 'spectacular' architectural form is associated with corporate image in order to create a new urban environment of pure imagery and brand names, without any real concern for the citizens and users who live and work there (see Ill. 10.1).

This has serious consequences not only for the urban environment, but also for the epistemology of urban planning. It also influences how serious critique of contemporary urban phenomena is formulated, since most of this critique is based upon formal and publicity-related parameters, and from that perspective, urban space is nothing more than merchandise. It is a degenerative process that relies principally on an unprecedented aesthetisation of space.

It may be argued that urban space has always been 'aesthetised', since public space has been used throughout history to convey meaning, rather than function; to represent economic or political power (see Ill. 10.2). However, urban space has always been a mismatch of social contradictions, but, if we accept the ideas of Theodor Adorno (1947, rep. Adorno & Horkheimer 1993) and Walter Benjamin (1936) on the culture industry and the loss of the 'aura', respectively, we may say that urban space has entered the realm of merchandise. It has become a cultural *and* commercial commodity. It is now tailored for consumption, not by the users of this space, but by a very vague set of market-dictated necessities. These 'necessities' arise more or less artificially, imposed, for instance, by the ideology of globalisation and the imagery of the global city, which are viewed by the main decision-makers as desirable and inexorable. This ideology is largely based on the aesthetisation of space, which we will analyse in the next section.

## The aesthetisation of space

Aesthetics plays a major role in the growing tendency to spectacularise urban forms. Appearance alone and 'spectacular' forms are becoming more important than discussions about the real material substance of public space and the social meaning of public actions in the city (see Debord, 1995). This creates an artificial social consensus among opinion-makers, which has little to do with the real material living conditions of the majority of citizens. Moreover, for developing cities in a world economy, those who are in some way 'excluded' from the main socio-economic streams are just considered a 'nuisance' for development. City competition to attract direct foreign investment rarely gives any thought to the wishes and needs of underprivileged sections of the population (see Ill. 10.3).

In J. Galard's (1999) opinion, reality is being increasingly diluted by aesthetics. Basing judgment and analysis in the urban field on aesthetic generalisations has generated a weakening of critical judgment, which has influenced both politics and decision-making processes. This has caused administrators and leading actors to opt for programmes that are mostly based on making urban forms spectacular in order to attract investment.

The triumph of aesthetics over citizens' real material needs is clearly advantageous for market forces, because 'commodity fetishism' leads to the programmed obsolescence of products, and also to the precocious obsolescence of spaces and buildings. Space is turned into a commodity; a form that lacks comprehensive social meaning, where social constructions even appear as inappropriate for explaining urban space.

Some researchers in the urban field are becoming increasingly convinced that space actually produces societal changes, rather than space itself being socially produced. An alternative line of thought, and

**Illustration 10.1:**
**Pudong, the new business district of Shanghai, the most dynamic city of China. Spectacular architecture is associated with brands and corporate image in order to create an image of modernity (photo by Frank de Graaf)**

**Illustration 10.2:**
**The Monument to Victor Emmanuel II, Rome. The enormous white marble monument at the Piazza Venezia was built as a tribute to the first King of a united Italy (photo by Roberto Rocco, 1996)**

**Illustration 10.3:**
**The development of a new business district along the River Pinheiros in São Paulo, Brazil, one of the countries with the largest inequalities (photo by Roberto Rocco, 2002)**

the one to which we subscribe, tries to reconcile both ideas by stating that space is indeed socially produced, but that it has the potential to yield or set off, dialectically, certain societal changes.

Deutsche (1996) considers spatial forms to be predominantly 'materialised social structures'. However, viewed from a functional angle, spatial order appears to be controlled by either mechanical or organic natural laws. In Deutsche's opinion, when space is divorced from its social production, it is consumed by 'commodity fetishism' and transformed by inversion. In other words, when space is represented as an independent object, it seems to exert control over the very people who produce and use it. The strong social potential of space is therefore 'neutralised', and once 'neutralised' in this way, space then becomes politically neutral – merely utilitarian. In Deutsche's view, the notion that the city 'speaks for itself' (i.e. that a city's form alone can explain the functioning of society) masks the identity of its citizens and users; those who speak *through* the city (Deutsche, 1996).

In effect, formalist discourses that instrumentalise space (that are based on form alone) result, more than anything, in public policies that aim 'to restore' public space. This does *not* mean restoring the public sphere as such, but rather creating a mock aesthetised type of public space.

## The neutralisation of space

The geographer M. Santos (1993), co-founder of *Nouvelle Géographie*, sees space as having rationality, but only the rationality that comes from the intentions behind the choice of objects that compose it. However, only rational social players can formulate the intentions behind the choice of objects.

Reducing space to a mere mathematical construction arises when reasoning in general is instrumentalised to act as a tool for the dominant ideology of competitiveness and productivity. In other words, an instrument is made of reasoning so that reality can be increasingly understood (and actions derived from this understanding) by means of indexes, numbers and values that can be isolated and played with. It can be seen as a time-saving short-cut, doing away with the need to undertake a more comprehensive analysis of reality. This means that the more space is reduced to mathematical forms, only comprehensible to 'specialists', the fewer people there will be who are able to 'act rationally' to space, because the increasing mathematisation of this space will render them 'incompetent'.

'Competence' is used here in the sense of having knowledge, understanding, and practical and thinking skills to perform effectively in meeting established standards, and in controlling the processes needed to arrive at those standards. Not everyone is 'competent' at everything all the time. 'Levels of competence' presuppose that people are competent at some things, but less so at others, so competence can be used as a tool for controlling the less able

Quantitative analytical methods can be manipulated very easily by those with power and 'competence'. This is especially true in the economic field. Santos (1993) concludes that this creates ideal conditions not only for intensifying profit, but also for divorcing change from the common man. By using competence in mathematics to rationalise space, the 'incompetent' are excluded. Ideological forces make use of the public's (alleged) 'incompetence' when judging what is best for the city. This allows the market to become 'tyrannical', and in the face of market necessities, even democratic and representative States have a tendency to become apparently 'powerless'. But then again, the State's 'lack of power' can also be interpreted as a new ideological construction. Recent studies show that, in many countries, the State continues to be a powerful ally of market forces, even helping to create spaces where capital can operate without constraints. Public places are spectacularised, and to help maintain the popularity of local administrations, public actions are given extreme visibility. Chomsky (1997) believes that publicity strategies such as these are fundamental in modern societies for building up political legitimacy and perpetuating political power.

The new phase of capitalism is largely based on entrepreneurial strategies of flexible growth. These include making labour regulations flexible and promoting a general lack of solidarity among different social groups. These trends are accompanied by the rapid decline of the public realm (Habermas, 1989; Sennett, 1992, 2000), which, as indicated by Sennett (1992), signifies a decline of solidarity and common welfare itself as valid parameters for social relations and decision-making. The parameters *en vogue*, as previously discussed, are competitiveness and productivity. And when reasoning is anchored onto market rules, then reasoning itself is superseded by the internal logic of the market. It is replaced by 'pragmatic reason', which justifies itself through the same ideological constructions – through publicity and building spectacular forms – that secure for the dominant classes their legitimacy to rule. Urban space has become a tool for these strategies, and at the same time is the result of them.

Under these conditions, urban space is assessed and understood through the filters of functionality (the 'well-managed city') or plasticity (the 'beautiful city'), without taking into account the social relations that are the constituents of urban space. It becomes a tool for development and progress, taking no account of *whose* progress and *whose* development it should convey.

In Deutsche's (1996) opinion, the ideology of function overshadows how, in practice, the uses and definitions made of cities are in conflict with each other, obscuring the very existence of groups who put the space to any other than the dominant uses. In defining the city as a product of social practices, Deutsche (1996) quotes Ledrut who disagrees with the technocratic view that the city is a product of 'specialists'. Ledrut insists that the city is not a spatial frame external to its users, but is produced by them. Deutsche also points out that, although Ledrut's formulations might seem disappointingly simple, they have profound implications, nonetheless. Ledrut not only explicitly recognises the participation of different social groups in producing urban space, he also argues against a concept of space that is imposed by public institutions and big corporations. These can only be guided by the need to make profit, and can only be legitimated by concepts such as efficiency and beauty. Such assertions coincide with aesthetising reality, a process described by Galard, and neutralising space, as described by Deutsche. And despite the cautionary voices of Deutsche and Ledrut, and the emphasis they place on the importance of the social element, the citizens and users of urban space, the ideas that have been used as pillars for large urban interventions in this age of globalisation have nonetheless been those derived from the 'beautiful city' and the 'efficient city'.

## 'Rescuing' public space

Instrumentalising urban space is the root of another fiction: 'rescuing' public space. The obsession about public space is centred on the wish to re-establish the social element that has been lost in most modern metropolises. Arantes (1999) sees a strong ideological bias in this. She sees in this trend an attempt to try and fill the void left when public space has been ruined. 'Public space' in these cases is no more than a simple marketing image; a scene of social life that has ceased to exist. For Arantes, a construction of this type is merely a managerial scenography of the city, something like the theatre of daily life, where the city's history is no more than the aesthetics of memory; a succession of representative tableaux of daily life (Arantes, 1999) (see Ill. 10.4).

According to contemporary philosophers such as Derrida and Habermas, the 'new public life' would take place through the metamorphosis of urban practice and theory. However, many philosophers of the Marxist tradition, Harvey and Marcuse for instance, disagree with them. Because there is strong evidence of a return of social separation and ghetto formation in the new city of flexible production and growth, they see new public life as being nothing more than 'urban decoration'.

It is here that the public sphere and public opinion appear to coincide, though more as a disjointed construction, full of 'gaps', that attempts to accommodate diverse aspects such as cultural industry, publicity and the manipulation of public opinion (see Adorno, 1993). Therefore, there is a construction behind the drive to rescue public space, that celebrates public space without allowing any parallel action to rescue the public sphere. It is revealing that big urban projects all over the world are only possible because of partnerships between the public and private sector. Private actors make strong demands for their brands and logos to appear prominently in the new spectacular spaces with their ultra-modern media facilities, designed by world-famous architects (Ill. 10.4).

In Kurz's (1996) opinion, 'modern irrationalism' does not just manifest itself through fanatical religious movements, but also through the rational façade of trendy political ideas and scientific knowledge. In this way, the ideas contained (and hidden) within the concept of partnership between public and private sectors can be identified as mere instruments for concealing conflicts, leading global cities towards an implausible 'happy ending' amidst the social barbarism of urban environments that is a feature of most big cities, whether in developed or developing countries.

## Governance as an ideological tool

Governance is a recurring term both in the media and in the academic world. The word itself is not new, but its meaning is being constantly refined by institutions and thinkers. For many of them (e.g. The British Council, The Centre d'Études en Gouvernance of the University of Ottawa, The Governance Working Group of the International Institute of Administrative Sciences in Brussels), the word conveys the idea of a 'process', a field for political action, rather than a fixed concept of public administration. Governance presupposes that there is 'positive tension' between institutional government, civil society (represented by old and new civil institutions, including non-governmental organisations) and enterprises (the 'private sector'). The general aim of this 'positive tension' is to create conditions that will trigger the optimisation of public administration, and generate welfare and social justice, through a better interaction among the three main players that steer society, mentioned in the decision-making processes above.

The concept of governance is based on the idea of balance among different social actors (the State, the private sector, and civil society). In formulating this concept, these actors appear to have the same level of importance and strength, working together to achieve common welfare. However, such ideal conditions do not correspond with reality. Different actors have very different weights in decision-making

**Illustration 10.4: A general perspective of the Piazza d'Italia Charles Moore in New Orleans, 1978. This is an early example of the contemporary trend to spectacularise space. It also demonstrates that the concept of public and private space is becoming increasingly blurred. (source: Johnson, J. & C. Moore, 1986, *Buildings and Projects 1949-86*, Rizzoli, New York, p. 88)**

processes. Among these, transnational corporations are becoming increasingly powerful in deciding where and when cities and regions will get investment and jobs. International organizations such as the International Monetary Fund and the World Bank are becoming more powerful than ever in deciding which countries will get loans and which projects will be funded, whereas the public sector (not to mention civil society) has increasingly become a mere spectator of decisions taken elsewhere. The strength of 'public opinion' is relative and, as analysed elsewhere in this text, is heavily influenced by ideological constructions. As a result, a dubious consensus is reached among urban actors who are, in actual fact, completely different in their capacity to make decisions and in their capacity to undertake real action, not to mention their divergent objectives and goals.

## The role of urban planning in triggering societal changes: The Porto Alegre model

The potential for change in the phenomena described earlier in this chapter can only be accomplished by strengthening the public sphere, using, for instance, direct participation in decision-making as a tool for consequent societal change.

There are very many well-studied cases of participatory decision-making processes. Porto Alegre (Brazil) is an acknowledged example of how participatory policies, combined with a strong legitimate local government can bring about real change through competent urban planning and by using innovative tools to administer the city. Societal changes regarding how urban space is conceived, planned and managed, not only by the public sector (in the form of local government), but also by society as a whole are thereby brought about. Porto Alegre, the largest industrial city in Southern Brazil, with 1.3 million inhabitants, has a local economy worth over US$ 7 billion, and a reputation of hosting a progressive civil society led by intellectuals and labour unions experienced in mobilising people to partake in public life and oppose authoritarianism (World Bank, 2003). The city is an emblem of the 'Urban Participatory' model (also known as the 'Brazilian model'), which is based on participatory public budgetary decision-making. This model expressively rejects urban competition, which it views as a tool for turning the city into a commodity, and for alienating its citizens from political life.

The 'Participatory City Budget' at Porto Alegre includes the following steps, all of which take place in one administrative year:

- Assessing the budget expenditure of previous years, and holding discussions to fix next year's priorities;
- Electing 'regional' delegates (to represent the 'regions' into which the municipality is divided);
- Holding 'regional' meetings;
- Defining priorities;
- Presenting the wishes of the community (NB. this strongly relies on how well citizens participate in 'regional' meetings);
- Reconciling technical/institutional needs and community demands;
- Preparing and submitting a budget proposal;
- Preparing regional and thematic investment plans;
- Getting legislative budgetary approval;
- Following-up meetings and monitoring them well;
- Electing regional delegates, and repeating the above steps again.

Despite its emphasis on economic liberalisation and market forces, the World Bank is also very interested in new experiments with governance. It considers the Porto Alegre case to be a success story. According to the World Bank, Porto Alegre

> "(...) has witnessed some spectacular achievements in recent years, credit for which has largely been given to the participatory budget process. Between 1989 and 1996, the number of households with access to water services rose from 80% to 98%;the percentage of the population served by the municipal sewage system rose from 46% to 85%; the number of children enrolled in public schools doubled; in the poorer neighbourhoods, 30 kilometres of roads were paved annually from 1989 onwards; and because transparency improved the motivation to pay taxes, revenue increased by nearly 50% (budget resources for investment only went up from US$ 54m in 1992 to US$ 70m in 1996). The Porto Alegre experiment also presents a strong example of democratic accountability, equity, and a redistribution of justice, with the participatory part guaranteeing legitimacy to make decisions, and an objective budgeting that ensures fairness in an otherwise arbitrary process of translating political decisions into distributed resources." (World Bank, 2003:3)

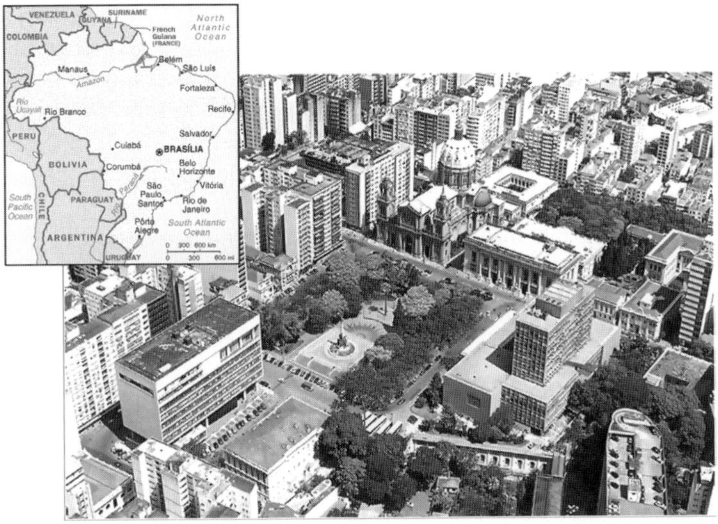


**Illustration 10.5: The Cathedral Square in downtown Porto Alegre and the location of Porto Alegre in South America (photo by E. Loch Terra, PUC RS, Brazil)**

The Porto Alegre 'Participatory Budgeting' model was nominated by the 1996 UN Summit on Human Settlements in Istanbul as an exemplary case of 'urban innovation'. It stood out because it demonstrated an efficient practice of democratic resource management.

The most renowned examples of participatory budgets, inspired in part or entirely by the Porto Alegre experiment, are: Montevideo, Caracas, some of the suburbs of Paris, and Toronto. German communities are currently testing the model in an attempt to revitalise citizens' will to participate, and to gain access to, and put to good use, their knowledge and experience. An example of this initiative is the 'Kommunaler Bürgerhaushalt' project instigated by the Ministry of the Interior Nordrheinwestfalen and the Bertelsmann Foundation (Goethe Institut, 2005).

# Conclusion

Our main conclusion is that new instruments must be used by decision- makers, managers and planners in order to come to resourceful and innovative ways of resisting the deep alienation of citizens from city life, city administration and urban planning itself. Urban planners should constantly keep in mind alternative participatory practices for managing the city, and include its users in decision-making processes. In other words, efforts should be made to include in the administration of the city those who would otherwise be considered 'incompetent' to manage it.

The opportunity for change in the Urban Planning field rests in its inherent trans-disciplinary character. It has the power to generate change by coordinating various fields of knowledge, technical and non-technical, to instigate political-institutional changes, participation (thus contributing to a true 'public sphere') and promote valuable spatial change.

There are alternatives to technocratic and formalist approaches to urban problems that allow public participation, help to build new modes of spatial intervention, and thus genuinely strengthen the public sphere. These are demonstrated by the experiences gained at Porto Alegre and other cities. Contrary to the belief that cities must concede to the top-down requirements of globalisation and relinquish equality as a valid parameter for policy-making, there are alternatives for sustaining homogeneous and egalitarian development in cities today, both in developed and developing countries.

In the interests of social stability, it is important that urban planners and other urban actors continue to experiment and prioritise these new alternative instruments for participatory urban management in the coming years.

## References


Adorno, T. & M. Horkheimer, 1993, The culture industry: Enlightment as mass deception; In: same authors, *Dialectic of Enlightenment*, Continuum, New York (first published 1947)

Arantes, O., 1999, *Urbanismo em fim de linha*, EDUSP, São Paulo

Ashworth, G., 1998, *The Transfer of Power: Decentralization in Central and Eastern Europe*, LGI, Budapest

Benjamin, W., 1936, The work of art in the age of mechanical reproduction; In: *Walter Benjamin: Selected writings*, Vol. 3, 1935-1938, Harvard University Press, Cambridge Mass.

Bottomore, T. (ed.), 1985, *A dictionary of Marxist thought*, Blackwell, Oxford

Chauí, M., 1981, *Cultura e Democracia*, Cortez, Sao Paulo

Chomsky, N., 1997, *Democracy in a neoliberal order: doctrines and reality*, University of Cape Town, Cape Town

Debord, G., 1995, *The society of the spectacle*, Zone Books, New York

Deutsche, R., 1996, *Evictions: Art and Spatial Politics*, MIT Press, Cambridge

Galard, J., 1999, *Arte & Cotidiano, Proceedings of Conference at Instituto Itaú Cultural*, São Paulo, April 29

Goethe Institut, *Participatory Budgeting in Porto Alegre*, http://www.goethe.de/br/poa/buerg/en/terra.htm (Jan 19, 2005)

Habermas, J., 1989, *The structural transformation of the public sphere: Inquiry into a category of bourgeois society*, Polity Press, Cambridge

Harvey, D., 1990, *The Condition of Postmodernity: An enquiry into the origins of cultural change*, Blackwell, Oxford

Hobbes, T., 1962 (1651), *LEVIATHAN, or the matter, form and power of a Commonwealth*, Clarendon, Oxford

Kurz, R., 1996, A biologização do social, *Folha de S. Paulo*, Cad. MAIS!, July 7, pp. 5-7

Marx, K. & F. Engels, reprint 1970, *The German ideology*, Lawrence & Wishart, London

Rousseau, J.-J., 1968 (1762), *The social contract*, Penguin, Harmondsworth

Santos, M., 1993, A aceleração contemporânea: tempo mundo e espaço mundo; In: Santos, M. *et al.* (eds), Fim de *Século e Globalização*, Hucitec, São Paulo

Sennett, R., 1992, *The Fall of Public Man*, W.W. Norton, New York

Sennett, R., 2000, *The Corrosion of Character: Personal Consequences of Work in the New Capitalism*, W.W. Norton, New York

Sennett, R., 2003, *Respect in a World of Inequality*, W.W. Norton, New York

Singer, P., 1999, *Globalização e Desemprego: diagnóstico e alternativas*, Contexto, São Paulo

World Bank, 2003, *Social Development Notes*, No. 71, World Bank Press, Washington D.C.

# 11 Spatial Developments in the Netherlands, 1975–2005

## Scale Increase, More Actors, More Disciplines

Joost Schrijnen

## Introduction

In the period from 1970 to the beginning of the 21st century, spatial planning in the Netherlands changed from spatial development planning, led and mainly financed by the government, to a multi-actor planning. In the same period, an explosive growth of urbanisation and mobility took place. Large parts of the Netherlands developed from territories with autonomous towns in an agrarian environment into areas with urban networks and network cities encircling the Green Heart. The official administrative organisations have not been able to adjust themselves to this situation. In the formal territorial administrative organs – State, Province and Municipality – elected administrators are responsible for assignments that will be supervised within the administrative networks. The management has found itself in a networking, supervising environment. Spatial development that was originally driven by collaborative interests such as public housing (in towns and cities), infrastructure and land consolidation (in agricultural areas), has grown into an integral spatial development assignment involving an increasing number of disciplines and many private parties (Wetenschappelijke Raad voor het Regeringsbeleid, 1998). In this chapter, these developments will be explained from the government's point of view, and suggestions will be made for a follow-up. Spatial and societal trends will be dealt with first: the acceleration of planning, the simultaneous globalisation and fragmentation of economy and space, and the opportunities and threats posed by a changing Europe. Then the dynamic developments from 1975 onwards will be reconstructed.[1] This will lead to a picture of the spatial planner's current field of activities.

## The acceleration of planning

In 1995, the manifestation '50 years of reconstruction, 50 years into the future' was held in the Laurenskerk in Rotterdam (Gemeente Rotterdam, 1995: Van Es' chapter). Held over five winter evenings to commemorate 50 years of reconstruction and the 50th anniversary of the 1945/46 Foundation Plan, these meetings were all about looking back and looking forward. This reconstruction plan to rebuild the inner city of Rotterdam after all the bombing that it had suffered during the Second World War, turned out to be a successful initiative that retained its usefulness over the years. The question remains open whether Van Traa, the city architect at that time, in formulating this 1945/46 plan had envisaged that it would remain useful for such a long time.

1 See on the spatal policies during this period also chapter 3 by Herman Roseboom.

Of course, the *Dienst Stedenbouw en Volkshuisvesting* [Municipal Office for Urban Design and Public Housing] knew that it would be an impossible task to make another 50-year plan. A detailed, formal plan covering such an extensive length of time is no longer relevant, nor are such plans even considered nowadays, despite the long-term impact of spatial plans that have already been carried out. The anniversary was more an invitation to reflect on trends and to analyse them. The public approach, with public debates about the city, design studies, and the exhibitions on the identity and spatial development of the city during this 50/50 manifestation were so successful that they were later imitated in many other Dutch cities that were considering their own (spatial) future. In the meantime, there has also been talk of carrying out a continual analysis of outlooks and trends to distil possible futures from them on every relevant level of scale. These developments followed in rapid succession, so that analysing and considering which interpretation would be the most relevant for daily work became a permanent undertaking: planning activities started to hyperventilate.

## Globalisation and fragmentation

The Port of Rotterdam was still the largest in the world in 1995, when the manifestation meetings were held. At one of these meetings, Henk Molenaar, emeritus professor in port economics at the Erasmus University Rotterdam and former director of the *Gemeentelijk Havenbedrijf Rotterdam* [Rotterdam Port Authority], sketched a scenario of possible extremes with regard to the port's future position in world trade (Municipality of Rotterdam, 1995: Molenaar's chapter). He talked about transport processes that were expanding to global scale, and the increasing capacity of means of transport whereby a spoke-and-hub structure was being created both in container and bulk transport and in the mass transport of people. Secondly, he mentioned the physical and institutional enlargement in scale of entrepreneurship and management. But, at the same time, he pointed to the trend towards disintegration, reduction and fragmentation of society and institutions.

In fact, it is this reduction in scale that releases the power of innovation and creativity. What now is the actual meaning of these trends for the spatial development of the Netherlands and the Randstad; and what is the relevance of Dutch territorial character, history and culture in that expanding world? The concentration of financial information is restricted to a few metropolises. At certain nodes, there are concentrations of trade flows and transhipment of goods, and at others, concentrations of people and information (Sassen, 2001). Institutional and progressive economic scale enlargement is a fact: in future this might lead to one European stock exchange and to enormous intercontinental transport firms, financed by international capital, and managed from just a few places around the globe.

## Everything is on the move

At the 1995 manifestation in Rotterdam, the changing environment was mainly looked at from the viewpoint of hardware and ICT. In the meantime, the social consequences of globalisation have also become more and more evident. Globally, flows of migrants on an annual basis are still rather small compared with the total world population, but because of their continuing size and force they lead to new worldwide racial, religious and cultural interactions. The only way to prevent large-scale migration from poor to rich countries is to sponsor local education and growth, and lower trading restrictions worldwide to allow third-world economies to further their own development. In fact, as far as migration patterns are concerned, the same processes are taking place in both the developed and the developing world. It is a movement from the periphery, from the countryside where employment prospects are decreasing, to central urban areas, where the employment opportunities offered by urban economies are generally better.

At the same time, the Europeans are working on a larger, internally borderless Europe, the size of which has not yet been established. This demands enormous investments in educating the population and the physical infrastructure. This offers chances for those who recognise the opportunities, and short-term threats for those who are unwilling to share some of the affluence of the old Europe. In addition to the destabilising effects of the unequal distribution of prosperity, coupled with a process of radicalisation, climate changes will also have a destabilising effect. These latter phenomena will not only have huge consequences for economic development, but will also have an enormous impact on urbanisation. After all, many intensively urbanised areas are located in deltas, which are also intensive food production areas. It is exactly these areas that will be threatened more by higher sea levels and the need to allow rivers the unrestrained use of more of their flood plains to drain off excess water from higher, and more intensive rainfall. It is a sacrifice that has to be made, to minimise or avoid even more widespread damage to human structures and populations.

These changes affect the mental map of our actions and the question of how to manage spatial transformations. They determine not only the strategic options, but also the actual processes. Management and strategy often fail to keep pace with actual developments. These developments are used as the context for the actions described in the following sections; actions such as looking for new ways of management that are more attuned to a new society, and looking for assignments and projects in line with European dimensions.

## Reconstructing the dynamic period, 1975-2005

### 1975-1985: Mainly public housing and production

At the end of the 70s, the implementation of the *Derde Nota over de Ruimtelijke Ordening* (Third Memorandum on Spatial Planning) (Ministerie van Volkshuisvesting en Ruimtelijke Ordening, 1976) came under attack. National housing production gradually stagnated. This led to a growing awareness that, also in environmental planning and public housing, the end of the reconstruction period was in sight. It took public housing many years to adapt to the culture change of 1968. For a long time, the main effort to meet the need for new houses would still be the industrial production of houses in extension areas. At the same time, demolition took place in old districts to help create mono-functional cities. In the early 70s, residents of the 19th century working-class districts demonstrated against the demolition of their districts and against their planned resettlement in urban extension areas, a forced move which they likened to deportation (Gemeente Rotterdam, 1988).

In order to stimulate the production of houses again, a number of large sources of financing were approached. First of all, money from the urban renewal fund was used to subsidise the costs of land bought by local authorities for exploiting their areas. Secondly, sources of financing were needed to renew and expand the existing stock of (social) housing. Through the ICOG process (ICOG: *Interdepartementale Coördinatie Overleg Groeikernen en Groeisteden* [Interdepartmental Committee on Development Towns]) location and infrastructure subsidies were also given to urban development towns to get production started. At national level, the negotiations were interdepartmentally and district oriented, whereby integral exploitation plans for large districts were tested against a system of standards for land use. Within this framework of standards, green areas, water, and facilities were all included in the exploitation as subsidisable planning items. In addition, there was a more detailed regulation to support the build up of a civil service and local administration apparatus in growth municipalities.

In consultation with the administrators of the development towns, the Minister of Housing and Spatial Planning intervened directly to force through agreement for these lines of action. The resulting money flows led to an immense production of houses in the 80s, both in the old districts and in the urban extension areas. An interesting detail is that once permission had been gained from the State, these municipalities were allowed great freedom at district level to decide for themselves how to develop the urban design in relation to available funding within the budgetary frameworks, while for housing subsidies, the run on State funds took place for each building plan. Subsidy applications for tens of thousands of houses were checked by the State every year.

## 1985-1995: Urban regions come into being

In the mid-80s, the availability of suitable locations in some development towns slowly came to an end. The first cracks became visible in the massive demolition programme and in the efforts to improve urban renewal, and social housing in particular. The middle class left the districts and towns, together with many firms that had been established in the old districts. In 1988, for instance, Rotterdam evaluated its 15-year urban renewal plan. The outcome was a desire for more social mixing, for the development of more expensive locations in the inner city, and the beginnings of an interest in urban economy (Gemeente Rotterdam, 1988) (see on urban renewal in Rotterdam also chapter 9 by Paul Stouten).

The division of urban functions in the 30s, originated by CIAM (Congres Internationaux d'Architecture Moderne), left deep marks during the reconstruction period, both in the development of urban extension and in urban renewal. Until the mid-80s, production was aimed at mainly old and new mono-functional districts with facilities linked to the district, and separate areas for employment and recreation. The flows of money that had been organised ensured that the house-building programme ran well. However, the economic decline of the mid-80s, led to a fundamental reconsideration of the approach. The Third Memorandum on Spatial Planning started to lose its effect, and thoughts began to turn towards formulating a Fourth Memorandum on Spatial Planning based on the aim of stimulating the economics of the towns themselves, partly by introducing a new form of integral spatial development.

This *Vierde Nota over de Ruimtelijke Ordening* [Fourth Memorandum on Spatial Planning] (Ministerie van Volkshuisvesting, Ruimtelijke Ordening en Milieubeheer, 1988) unfolded a main-port strategy (incidentally, this is a notion that is only used in the Netherlands). The continued growth of the Port of Rotterdam and the airport of Amsterdam at Schiphol was expected to act as a support to guarantee the Dutch position in Europe. The idea arose to connect up with the French high-speed passenger rail network. To increase the small share of freight transport by rail from Rotterdam (compared with Antwerp and Hamburg) the construction of the Betuwe freight rail link to Germany was started. The plan was then to strengthen the urban economies, at the same time, by starting 'large city projects', which would be developed by public–private collaboration. The intention was that these would become top locations for internationalising urban economic functions. The locations of this first generation of Key Urban Projects (Ministerie van Volkshuisvesting, Ruimtelijke Ordening en Milieubeheer, 1991) included the 'Kop van Zuid' in Rotterdam, the banks of *het IJ* [Lake IJ] in Amsterdam, and 'Den Haag Centraal' in The Hague. Further demands for urbanisation were met by switching from the 'development' strategy of the Third Memorandum to an urban approach: first within, then bordering, and only then, outside the city.

These spatial-functional strategies were followed by wishes for higher administrative levels of scale, because this form of urbanisation led to urban regions that had to meet citizens' spatial needs. Eventually, in 1994, after both the Fourth Memorandum and the *Vierde Nota over de Ruimtelijke Ordening Extra* [Fourth Memorandum on Spatial Planning-Plus] (Ministerie van Volkshuisvesting, Ruimtelijke Ordening en Milieubeheer, 1996), the regions and the State made development contracts for the urban regions. A further emancipation that would have turned the urban regions into urban provinces, with far-reaching decentralised and supermunicipal authority, was cancelled, because of resistance from the population. The old toolkit for current urban renewal remained in use, because it was thought that it would be ready some time in the mid-90s. On a location level, negotiations took place concerning the so-called 'large building locations', such as Nieuw Sloten in Amsterdam, Prinsenland in Rotterdam, followed by key projects such as 'Kop Van Zuid' in Rotterdam and 'Sfinx Ceramiek' in Maasticht. Urban renewal did not really come to an end, but the funding system did, and so did the funding system for social housing. The social housing corporations were privatised and were left to fund the urban renewal themselves. Some money flows remained open to subsidise the costs of buying land for urban-function change. It is interesting that contracts between the State and the regions led to intermunicipal equalisation agreements for funding local amenities.

As a conclusion, it can be stated that the period 1985-1995 was characterised by:

- A moderate scaling up of municipalities to urban regions;
- Integral realisation contracts for the urbanising between the urban regions and the State;
- Regional internal equalisation;
- A certain level of integral funding by the State;
- Key projects that received special attention and supervision by the State;
- An appointed co-ordinating minister, a system that had already been thought of in the 80s; and
- The main-port approach for the Port of Rotterdam and Schiphol Airport, as national projects.

## 1995-2005: The urban-regional level is realised; preparing to up-scale to Wings

The actual operation to strengthen the position of the Netherlands in Europe took place on an urban-regional level and in the main-port projects. However, the consequences were also noticeable in other parts of the Netherlands. The growth of mobility and urbanisation, smaller families and higher levels of education, the increase in the number of working members in each household, and with that, the growing spending capacity of households, together led to a much larger spatial dynamism than had been anticipated in the Fourth Memorandum and the Fourth Memorandum on Spatial Planning-Plus of the 80s and 90s. Managerial thinking about spatial developments was focused on the urban-regional level, but the dynamism itself also extended much further than this level. The administrative organisation and developments in the infrastructure were not prepared for this.

At the end of the 80s and beginning of the 90s, however, various national and interprovincial policy documents appeared on developing the Randstad (Ministerie van Volkshuisvesting, Ruimtelijke Ordening en Milieubeheer, 1990). They were mainly used to plan urban-regional assignments or for the key projects and the main ports. A further integration of the Randstad, or its 'wings', to form a metropole had not yet come up for discussion. It is interesting that, in the mid-90s, the Minister of Spatial Planning and Environment, Mrs De Boer, herself led the consultations about spatial development at regional level. For this, the State divided the country into four administrative parts: North, East, South and West/Randstad. The minister did that based on her experience as provincial administrator for the province of North-Holland. During that period she had drawn up a structural outlook for the Randstad in collaboration with an inter-provincial office.

Along the South Wing of the Randstad, the province of Zuid-Holland put more highly integrated developments on its agenda, but the towns and cities were not prepared to accept them at this stage. A similar trend had begun along in the North Wing, where the proposal to develop Amsterdam and Almere into a 'double city' was rejected by Almere in 1994 to prevent the disintegration of the province of Flevoland, a move that would have forced Almere to participate in developing the Amsterdam area.

While the work on the first generation of Key Projects (Ministerie van Volkshuisvesting, Ruimtelijke Ordening en Milieubeheer, 1991) such as the 'Kop van Zuid' in Rotterdam, the centre of The Hague and Amsterdam Nieuw Oost was in full progress, in formulating the actualisation of the Fourth Memorandum (VINAC; Ministerie van Volkshuisvesting, Ruimtelijke Ordening en Milieubeheer, 1998), mention was made of a possible second generation of key projects, still called 'strategic interventions' at that time. At first, the State did not want to collaborate with this, but the regions enforced this integral project-wise approach. They, in fact, had a much larger number of projects in mind than did the State (Stuurgroep Randstad, 1996). Eventually, six new (second generation) Key Projects were named in this VINAC: the railway stations along the new high-speed lines, two of which (the Central Station of The Hague and the Utrecht City Project) had already been included in area projects listed among the first-generation key projects.

## From ICOG to ICES and back: Interdepartmental super co-ordination

The negotiations about implementing the urbanising projects took place in the 80s within the ICOG (Interdepartmental Committee on Development Towns). These consultations were focused on locations, subsidies and production. There was a direct relation between municipalities and the State under the supervision of the Minister of Housing, Spatial Planning and Environment, and the building assignments were funded interdepartmentally.

In the late 80s and the early 90s, the State began to realise that the national investments for infrastructure had collapsed to a historic low in comparison with surrounding European states, so funds were released to tackle large new infrastructure projects, such as the Betuwe freight rail link, the high-speed railway line and large-city projects. Interdepartmentally, on the initiative of the Secretaries General of the Ministeries of Economic Affairs; Housing, Spatial Planning and Environment; Transport, Public Works and Water Management, this led to the foundation of ICES (*Interdepartementale Commisie Economische Structuurversterking* [Interdepartmental Commission for Strengthening Economic Structures]). By co-ordinating their efforts interdepartmentally, their aim was to strengthen the economic structure and propose an integral implementation programme to the Cabinet. Because of increasing prosperity, extra funding was available. So, in a sense, this led to super co-ordination and integration at national level, especially for spatial investments, coupled with a spatial and mobility strategy, strongly focused on internationalisation. There was growing orientation towards Europe at this time.

The interdepartmental co-ordination, brought about by ICES intervention, also stimulated the national administrative regions and sub-regions to bring forward their claims on the funds held by ICES. Although the regions also formulated programmes, in the end, they just amounted to endless project lists submitted to central government without any internal priorities having been made. These lists were then examined within assessment frameworks drawn up by the planning offices. This all-embracing approach, together with other factors, proved too much for the ICES, which suddenly disintegrated early in 2000. Following the collapse of the ICES, the central government withdrew from the regional and local policies in the *Nota Ruimte: Ruimte voor Ontwikkeling* [National Spatial Strategy: Space for Development], and also from investments. Nevertheless, informal interdepartmental consultations were held during the post-ICES era, and priorities for investments were set, after all.

The central government formulates project packets of its own projects and offers them to the regions. Agreements have already been made with four regions (the North Wing Randstad/Utrecht, the South Wing Randstad, the Green Heart and Brabant City) about the implementation of those packets, in addition to agreements mainly aimed at larger projects, about which the State negotiates directly with cities or regions. With the specification of the National Spatial Strategy, the scaling up of the 1994 urban-regional implementation agreements to Randstad-wing level became a fact (Ministerie van Volkshuisvesting, Ruimtelijke Ordening en Milieubeheer, Ministerie van Landbouw, Natuur en Voedselkwaliteit, Ministerie van Verkeer en Waterstaat & Ministerie van Economische Zaken, 2005).

## 2005 and beyond: Growing towards an overall organisation for the Randstad?

The *Nota Ruimte, Ruimte voor ontwikkeling* (Ministerie van Volkshuisvesting, Ruimtelijke Ordening en Milieubeheer, 2004) established in 2005 as the operational successor of the Fifth Memorandum on Spatial Planning, was never approved by Parliament. It takes as its point of departure a new philosophy for managing the spatial development of the Netherlands. Under this strategy, the State will restrict its efforts to an agenda and to projects that it considers important, and the lower authorities will implement their own urbanisation policy and assignments within the more general frameworks laid down by central government. To make that possible, a proposal has been put to Parliament to adapt the legislation by passing a new Spatial Planning Act and with new land-use policy instruments. The power to implement this Act will also be linked to the regional level. The State effort to create a network society, and collaboration between towns and administrative bodies in urban networks, will become operational to the extent that the networks themselves start working on it. Central government is still undecided as to what action to take regarding the Randstad, because this level is higher than that of the provinces.

The State confers with the partners in the Randstad, in consultation with all the important departments with which the provinces, urban regions and large cities have dealings. The various governing bodies of the Randstad have a co-ordination office, 'Randstad' in Utrecht, which is also the secretariat of the *Bestuurscommissie Randstad (BCR)* [Administrative Committee for the Randstad]. The Randstad partners produce preparatory policy documents for the central government, which are used as input for national policy. This in itself integrates the policies of the lower authorities, and slowly but surely, this collaboration is helping to bridge the animosity that exists between the various administrative levels. This has only been made possible by the personal efforts of the various administrators. However, so far, this collaboration has only been successful with regard to policy. On the level of the Randstad, no priorities have been set as yet for investment.

In the same period, the *Vereniging Deltametropool* [Delta Metropolis Society] (Vereniging Deltametropool, 1998) was founded, originally on the initiative of the four aldermen responsible for spatial planning from each of the four large cities of the Netherlands (Amsterdam The Hague, Rotterdam and Utrecht), together with larger municipalities in the Randstad and non-governmental non-profit organisations. The Society is an open thinktank in which a strategy for the Randstad can be deliberated without hinderance from consultations between the large societal-interest organisations and the larger cities and provinces.

So, in the course of thirty years, we have scaled up in the Netherlands from municipal, to urban-regional and now to Randstad-wing level in making agreements on how to implement urbanisation. At the same time, there are integration mechanisms to achieve policy development and integration on central-government-administrative-division level in an interaction between the State and these administrative divisions. It is interesting that, in addition to the club-like structure of the Delta Metropolis Society, the regulation of the Randstad also has a legal basis, while the South Wing level is an informal arrangement, that nevertheless has its own co-ordination office (Bestuurlijk Platform Zuidvleugel, 2004). The North Wing does not have an official co-ordination centre yet, but there is an administrative platform. In the region of Utrecht, the form of collaboration has still to be decided, and Utrecht has now applied to negiotiate with the State under the umbrella of the North Wing.

Anno 2005, the picture is as follows:

- The *Vereniging Deltametropool* [Delta Metropolis Society] is a strategic thinktank focused on long-term strategies;
- The *Gemeenschappelijke Regeling Randstad* [Collective Regulations for the Randstad] is focused on strategy and on preparing policy between the State and the regions;
- The *Bestuurscommissie Randstad* [Administrative Committee for the Randstad] is the negotiating platform between the State and the regions for integral strategic spatial development;
- The North and South Wings of the Randstad each have their own bodies to help the areas attune to the integral urbanisation strategy and to agreements with the State on project priorities;
- The progress of the *Projecten* [projects] themselves is in the hands of various project owners: ministries, provinces, regions or municipalities.

Within the span of thirty years, the centrally administered planning in a relatively protected environment in the Netherlands has changed to a multi-layered governmental management that interacts with many more actors and disciplines. It started off from a closed national and West European context, moved to a larger and more open European territory, with many more actors, before moving again into an open world economy. During these thirty years, an enormous scaling up has taken place in the administration of the Netherlands. At the same time, the formal division into State, Province and Municipality has remained almost completely intact, and within this, a way of working has emerged that, despite the hierarchical organisation of the administrative bodies, they can present themselves as equal negotiating partners. Illustration 11.1 gives an overview of the situation in 1990, showing the

| Situation in 1990 | Vision and strategy | Official plan | Priorities | Funding |
|---|---|---|---|---|
| The Netherlands | Yes | Yes | Yes | Yes |
| province | Yes | Yes | Yes | Yes |
| municipality | Yes | Yes | Yes | Yes |

| Situation in 2005 | Vision and strategy | Official plan | Priorities | Funding |
|---|---|---|---|---|
| Europe | Yes | | Yes | Yes |
| European region | Yes | | | |
| The Netherlands | Yes | Yes | Yes | Yes |
| Randstad region | Yes | | | |
| Province | Yes | Yes | Yes | Yes |
| Randstad Wings | Yes | | Yes | |
| Urban region plus | Yes | Yes | Yes | Yes |
| Region | Yes | | Yes | |
| Municipality | Yes | Yes | Yes | Yes |
| Urban district | Yes | Yes | Yes | |

**Illustration 11.1: Administrative development 1990-2005**

three original official administrative organs: the State, the provinces and the municipalities. Now, in 2005, there is a multitude of old and new organs, democratically ranked in various ways (Teisman, 1992). Together, they form a network of administrative organs for different sorts of tasks on different scale levels.

## Finally

The discipline of the urban designer and the spatial planner must adapt itself to the complex context described above. Only then can it practice its profession in a relevant way (Schrijnen, 2005). The designer must make his visions and images accessible to the many actors in the field; and the spatial planner, given his role as a bearer of knowledge about the processes of spatial change and the decision processes associated with them, must uncover the relations between the administration and the planning product. In the course of these thirty years, the designer has changed from being an elaborator of a given assignment, to a co-creator of the assignment, by putting himself, with his force of imagination, into the position of his multiple customers and, if possible, by co-ordinating (as far as possible) the variety of interests. The town and country planner has developed from a expert in a stable geography to a process engineer of spatial plans in a permanently changing environment.

## References


Bestuurlijk Platvorm Zuidvleugel, 2004, *De Zuidvleugel van de Randstad, Netwerk van bestuur & recht, kennis en logistiek*, Den Haag

Gemeente Rotterdam, 1988, *Vernieuwing van de Stadsvernieuwing*, Rotterdam

Gemeente Rotterdam, 1995, *Rotterdam 2045. Visies op de toekomst van stad, haven en regio*, Rotterdam

Ministerie van Volkshuisvesting en Ruimtelijke Ordening, 1976, *Derde Nota over de Ruimtelijke ordening. Nota landelijke gebieden*, Den Haag

Ministerie van Volkshuisvesting, Ruimtelijke Ordening en Milieubeheer, 1988, *Vierde Nota over de Ruimtelijke Ordening, deel a: Beleidsvoornemen. Op weg naar 2015*, Den Haag

Ministerie van Volkshuisvesting, Ruimtelijke Ordening en Milieubeheer, 1991, *Sleutelprojecten ruimtelijke inrichting tussenstand 1991*, Den Haag

Ministerie van Volkshuisvesting, Ruimtelijke Ordening en Milieubeheer, 1996, *Vierde Nota over de Ruimtelijke Ordening Extra, deel 1*, Den Haag

Ministerie van Volkshuisvesting, Ruimtelijke Ordening en Milieubeheer, 1990, *De Randstad op weg naar 2015*, Den Haag

Ministerie van Volkshuisvesting, Ruimtelijke Ordening en Milieubeheer, 1998, *Actualisering Vierde Nota over de ruimtelijke ordening Extra (VINAC)*, Den Haag

Ministerie van Volkshuisvesting, Ruimtelijke Ordening en Milieubeheer, 2001, *Ruimte maken, ruimte delen. Vijfde Nota over de Ruimtelijke Ordening 2000/2020*, Den Haag

Ministerie van Volkshuisvesting, Ruimtelijke Ordening en Milieubeheer, 2004, *Nota Ruimte: Ruimte voor ontwikkeling*, Ministerie van Volkshuisvesting, Ruimtelijke Ordening en Milieubeheer, Den Haag

Sassen, S., 2001, *The Global City*, Oxford

Schrijnen, J., 2005, *Land en Stad, de creatie van een opgave*, Inaugural Lecture, Technische Universiteit Delft

Stuurgroep Randstad, 1996, *Gespreksnotitie II*, PKB-Vinexact, Den Haag

Teisman, G., 1992, *Complexe besluitvorming, een policentrisch perspectief op besluitvorming over ruimtelijke investeringen*, Rotterdam

Vereniging Deltametropool, 1998, *Deltametropool in Europees perspectief*, Technische Universiteit Delft

Wetenschappelijk Raad voor het Regeringsbeleid, 1998, *Ruimtelijke ontwikkelingspolitiek*, Den Haag

# Take a Break

Iwan Kriens

A lot is happening in the field of spatial planning, so it is about time that an inventory is made of these trends. On the 18th June 2003, as a 'farewell present' for one of their retiring senior planning consultants, Royal Haskoning organised a conference titled 'Time to Move'. More than 150 colleagues and related specialists from all over the Netherlands gathered to discuss the movements in the field. Why are these movements taking place? What exactly is going on at the present time and where is it leading?

Five themes were formulated, touched upon by a number of keynote introductions:

*Sustainable development*

Is this a transient phenomenon or does it appear to be a permanent change? Should spatial planning return to its roots or, on the contrary, should it broaden its scope by including many more aspects of sustainable development than just water and nature?

*Design on a regional scale*

Has taking a long-term view, with a vision that extends beyond local boundaries, become a matter of urgency again within conceptual designing and planning? And who should be doing something about it?

*Actors and roles in spatial planning*

How can the process of spatial planning be organised in such a way that all actors, including users, have a chance to express their ideas fully? How does one deal with multiple roles, and what contribution can the planners make to improve effectivity?

*The time factor*

In spatial planning, shouldn't the time factor be considered differently in relation to settlement patterns, consumer mobility and other spatial behaviour? And what would be the implications for the field of study?

*The relation between urban design and traffic*

Is it possible to have spatial quality, optimal mobility and traffic safety as collective targets? Shouldn't there be a more rapid fusion, on all scale levels, between the different worlds of urban designers and traffic engineers?

Around these themes, all touching upon relevant questions, guaranteeing beforehand that the conference would end with just as many (other) questions, heated group discussions took place.

In one of the discussion groups, the theorem was more contentious: the obsolescence of spatial planning can be traced back to the lack of a theoretical framework about how to deal with societal changes. The keynote speaker asserted that, in the Netherlands, as there is no cultural identity any more, people will no longer desire to belong to a group. The mainspring will then be individuality, singularity even. The result is that today's politics cannot function anymore, because 'representing' and 'promoting' people's interests no longer matter. The city becomes a camp. This word 'camp' was used metaphorically by Giorgio Agamben (1993, 1998) to describe an unfriendly place for collective living, aimed at destroying life and identity. The state of 'just existing' implies that morals have disappeared. This means disengagement.

The individualisation that characterises today's society tends to negate the need for care and solidarity. And if solidarity is a condition for positive development, then the question that has to be asked is: how can a new form of solidarity be created? Spatial planning can play a role in this, by shaping public space differently. However, this poses the question whether the social component of solidarity can be approached in this way. And how can this component be found, if there is a lack of culture and cultural identity? The lack of 'personality' and 'cultural identity' implies, then, that there is no collective principle from which to design. It will then also become more difficult to build and plan using the principle of social models. This, in turn, will give private clients more scope, and they, unfortunately, pay more attention to architecture than to planning problems.

The majority of the present voters can be described as a collection of individual, singular entities that has gone adrift. Politics, therefore, needs sources of legitimation that harmonise with individuality and singularity. In the past, spatial planning was one such source of legitimation. So, in order to retain this position, should spatial planning make a turnaround in order to retain this position? A condition for this is that individualities and singularities should offer spatial planners opportunities of taking up a new position. In turn, these planners should then be prepared to focus on the wishes of those individualities and singularities, and stop trying to look for arguments in the market to strengthen their old positions.

These considerations made the specialists' heads spin. But in the end the conclusions were:
- The discipline of spatial planning is not renewing itself enough and is contaminated by politics and the market;
- The problem is not so much the lack of a theoretical framework, but the incapacity of many professional planners to pose societal and philosophical questions and look for answers.

And with these conclusions everything was back to normal. As a participant I regret this. As there is no account of the discussion, an opportunity is vanished to formulate a sixth theme: innovating spatial planning, especially by contributions from other disciplines. To start with the changes in cultural identity, declining solidarity, increasing individualisation and privatisation in their consequences for the design and planning task for urban public space.

Take a break is motivated by the need to continue the debate.

## References

Symposium *Time to Move*, farewell to Peter Jonquière, Utrecht, 18 june 2003, Royal Haskoning Division Spatial Development, the Netherlands
Agamben, G., 1993, *The coming community*, University of Minneapolis Press, Minneapolis
Agamben, G. & D. Heller-Roazen, 1998, *Homo Sacer: Sovereign power and bare life* (translation), Stanford University Press, Stanford

# Part III

## Networks

# Introduction to Part III

The central theme of Part III is *Networks*. Spatial planning from the perspective of urban networks implies not just planning in terms of connecting cities but focusing on urban processes that have a spatial dimension, i.e. processes involving goods, energy, water, information and, not least, people. This means linking the notion 'space' with that of 'time', by means of the notion 'use'. 'Time' is associated here with cyclical daily, weekly, seasonal and yearly processes, rather than with the transformation of the urban environment throughout the years. Moreover, as a healthy and safe environment is a necessary condition for urban processes, urban designers and planners should not just seek to relate the physical urban system (*urbs*) with processes in the societal urban system (*civitas*), but should look for ways of accommodating natural processes as well.

This part of the book is structured from the general to the specific and from the abstract towards the concrete. Remon Rooij with *The Urbanism of Networks* and Riet Moens with *Layers, Patterns and Networks in the Landscape* first introduce some basic terminology. Ina Klaasen, in *Putting Time into the Picture: The Relation between Space and Time in Urban Planning* then explores the many ways in which 'time' is related to 'space' within the context of design and planning, and the problems that this causes due to the difficulty of representing time in designs and spatial plans. In *Time in Urban Planning and Design in the ICT age* Paul Drewe pursues the matter of what 'time' means for social, cultural and economic activities in cities. He stresses how the development of information and communication technology (ICT) has changed the time dimension in spatial planning, but cautions us not to lose touch with the 'classics' in our field of study.

In the following four chapters, the focus is on specific networks. Ana Maria Fernandez-Maldonado concludes in *ICT Infrastructure Networks as Supports for New Urban Processes* that new trends in the ICT sector, based on market principles, are contributing to an unequal development of cities and regions around the world, a development that favours the 'rich-gets richer'. Ernesto Philibert Petit, in *Metamobility: In Search of Connections within the Networks of Mobility*, makes a plea for increasing connectivity in cities, using the concept of 'mobility' to develop new concepts of what, where and how to connect. In *Small but Vital: The Influence of Small-Scale Mobility on Sustainable Urban Functioning*, Remon Rooij draws our attention to the lowest scale levels in the city, i.e. the dwelling environment and the neighbourhood. He stipulates that, although a lot of scientific and practical knowledge is available about the groups that use this level of the city, most of this knowledge is unavailable in the form of design tools. He goes on to suggest ways of developing such tools. The relation between the physical urban systems and their networks is then put in a theoretical light by Jeroen van Schaick. In *Integrating Social and Spatial Aspects of the City: Comparing the Models of Heeling, Dupuy, Castells and Lefebvre* he explores how the structure of societal urban systems can – or cannot – be related to that of the physical systems, and concludes that, for urban designers and planners to be able to deal with the dynamics of socio-spatial systems, a better understanding has to be developed of issues such as the languages and representation methods used, and how time is related to different scale levels.

In the last two chapters 'connectivity' and 'process-thinking' are applied in concrete situations. In *Towards a New Urban Philosophy; The Case of Athens*, Nikos Salingaros unfolds his ideas about how to repair Europe's damaged urban fabric. He formulates a detailed New Urban Philosophy, stressing the importance of the relationship of humankind to nature and the environment. He then applies this to Athens. Finally, in *Finding a New Meaning For Public Spaces In Postmodernity; The Raval District in Barcelona,* Francesc Magrinyà shows us, using examples from Barcelona, how the traditional idea of public space has lost its meaning in an age characterised by complex and fragmented territory and the co-existence of different mobilities.

# 12 The Urbanism of Networks

« The design ideas and the instruments of a lot of today's urbanists still result from a traditional and geographical concept of space and time. Therefore, they are hardly suitable to play along with the fundamental fleetingness and infiniteness of the network society. »
Boelens (2000), *The Netherlands as Country of Networks*

Remon Rooij

## Introduction

In spatial planning, networks have often been defined according to a geographical point of view (Caso, 1999). Networks of infrastructures are channels used for many purposes, that interconnect places along lines and pathways. Transportation networks support the mobility of people and goods, telephone networks provide people and firms with voice and (other kinds of) data transmission.This conception of networks is focused on a geography of places and may lead one to underestimate the centrality of the human element of networking. Indeed, networks interconnect people rather than places; infrastructures of cables, electric wires, highways, railways are planned and designed in order to serve human settlements. In this view, elements of an urbanism of networks should consider the centrality of people in networking, by looking at how networks are used by people to reach (their) goals.

## Theory

A theory of networks proposed by Dupuy (1991), elaborating on previous investigations (see Fishman, 1990; Wright, 1943), recognises three levels of 'operators' of networks who (re)organise urban space (Ill. 12.1) (see also chapter 15 by Paul Drewe and chapter 19 by Jeroen van Schaick). On the first level, there are the suppliers of technical networks, such as streets, highways, cables, wires, sewerage, and so on. They are in charge of providing the physical elements of the networks (infrastructure management) and the services on those networks (exploiting the infrastructure). On the second level there are the suppliers of functional networks. They use the level immediately below to provide services, such as production, consumption, distribution, to the upper level. On this third level, the operators are people in their daily life. They make use of the first two levels to create their personal networks by interpreting possibilities and linking activity places, spaces, services, desires and needs in a single personal (or household) behaviour. In this way, people create their own virtual cities. As the first two levels are still characterised by a certain degree of 'objectivity', the third level is mainly a 'subjective' environment where personal, household, or company choices are made, even if conditioned by the lower two levels.

This distinction in levels reflects the articulation in physical, functional, social, and personal components that is inherent to the Network City. If we try to understand the city as a technology (Zeleny, 1987) we can identify these components with the hardware, software, and brainware of the Network City. The relationships between the software and the brainware can be expressed in terms of activities. Activities

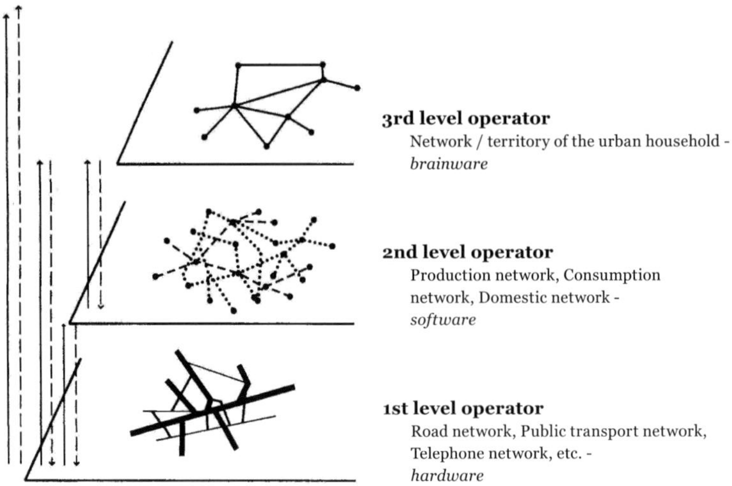


**Illustration 12.1: The urbanism of networks (Dupuy, 1991)**

are the way in which people (on the third level) make use of services or interact with other people (on the second level). The relationships between the hardware and the software can be expressed in terms of a programme. The programme defines the requirements for realising or adapting the physical environment (the first level) in relation to a certain set of objectives, while, in turn, the physical environment conditions the range of possibilities regarding its availability for functional or social uses (the second level).

It is evident that planned spatial actions cannot be placed in the third level, but are confined to the first and second levels. However, knowledge of the characteristics of the third level is very important in order to shape better a respondent environment supporting a real possibility of choice for people.

## *Urbs* and *civitas*

Under the influence of the rise of the network society - which refers to a society, in which social, economic and cultural structures are not solely determined by the shared use of a certain space, but also (primarily) by the connections that an individual actor (person, company, institution) has with places, persons, or activities elsewhere – a situation develops, in which the identity of the *urbs*, that is physical dimension of the city, and the *civitas*, that is the social dimension of the city, are no longer self-evident. The *urbs* and the *civitas* always used to go together, but that is no longer the case anymore; the *civitas* can take place anywhere: in places with a high or low density, in the train, on the Internet, etc. However, it is wrong to think that because the *urbs* and *civitas* are separated, places such as historic city centres will disappear as snow in summer. The Dutch Advisory Council for Housing, Spatial Planning and Environment [*VROM-raad*] states that "...different forms of urbanity exist..." and that "...the importance of the form of urbanity that is based on face-to-face contact and ... on activities in an urban setting with a relatively high density, has definitely not decreased" (VROM-raad, 1999:8). In the future, both an increase and decrease of peoples' action spaces can be expected: both a concentration and deconcentration of urban activities and/or the built environment. This is referred to in the literature (Boelens, 2000) as the 'double movement.' The technology that makes it easier to fly around the world, both physically and electronically, is also making it easier to bring the world together. That is why 'glocalisation' is a good description for what is happening in the world today. It is not just making things bigger, it is also making them a lot smaller. Peace and prosperity are much more likely when we all feel part of the same neighbourhood.

So, the Network City stands for multiple options instead of one uniform spatial-functional concept for cities. Therefore, it becomes more and more important not to discuss urban trends in general terms, but to be very precise and accurate about which sector of the employment, about which part of the housing demand, about which elements of the apparatus of facilities, one speaks. In fact, what we are talking about are several different networks, each with their own nodes and hierarchies. There is no single node at the centre for everything and everyone anymore (such as the market square, the traditional economic, societal, political and religious centre of towns and villages throughout history). For spatial planning, the good old one-dimensional Christaller scheme has become outdated. Now, other Christaller schemes need to be defined to explain different urban functions, both physical and electronic, on different levels of scale, each with their own conditions, rules, and order (see Ill. 12.2).

Within the hierarchies of these various networks, transport nodes, the places where (different) hierarchies and levels of scale of infrastructures come together, are (potentially) among the most interesting locations within the Network City. The nodes of the network, such as public squares, stations, park-and-ride facilities, motorway access and exit points etc., are situated on the field of tension between 'place' and 'flow', where the 'space occupied by flows' meets the 'space occupied by places'. In and around the transport nodes, the network is linked to the geographical surface and environment. Here, new development potentials, new possibilities, but also new threats arise.

## Conclusion: networks as a central theme

"Throughout the 20th century, cities have been planned, and in doing so, networks for the transport, not only of passengers and goods, but also of water, energy and information, have played an ever increasing role. However, seldom has urbanism been (re)thought in terms of networks: their topology, nodes, connectivity, capillarity or similarities. Nevertheless, some urbanists understood a long time ago already, at

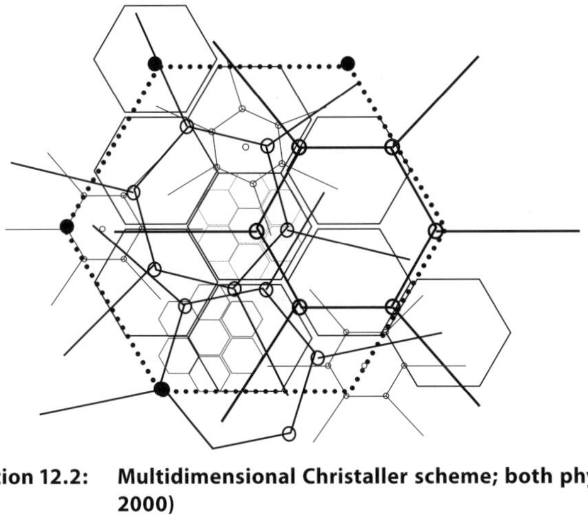

**Illustration 12.2: Multidimensional Christaller scheme; both physical and virtual hierarchies (Boelens, 2000)**

least in part, the meaning of networks to cities, but this was later marginalised by the mainstream zonal urbanism of the Athens Charter of 1933" (Drewe, 2002).

The network in its modern meaning is characterised by three principal criteria – the definitions of which are based on Caso (1999) and Drewe (2002):

- *The topological criterion*: topology refers to the geometrical or physical configuration of a network; to the way in which the nodes of a network are physically connected. Networks are not abstract entities, they are related to the spatial dimension by connecting links, via nodes, in space. This involves discontinuity and heterogeneity. The topology of a network is open and united. There are no separations such as city/countryside, centre/periphery and zoning in this configuration.
- *The kinetic criterion*: kinetics pertains to movement and communication between nodes. It is basically a relationship between space and time: speed. The rapidity of the connections within a network is a measure of the quality of the network itself. So, instantaneousness, homogeneity of speeds, the interest for rapid transfers and transits without losses of time or interruptions makes the network apt in accommodating movement and defines the kinetic criterion.
- *The adaptive criterion*: adaptability concerns the capacity of networks to evolve over time and space. On the one hand, a network should be able to modify its structure to welcome new systems or to extend the applications of existing ones. On the other hand, it should be able to adapt itself to the needs and desires of its users by offering them a range of choices to help them reach their goals.

In visualising the city of the 21st. century, The European Council of Town Planners (2003) focuses on the *Connected City*. The Council suggests that "…the connected city is comprised of a variety of connective mechanisms acting on different scales. These include tactile and visual connections with the built environment, as well as connections between a diversity of urban functions, infrastructure networks, and information and communication technologies." With respect to 'movement and mobility', the Council states that "…within city networks, mobility will be improved by interchange facilities between the various modes of transport". To do so "…the spatial organisation of the connected city will include a full integration of transportation and town planning policies. They will be complemented by more imaginative urban design and easier access to information…" and "… ease of movement and access will be a critical element of city living, together with greater choice in mode of transport."

## References


Boelens, L., 2000, *Van de Rand en de Stad*, Essay in het kader van het project 'Binnenstad en Stadsrandcentra' van de werkgroep Mobiliteit NIROV, Holland Railconsult, Utrecht

Caso, O., 1999, *The City, The Elderly, and Telematics. Design aspects of telematic applications in a residential neighbourhood*, PhD Thesis Delft University of Technology, Spatial Planning Group, Transformations no. 2, Delft University Press, Delft

Drewe, P., 2002, The Network City – from Utopia to new paradigm, *Atlantis*, 14.1, pp. 24-29

Dupuy, G., 1991, *L'urbanisme des réseaux. Théories et méthodes,* Armand Colin, Paris

European Council of Town Planners, 2003, *The New Charter of Athens 2003, The European Council of Town Planners' Vision for Cities in the 21st Century*, Lisbon, http://www.ceu-ectp.org/e/athens

Fishman, R., 1990, Metropolis Unbound: the New City of the Twentieth Century, *Flux*, International Scientific Quarterly on Networks and Territories, no. 1

VROM-raad, 1999, *Stad en wijk:verschillen maken kwaliteit, Visie op de stad*, Advies no. 13

Wright, F.L., 1943, *Broadacre City. Box six of an autobiography*, Spring Green, Wisconsin

Zeleny, M., 1987, La gestione a tecnologia superiore e la gestione della tecnologia superiore; In: Bocchi, G. & M. Cerutti (eds), *La sfida della complessita*, Feltrinelli, Milano

# 13 Layers, Patterns and Networks in the Landscape

Riet Moens-Gigengack

## Introduction

When making a spatial analysis of a region, landscape architects consider the built-up area to be part of the urban layer. This layer is influenced by other layers and especially by the underlying cultural and natural layers. In landscape architecture, for analysing the landscape, the concept of patterns and networks is used. The term 'pattern' is used to describe specific regular and repeating forms in the landscape that have one common denominator. The term 'network' anticipates exchanges and processes in the landscape along specific paths.

## The layers of the landscape

A well-known method for analysing landscape is to describe and to map the information. The landscape is analysed from different aspects, such as soil, water, morphology, plant cover, etc., which can then be combined or classified in various ways. A common classification is to subdivide the analysis of the landscape into three separate layers: the 'natural landscape', the 'cultural landscape' and the 'urban landscape'.

### Layers

The landscape can be analysed as though it were a building consisting of various systems (Ill. 13.1). Any one system is laid out over an earlier system of the same region, and both of them interact with each other over time and space. The mutual effects are the result of a series of morphological and functional transformations. For instance, the agricultural landscape is the result of reclaiming and cultivating the natural landscape, and the urban landscape is the result of civil engineering and urban development carried out on both the natural and the agricultural landscape. In addition to these reclamation, farming, engineering, and building techniques, the urban landscape also reflects the history of natural geomorphological and pedological processes that took place in the past. The overall structure of the urban landscape is thus a combination of its history and how the various current systems fit in with each other.

### Transformations

Transformation in a landscape is a change of form within the framework of the original landscape by using its possibilities. The transformation is determined by what was previously present in the landscape, and is brought about by different processes, which can be of natural or cultural origin.

The transformations that originate naturally are for example those shaped by tectonic movements and volcanic eruptions. These result in new structural forms, which, in turn, are then changed by the action of agents such as water, wind, ice and gravity. Changes also take place when flora and fauna invade the land and sea, and the joint effects of both these and the natural geomorphological forces result in yet more changes. For instance, rocks become weathered, and vegetation zoned by selection through climate and altitude. Further patterns can be created by the grazing habits of large mammals. Nevertheless, it is probably mankind who has brought about some of the largest changes on earth, especially in recent times; and most of these have been to a greater or lesser degree harmful to the natural landscape. When people lived in a natural relationship with their surroundings, in harmony with their habitats, the natural balance remained largely undisturbed, but as soon as we became *homo technicus*, we started to alter the earth to suit our own requirements, and from that time onwards, our influence has dominated that of the vegetation and fauna.

Nonetheless, natural forces will always continue to exert an influence. In the case of natural disasters, such as volcanic eruptions, earthquakes and tsunamis they will always prevail over anything that *homo technicus* has tried to achieve. Processes that are caused by such forces will continue to reoccur, until such time as the natural force no longer exists.

So landscape is the result of a series of natural, cultural, functional and technical transformations. The morphology of the landscape is determined by the various processes that lead to these transformations. However, their impact will not be the same everywhere, nor will the changes take place at the same time. These processes can be started anywhere, at any moment, by natural forces, or by human influences, and the velocity of change will also vary from place to place.

On closer examination, the structure of the landscape reveals a kind of stratification within its own morphology. This is the result of the influence and manipulation of the various forces that were active during its formation. In short, the form is determined by the native flora and fauna, land reclamation and cultivation for farming, and by urban land use. The result is a patchwork of land enclosures, woods, water bodies, road infrastructure and buildings.

## Describing the different layers of landscape

The natural landscape (i.e. landscape not influenced by man) reflects its geological and geomorphological history; its forms are caused by internal and external natural processes. In this natural landscape, flora and fauna establish themselves by finding their own niche – a place that totally fits the needs of that species with respect to climate and substratum.

By contrast, the agricultural landscape is the result of land reclamation and the colonisation of the natural landscape. It is the cultivation pattern that becomes the most striking characteristic of this landscape, but the patterns and land use of the agricultural landscape do not always match the substratum, although they are largely determined by it.

As soon as man consciously determines the form of the landscape, it changes into an urban or architectonic form. This includes not only the built-up area, but also the technically determined infrastructure and the more technically advanced agricultural landscape. Natural landscapes, such as forests, water, barren land, tundra and areas of perennial snow and ice, untouched by man, form no part of the man-made landscape, although they are part of the total landscape (Steenbergen & Reh, 1996).

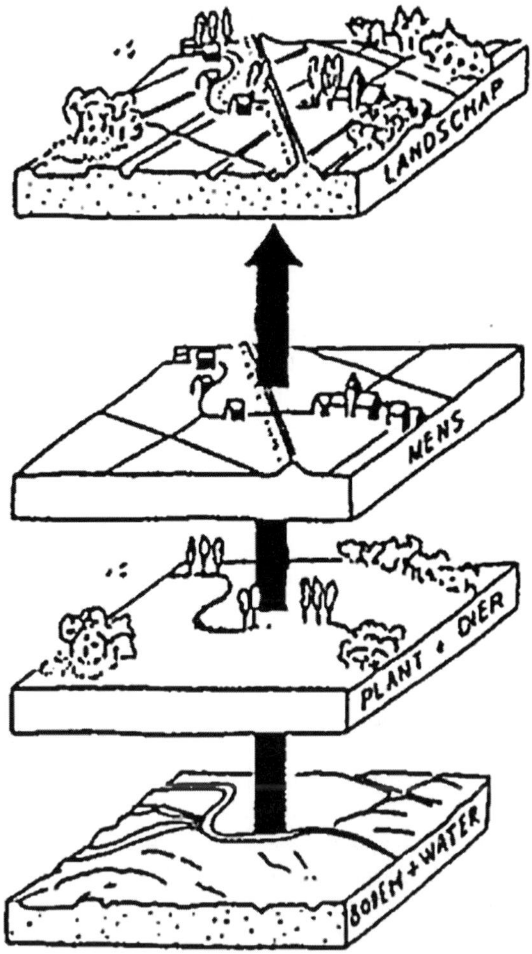


**Illustration 13.1:**
**From the bottom, upwards: the natural landscape, cultural landscape, urban landscape and architectonic landscape. Each layer of the landscape is a transformation of the layer that underlies it**

# Physical patterns and networks in the landscape

Since this chapter deals only with purely physical and substantial systems, the definitions that are used, and the examples given, will be restricted to these.

## The physical pattern

A physical pattern supposes equality in a set or a collection of forms. A pattern arises when an undefined agent causes a set of discernable uniformities (De Jong, 1992). Generally speaking, there can be a pattern on every level of scale, that can be extended in every direction. This physical pattern is determined by the area and the subsoil (or base). These patterns can be transformed or even halted by physical constraints such as tectonic forms (e.g. mountains) or large natural phenomena, such as water.

In describing, analysing and designing landscape, elements are indicated by points, lines and surfaces. When equal or uniform elements form a set or a collection, then, using De Jong's definition, given above, then they can be considered to be a pattern.

In landscape analyses, especially when analysing reclamation systems, the term 'pattern' is used to describe a system of ditches in the agricultural layer. The same term is also used to describe the organisation of plots of land (Ill. 13.2a). In the natural layer, the term is used to describe a river system including all tributaries; in soils patterns can be determined as well: soils formed in river deposits usually show a characteristic pattern. Or on an even lower scale a pattern in a frozen soil is formed by frost cracks (De Jong, Moens, Van den Akker & Steenbergen, 2003). Trees in the natural layer, for example, in the tropical rain forest, also form patterns, as do trees and plants in an orchard, plantation or vineyard. In the rainforest, this pattern will be random, more dictated by habitat, the amount of light and the distance to the same species of tree, while in the vineyard and orchard, the system will be linear to maximise the growing conditions of all the plants in the vineyard, and to make harvesting easier.

The patterns do not typically indicate direction, connection or intersection. If these elements are significant features of a certain pattern, then this will be indicated in the form of an additional adjective to describe the system, for example, a *north–south* ditch pattern, an *elongated* pattern of plots of land, or a *hexagonal* pattern of frost cracks.

As long as the systems or collections of discernable uniformities describe just their form, without indicating their function, then they will be classified under the term 'pattern'. But, for example, as soon as a pattern of ditches in a polder is used for drainage, it is no longer called a pattern, but a network used for draining the polder and keeping it dry. As soon as a function is attributed to the pattern, then the 'pattern' concept is extended into a 'network' (Ill. 13.2b).

## Physical network

A physical network is a system that is determined by topology, flows and adaptation (Dupuy, 1991). Or according to van Leeuwen (1973) a (physical) network is a pattern linked to functions and time. The form of the network is not only determined by equality within a set or collection, but also by function or potential use and by flow. There is always a flow along the links between two points or junctions. The flow can be in one direction only, or in both directions. A river with all its tributaries is then not just a pattern, but a (natural) network.

A network is a collection of more or less equal links and junctions. That the function within a network is equal is inherent to the system itself. For example, in the case of road infrastructure, function is built up with one purpose in mind, i.e. to connect the meeting points, nodes or junctions, in order to settle the traffic as quickly and safely as possible. Since links and junctions are characteristics of a network, these will develop in relationship to each other within the system. Whether one starts to built a network with the links or the junctions is irrelevant.

Nowadays, the term 'network' is also used for infrastructure, waterways, cities, and human contacts, etc. A network can be both a distribution and a contribution system at the same time; the direction of the flow is becoming less important. The focal points in the system are the connections in the network, the junctions, and system's function. These modern, man-made networks can be extended and changed on every scale, sometimes at any moment, and in any direction unlike in natural networks.

**Illustration 13.2a:**
**The Dutch polder 'de Schermer' showing its pattern of plots of land and ditches (Source: Steenbergen, Reh & Van der Kooij, 2002)**

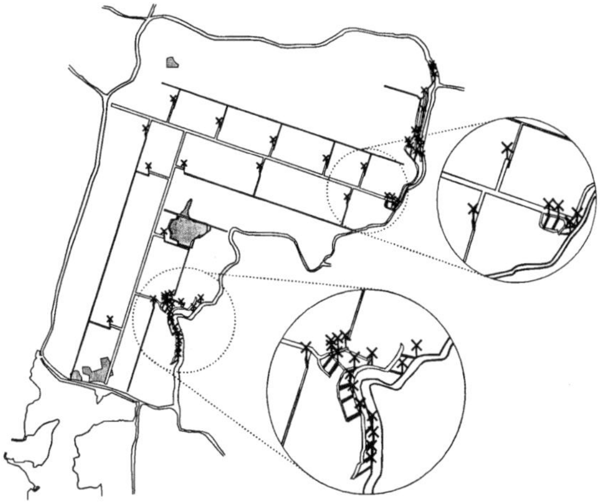

**Illustration 13.2b:**
**The Dutch polder 'de Schermer' showing its network of drainage systems (*ibid.*)**

## Conclusion

The terms 'pattern' and 'network' are used in both landscape analysis and landscape design. Depending on the context, the same physical configuration can belong to either patterns or networks. The difference is that 'pattern' is associated more with form, and 'network' with use. To use Van Leeuwen's (1973) definition: 'pattern' indicates a situation, while 'network' indicates a situation in which processes can take place.

## References


De Jong, T.M., 1992, *Kleine methodologie voor ontwerpend onderzoek*, Boom, Meppel (Neth.)

De Jong, T.M., M.J. Moens, C. van den Akker & C.M. Steenbergen, 2003, *Sun, wind, water, earth, life and environment: legends for design*, Publicatieburo Bouwkunde, Technische Universiteit Delft, Delft

Dupuy, G., 1991, *L'Urbanisme des Réseaux - Théories et Méthodes*, Armand Colin Editeurs, Paris

Steenbergen, C.M. & W. Reh, 1996, *Architecture and Landscape*, Thoth Publishers, Bussum, (Neth.)

Steenbergen, C.M., W. Reh & E. van der Kooij, 2002, *Experimenteel ontwerponderzoek de Schermer*, Publicatieburo Bouwkunde, Technische Universiteit Delft, Delft

Van Leeuwen, Chr.G., 1973, *Ekologie*, Faculty of Architecture, Technische Universiteit Delft, Delft

# 14 Putting Time into the Picture
## The Relation between Space and Time in Urban Design and Planning

« Space is in its very nature temporal and time spatial »
(S. Alexander (1920) - cited in Parkes & Thrift, 1980:12)

Ina Klaasen

## Introduction

In urban design and planning the concepts 'space' and 'time' do not need a profound philosophical approach. Neither do they need an Einstein: Newton suffices. 'Space', as we learned in school, is three-dimensional; 'time' can be considered to be a fourth dimension. However, this simple approach to these two notions does not mean that the relation between space and time in urban design and planning is unproblematic. This is largely due to the fact that the reflection on, and communication about, urban design and planning is usually undertaken with the help of spatial models. 'Model' is used here in the sense of a simplified representation of the real world, based on a conscious interpretation of that (past, present or future) reality and not as 'an example to follow'. As it is a two- or three-dimensional representation of a situation at a certain moment in time, the spatial model is inherently static in character, and the inability to depict the temporal dimension means that 'time', and therefore 'process', can only be indicated indirectly. As a lot of processes in urbanised systems are linked in some way to the physical urban system - which is the object of urban design and planning - designers and planners should be very much aware of this limitation of spatial models with regard to depicting 'time' and 'processes'. This chapter aims to contribute to this awareness, as unfortunately, it is sometimes lacking.

## Spatial models and their limitations

In daily life as well as in science, people use models to reflect on, and communicate about, reality; either verbal models, or mathematical, spatial or mechanical ones. They all have their limitations, because reality is always reduced in a model, by omitting elements and attributes to them that are considered irrelevant, and by categorising elements and their attributes. Illustration 14.1 shows the relation between reality and the model. This simplification (reduction) of reality is partly unconscious, and is based both on the scope of the natural human powers of perception and thinking and on cultural assumptions (which may also be affected by individual differences) (Klaasen, 2002). Depending on the perspective from which they are made they are also consciously simplified. Models are thus neither value-free, nor should one try to make them so (see also Bailer-Jones, 1999:29; Giere, 1999:46). This implies that before we start using a model, we have to think about its area of applicability. Venturing outside the area of applicability is known as over-extending the model.

In the case of spatial models there are specific limitations with respect to reality being scaled down (or in some disciplines scaled up) and to the difficulty of depicting non-visible components such as the components 'time' and 'process' (Ill.14.2). If we want to include the time dimension directly, then we need a mechanical model; a planetarium, for instance, a dynamic model of our solar system. Computers can be used to simulate a mechanical model compare with a film, a rapid succession of static images.

When urban designers/planners analyse an existing situation, they generally use (descriptive, conceptual or concrete) spatial models (Ill.14.3). When analysing probable developments within this situation, they use predictive models, and when analysing possible developments, potential-projective ones or intentional-projective ones (Ill.14.4) (see Klaasen, 2002).

Non-visible components are often, literally, left out of the picture. People do not just experience a city visually, however – they hear, smell and feel it. Those who cannot see, nonetheless experience the city. There are also urban elements underground that may have above-ground effects: cables, pipes, tree-root systems. Societal and natural processes, however, remain invisible: "The disadvantage of designing by drawing is that problems which are not visually apparent tend not to come to the designer's attention. Architects could not 'see' the social problems associated with new forms of housing by looking at their drawings."(Lawson, 1990: 18-19).

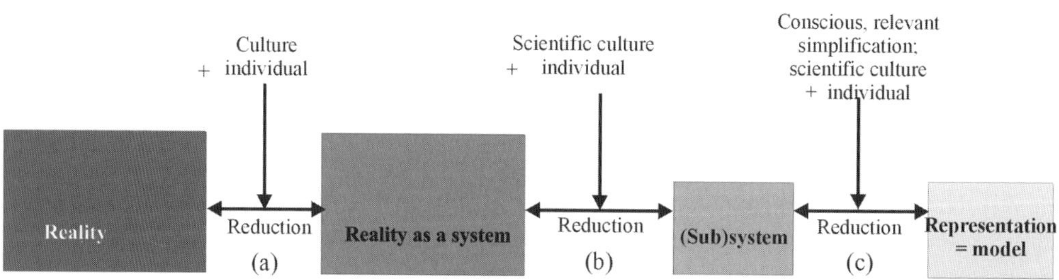


**Illustration 14.1: We can conceive of reality as a system or a complex of systems (a). We can extract selected systems and/or subsystems (b) from this reality on the basis of intentional considerations and scientific assumptions (which may also differ per individual), and then represent these as models (c)**

| Static spatial model | versus | Dynamic reality |
|---|---|---|
| Space | | Space and time |
| Visible phenomena | | Visible and invisible phenomena |
| Objects and spatial relations: patterns | | Objects and spatial and temporal relations: patterns and processes |

**Illustration 14 .2: The essential differences between the real situation and a spatial model of this situation**

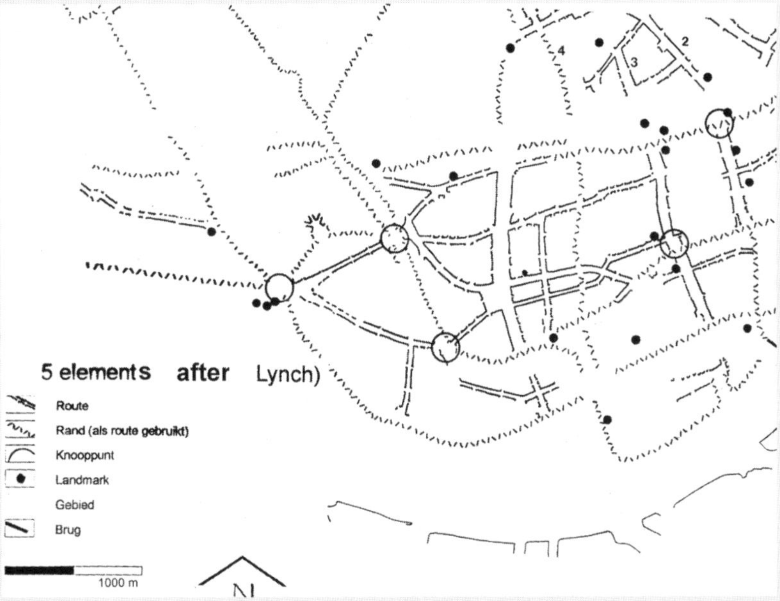


**Illustration 14.3: A descriptive spatial model: an analysis of the north-western part of Rotterdam according to 'Lynch's method' (Wijnen, 2002:38)**

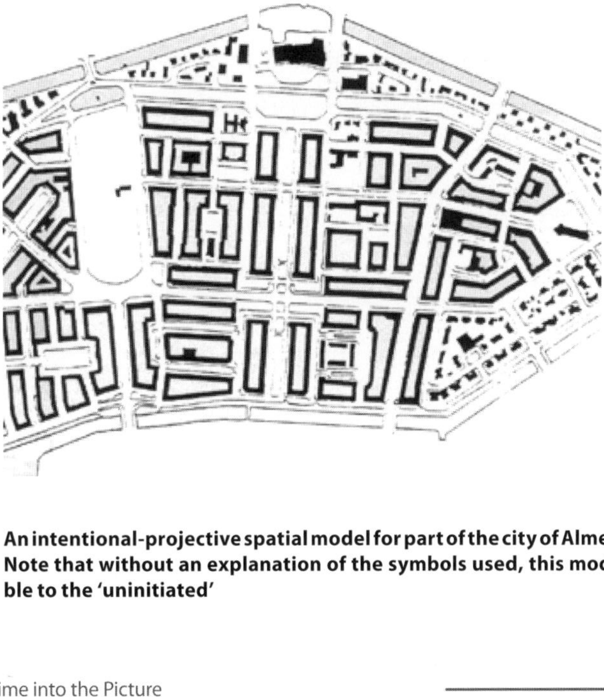

**Illustration 14.4: An intentional-projective spatial model for part of the city of Almere (Thüsh, 1993:140). Note that without an explanation of the symbols used, this model is incomprehensible to the 'uninitiated'**

When one is aware of the time dimension, and one does wish to visualise it, there do exist ways of doing this, for example, by presenting a series of spatial models at different points in time, by adding symbols to the visual language used in spatial models and/or by using measures, such as isochrones, to indicate travelling time (Illustrations 14.5-8).

## Time and processes

Systems including, of course, urbanised systems – from the scale of the neighbourhood to that of the region, change in time because of changing relations (interactions) between the elements and their attributes, and changing relations between the system and its environment: flows of information, energy, material and living creatures – i.e. processes. Whether or not we register temporal changes, depends on the temporal grain of the observation, the same way as registering spatial changes depends on the spatial grain of the observation. "The smallest difference that we wish to, or are able to perceive, conceive or represent while not yet designating it as 'equality' is ... the 'grain' of the perception, concept or representation," (De Jong, 1992:16). The concept 'spatial grain' is related to the concept 'spatial scale', and the 'temporal grain' to the 'timescale'.

A process is said to exist in the case of a prolonged, regular action or succession of actions that take place or are carried out in a certain manner. Processes may be divided into cyclical and linear processes (Ill. 14.9), and cyclical processes may themselves have a linear component (Ill. 14.10).

If a system has an internal structure with only spatial dimensions or only a temporal one (e.g. a piece of music), we can call this a composition or a pattern. The use of spatial models by designers/planners when reflecting and communicating, easily leads to regarding physical systems – i.e. configurations of spatial elements like buildings, streets parks, sewers – as autonomous objects characterised by a certain spatial composition, a certain pattern. The fact that the main Chair of Urban Design at the Faculty of Architecture of Delft University of Technology is named 'Urban Composition', underlines this.

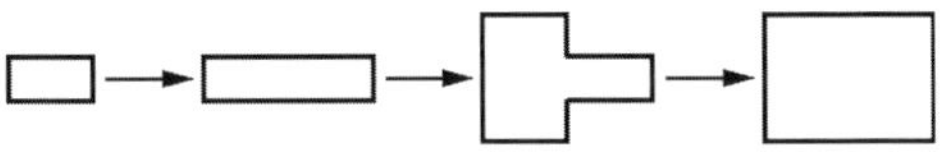

**Illustration 14.5:**
**An example of how arrows can indicate a development process: a conceptual spatial model of the development of a city**

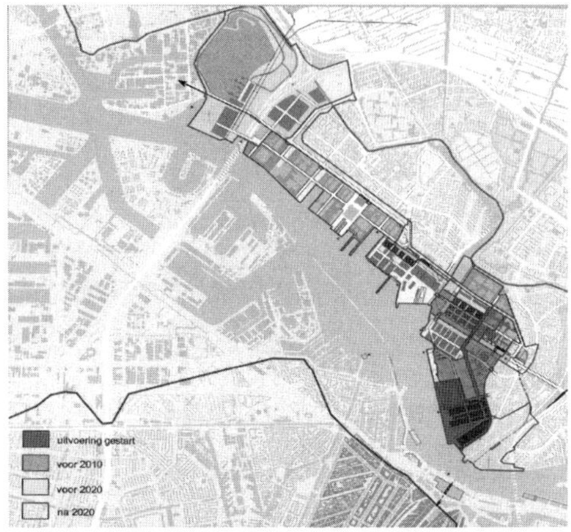


**Illustration 14.6:**
**An example of using visual language symbols to indicate the phasing of a spatial development(Stadsdeel Amsterdam-Noord, 2002)**

**Illustration 14.7:**
**An example of visualising travel time by using indicative isochrones. A spatial model for a planned eastwards expansion of Amsterdam into the IJsselmeer [Lake IJssel] accessibility isochrones of pre-transport for a prospective rail link**

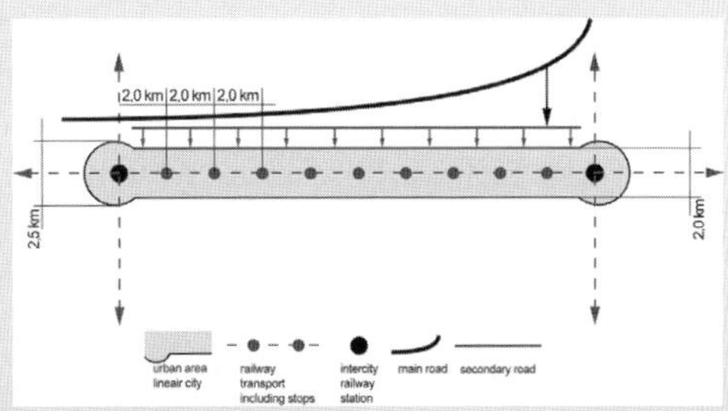


**Illustration 14.8:**
**An example of visualising travel time by inserting measures in a spatial analysis or design: a quantitative theoretical spatial model of a linear city**

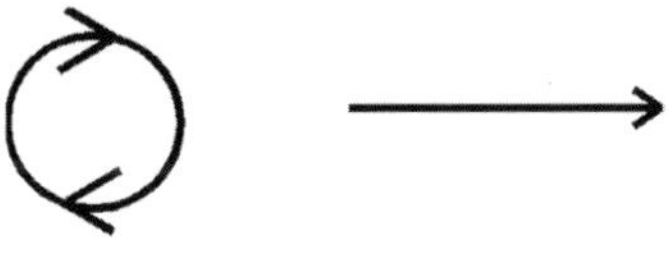

**Illustration 14.9:**
**Representations of a cyclical process (left) and a linear process (right)**

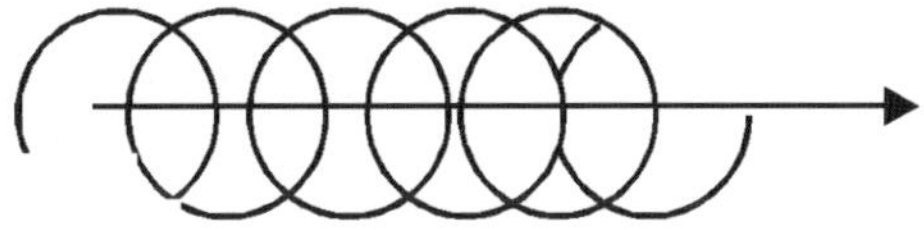

**Illustration 14.10:**
**Representation of a cyclical process with linear component**

## Relevant processes in urban and regional design and planning

Physical urban systems are generated by processes in urbanised systems on the one hand (initiated by designers, planners, decision-makers, builders, etc.),while on the other hand processes in urbanised systems need physical systems in a conditional sense (Ill. 14.11). When processes in urbanised systems have a spatial component, they are relevant for urban and regional designers/planners. Physical urban systems comprise urban spatial elements/objects in mutually coherent combinations and in coherent combinations with natural (and possibly rural) spatial elements/objects.

Generating – changes in – physical urban systems involves planning, a design process, a decision process and a construction process. Generally, the more complicated these processes are organisationally, financially, spatially, etc., the more time they will take. Planning and design processes may have cyclical aspects, but their main character, like that of construction processes, is linear. There is usually a relation between the orders of magnitude of the temporal and spatial grains, and thus between the spatial and temporal scales (Ill. 14.12). For instance, the first design for Amsterdam's expansion eastwards into the IJsselmeer [Lake IJssel] (see Ill. 14.7), now in its realisation stage, dates from 1965: Van den Broek's and Bakema's so-called 'Pampusplan'. As some of the elements of the physical urban system are part of the natural system; for instance, trees, but also hydrological elements, the notion 'construction' does not cover all the processes that are being implemented. Relevant processes in the natural system can have temporal grains that differ from those of construction processes.

Once brought about, elements of the physical urban system undergo reversible (cyclical, variable temporal grains) and irreversible (linear, large temporal grains) wear and tear under the influence of both societal and ecological processes. At some stage, the system as a whole will be replaced, because, functionally, it falls short of requirements, leading to an urban transformation of the area. The temporal grain will be relatively large, varying in relation to the spatial grain (from years to decades); the process of course, is linear. Examples of well-known urban transformations taking place at the present time are the redevelopments of former harbour areas.

Physical urban systems are not just generated by society, they for their part support urban processes-with-a-spatial-dimension. These processes are cyclical in character and have a small to very small temporal grain compared to the temporal grain of the linear urban development and transformation processes (Ill. 14.13). Examples of these small-grained temporal cyclical processes are the daily journey to work or school, routine shopping, retail stock deliveries, the supplies of water and gas, regular waste collection, weekly trips to football stadiums, etc.

These processes require 'activity carriers', as well as elements of the urban system sited in relation to each other on the grounds of their functional attributes, i.e. a system structure that takes into account that time is a constraining factor in most of these processes, certainly when people are involved. The primary constraint is the daily hormonally/astronomically based human biorhythm. The time people have to spare for mobility is, after all, limited; it has been empirically set at 2 x 45 min. = 90 min. daily (see studies by Vance, 1977, and by Schafer & Victor, 1997, both cited in Jacobs, 2000:45,57). Of secondary importance is the culture-based weekly rhythm (Bible, Koran), although this has declined in significance in the post-industrial era, and a seasonal rhythm, particularly where there is a well-marked climatic variation between seasons. With regard to experiencing the urban environment it's relevant to

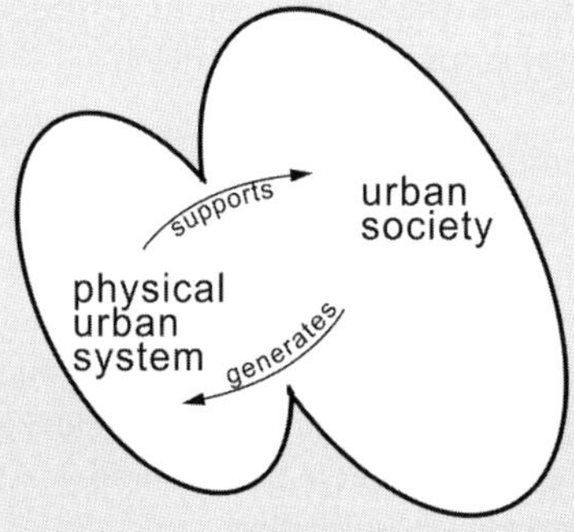


**Illustration 14.11:** **The physical urban system makes societal processes possible in a conditional sense. Conversely, societal processes lead to changes in and extensions to this system, and provide control over it**

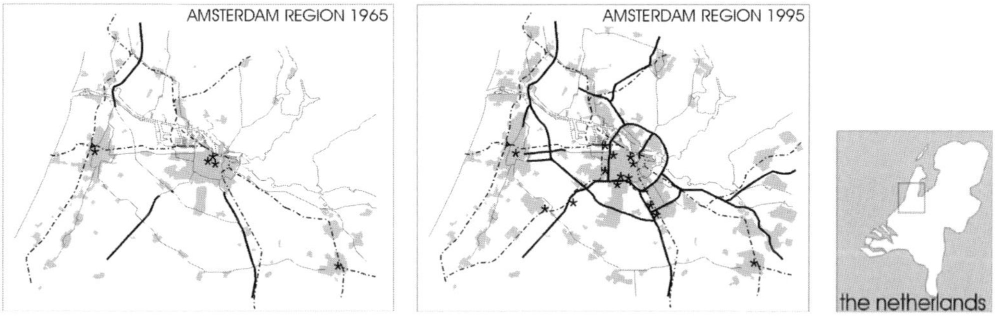


**Illustration 14.12:** **An example of a transformation process at a regional scale. The maps depict two snapshots of the (linear) spatial development of the Amsterdam region (Klaasen & Jacobs, 1999:22/23)**

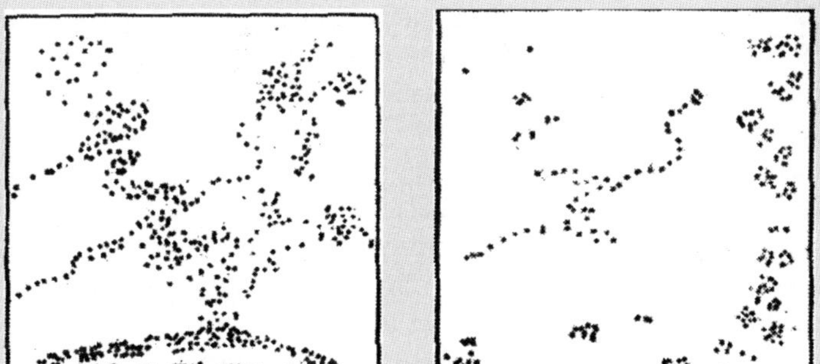

**Illustration 14.13:** **The large difference in intensity of use (represented by dots) of the same area at differing times of day (Doxiadis, 1968:325)**

realise that the information that reaches the user of this environment depends on his or her standpoint and the velocity at which that point may in some cases be moving.

Finally we have to take into account processes in the natural system that influence processes in urban systems. Healthy and safe living-conditions relate to the sustainable preservation of processes necessary and desirable to society in the natural system, and also to protection from undesirable processes. Examples of the former category are cyclical atmospheric currents (wind, temperature), the recycling of materials and the growing of plants and trees, etc. Apart from being functional, these processes usually have an experiential value as well: they are visual, auditory, olfactory and/or tactile in character. An example of the latter category, particularly relevant in the geographical situation of the Netherlands, but of course in other parts of the world as well, is flooding.

## Relating space to time: problems and some solutions

Problems regarding the relation between space and time in urban and regional design and planning can chiefly be traced back to the fact that the main tools to aid reflection and communication are models that only allow time to be shown indirectly. Evidently this makes urban and regional designers/planners less 'time-conscious'. A design for a building, a bridge or a park, where the size of the physical system, and hence the relatively short realisation time, make it possible to get by with a blueprint design. This type of design, however, is inappropriate for a physical urban system, as the implementation of a design for a physical urban system, particularly when at a relatively large level of scale, is beset with uncertainties owing to an extended construction phase and the large number of actors involved (Faludi, 1973; Barnett, 1982; Nio & Reijndorp, 1997:238). As Broadbent wrote, "... there may be real dangers in presenting a model of the city in physical terms as a three-dimensional spatial model. Planning committees and the public at large, not to mention the planners themselves, become notoriously attached, or antagonistic, to the actual forms which such models present; they tend to 'freeze' development of the city in a particular state." (1988:94).

During the years that it takes to implement proposed developments, almost unavoidably, changes will be made with regard to the original blueprint. As a blueprint design does not differentiate between essential and non-essential spatial interventions, the spatial quality aimed at by the designer may be lost. A solution to this problem is to indicate the essentials in a design and to make suggestions – give alternatives even – for non-essential aspects of the proposed interventions (Ill. 14.14). Essentials could be to indicate the direction that a certain urban development should take, if necessary also stating the spatial limitation of the development. Also alternative development routes, for instance depending on available finances, could be indicated.

The non-temporal, but exclusively spatial, character of models could also be the reason why some proposals for spatial interventions in physical urban systems reveal that little, if any, attention is paid either to maintenance and its associated costs, or, in the case of the example given as Illustration 14.15, to the dynamics of the urban situation. It follows from this example that it is also very advisable to differentiate between essentials and non-essentials when designing public places in dynamic urban situations like city centres.

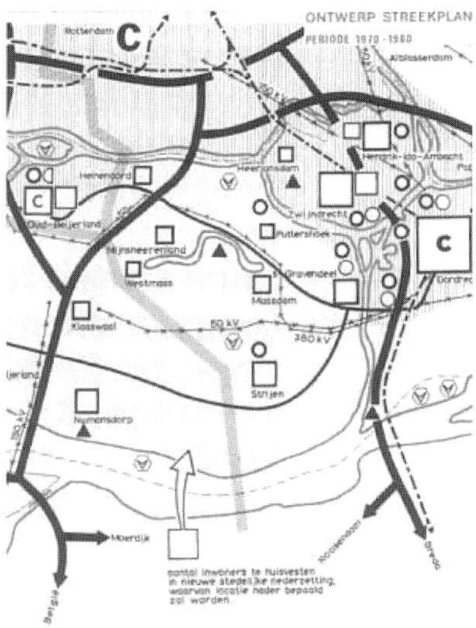


**Illustration 14.14:**
**In this fragment of a regional plan, certain potential urban developments are indicated by symbols only (squares of different sizes, arrows) (De Boer, 2005:31)**

**Illustration 14.15:**
**The Nieuwmarkt in Amsterdam. The dynamics of the metropolis have disrupted the original composition of the square, dating from 1990.**

**The long lines of lighter grey paving tiles were laid along the brownish yellow brickwork pavement to demarcate an underground waterway. The repaving of the street since then has disrupted these lines. Following a decision by the municipal council to assign part of the public space to a parking function, white paint was used to mark out the parking spaces. The same paint was used to demarcate the cycle path with the aim of drawing the attention of cyclists and motorists to this path (see picture to the right).**

**The part of the square not shown here is now planted with trees, despite the designers having successfully raised objections to these trees in the initial decision phase.**

**The fence around the centre of the square, which you can see in the picture to the left proved to be very vulnerable. Cars quite often damage this fence, a costly affair to repair.**

Another side to the dynamics of urbanised systems should be taken into account – also with regard to the relation between spatial and temporal grains. If large-grained spatial transformations of physical systems take place within a short period, i.e., within just a couple of years – as when, for instance, 'new towns' are developed in large building streams – then 'time', in a historical sense, cannot be experienced. In developments such as these, 'history' has no time to leave its mark on the built environment, and this reduces its experiential value. The slower the changes in the physical urban system can take place, the more the users of an urban area will be able to experience the passing of time. Preserving functionally out-of-date system elements, or complete systems (monuments), will add to the cultural-historical experiential value.

Regarding physical urban systems as autonomous systems, and designs as compositions, focuses on the generation of these physical systems by societal processes (see the arrow pointing to the left in Illustration 14.11). It is easy to reflect on, and communicate about these large-grained linear urban transformation processes using spatial models (Ill. 14.12 is an example). A problem that arises, though, is that the physical urban system consists not just of urban spatial elements, but also of natural ones, and that transformation processes in natural systems often have a different time scale from those made by people. Changes in ground-water level may take much longer to effect, than redeveloping a city neighbourhood (Klaasen, 1993:87 ff.); trees may take much longer to mature, than the time it takes to build houses (Ill. 14.16).

Unfortunately in a design and planning approach directed towards transformations of physical urban systems, the small-grained cyclical urban processes are not 'in the picture'. One of the consequences is that designers almost always opt to convey impressions of physical urban systems using a 'snapshot' of the situation at a given moment during the day, in summer, with the sun shining on fully-grown trees in leaf (Ill. 14.16). This representation bypasses the fact that people's experience of a city at night, in the winter, and in a gale or heavy rain, is different from that in the summer sunshine. Putting oneself into the position of the users of urban areas might lead to planning appropriate protection against unpleasant climatic conditions (depending on the climatic zone the measures may differ), and, from the viewpoint of public safety, to consider creating different routes for day-and-night use by slow traffic (pedestrians and cyclists) (Ill. 14.17). On the other hand people may actually enjoy experiencing – changes in – some climatic phenomena. Furthermore, as mentioned in the previous section, the way people experience the built environment depends on their mode of transport and the velocity at which they move. Given the static character of spatial models, designers tend to overlook such factors.

Focusing on small-grained cyclical processes demands another approach than transformation-directed design. This approach could be called process-directed design. A mainly quantitative schedule of design requirements (numbers of dwellings, surface areas of industry sites, traffic volumes to be accommodated, square metres of park per household, etc.) is insufficient, because it neglects the bio-rhythmical and culture-rhythmical constraints of societal activities. The 'activity carriers' need to be spatially structured in such a way that desirable daily and weekly temporal-spatial activity patterns are indeed accommodated. An example of this type of activity pattern is shown in Illustration 14.18.

Functional-spatial structures to accommodate temporal-spatial activity patterns should be supported by a relevant formal design to assist users in identifying and defining direction and ensuring pleasant surroundings. In this design approach, the physical urban system is seen explicitly as a subsystem of the urban system as a whole. Examples of building blocks for the spatial-organisational functionality of urban area designs and plans are shown in Illustrations 14.19-23.

**Illustration 14.16:** **A very clean urban street, on a summer day. The sun is shining and the fully-grown trees are in leaf. Now picture this in winter, during a gale, with plenty of moving cars. (source: Woningkrant IJburg Gemeente Amsterdam, November 2002)**

**Illustration 14.17:**
**'Cycling through Schiedam at night' is the title of a pamphlet published by the Municipality of Schiedam (Neth.), which is the source of this map (Undated)**

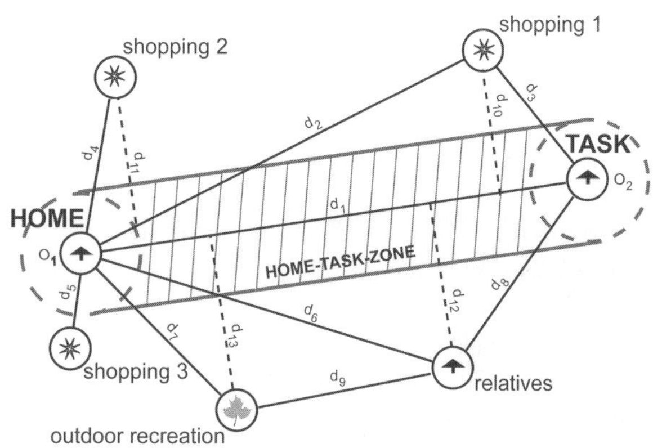


**Illustration 14.18:**
**Example of a temperal-spatial activity pattern (after Vidaković, 1988:122)**

Designers and planners should be aware that knowledge about temporal-spatial activity patterns is usually provided by empirical research into the relation between space and time in different categories of households (e.g., Mey, 1994, 1996; Dijst, 1995; Mey & Ter Heide, 1997). Design though, belonging to the domain of practical sciences (Klaasen, 2004), should not just be directed towards existing activity patterns, but towards accommodating possible (desirable) activity patterns.

In this respect, an interesting example of the relation between space and time is the 'Compact City' designed by two operational researchers (Dantzig & Saaty, 1973). George Dantzig and Thomas Saaty designed a 'four-dimensional city' with an 'around-the-clock' use of all urban facilities. As an example, they sketched the following scenario: 24 people having lunch in a restaurant at the same time would need 24 times the space that they would need if they had gone to lunch one after the other, making use of the 24 hours in the day. In this way, a city of 250,000 inhabitants needs a building of only 10 layers, each of them 10 m in height, with an average cross-section of 1.5 km, and includes everything from houses with outside space to specialised educational facilities and even a park (on the top layer) (Ill. 14.24). This would be a very time-efficient spatial organisation.

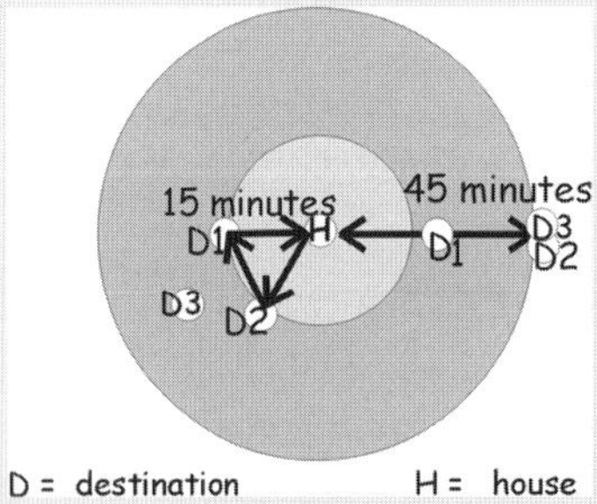


**Illustration 14.19:**
**The principle of time-space activity zones. The number of activities that can be carried out by an individual in the course of a single day increases along with the degree to which their 'carriers' are located in one another's vicinity, or on the route from one to another. With a maximum amount of available travel time of 90 minutes per day, point D3 is inaccessible if criss-cross trips are called for**

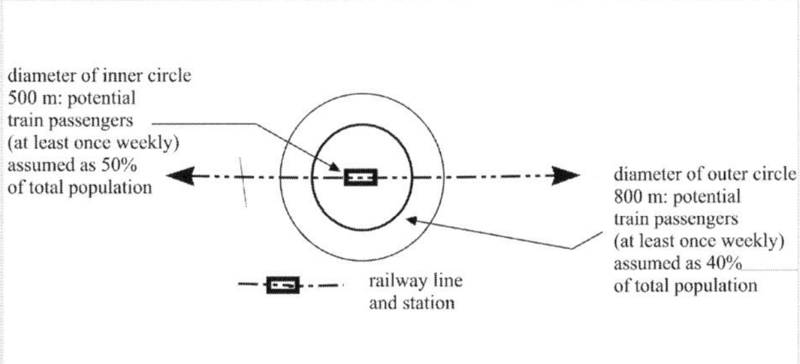


**Illustration 14.20:**
**Principle of a potentially sufficient user base for a collective function, taking a train station as an example. The norms applied here are based on assumptions. Spatially, the number of stops a train makes cannot be extended beyond a certain limit**

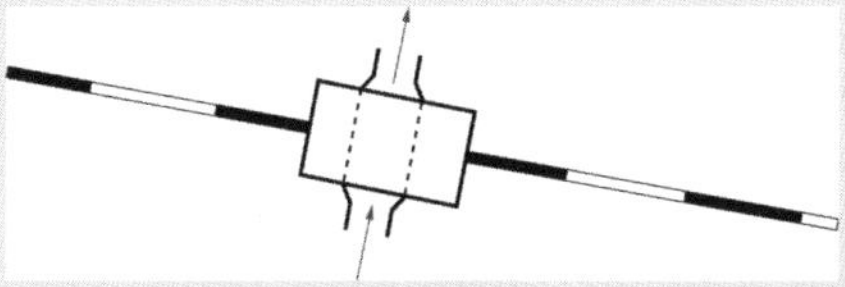

**Illustration 14.21:**
**Connections on a certain scale often form a .barrier on a lower level of scale. When the pedestrian/cycle routes between city neighbourhoods located on either side of a railway line converge with access to the station and its associated activities (ticket sales, kiosk, etc.), this improves public safety for all users**

**Illustration 14.22:**
**An example of how formal design can make it easier to find one's way. "Paths may not be identifiable and continuous but have directional quality as well: one direction along a line can easily be distinguished from the reverse. This can be done by a gradient, a regular change in some quality which is cumulative in one direction." (Lynch, 1960:54)**

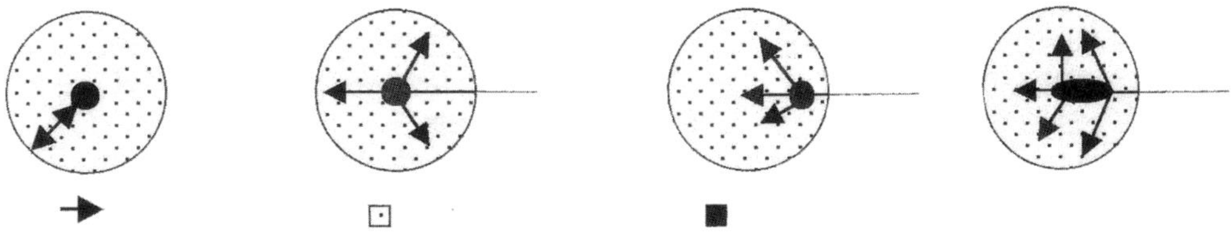

**Illustration 14.23:**
**Organisation principles for the location of collective functions at neighbourhood level under different conditions. (a) In a homogeneous situation, the centre of a circle is generally the most easily accessible point, and thus is appropriate for locating collective functions (which make an intensive use of space). The radius of the circle is a criterion for the functional spatial quality – the time and energy required for assumedly equivalent movement options (walking, cycling). The residential density and the surface area of the circle are joint criteria for the potential quality of the collective functions. In (b) and (c), the residential area is linked to the outside world. All residents and visitors pass through a single entry point. In (b) the point is a bus or metro station, one of the collective activities that are situated in the central zone. In (c), the entry point for example, is a town-centre parking garage located on the edge of a 'pedestrian-only' residential area. Example (d) consists of a combination of a bus or metro station and a parking garage. For collective functions, the zone between the centrally sited station and the eccentrically sited parking garage now has the highest location value**

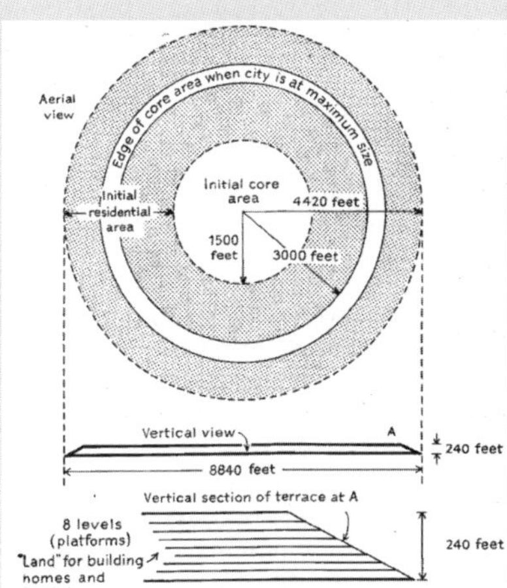


**Illustration 14.24:**
**Top and side view of a Compact City accommodating 250,000 inhabitants (Dantzig & Saaty, 1973:37)**

The top and side view of Compact City, as depicted in Illustration 14.24, is again an example of measures being needed to communicate indirectly about the relation between space and time. This puts forward the question of whether new legend keys can, and should, be developed to complement those already in use: for instance, a legend key to indicate how housing areas differ in possible temporal-spatial activity patterns, or to indicate which accessibility profile characterises a specific urban area (Ill. 14.25).

## Conclusions

In urban and regional design and planning, spatial models are the designated tool for reflection and communication. However, spatial models have their limitations, one of these being that the relation between space and time, and therefore spatial urban processes can only be visualised indirectly. As a result, urban processes tend to be given insufficient attention. This concerns planning, decision and construction processes as well as wear and tear of implemented plans, and certainly also small-grained cyclical societal processes, including the way the built environment is experienced depending on the mode of transportation used. To avoid problems arising because of the time that elapses between developing and implementing plans – and, at a later stage, adapting and/or maintaining implemented plans – making blue prints should be avoided. At the very least, the essential and non-essential components of plans should be differentiated.

Designers and planners should also become more aware of daily and weekly cyclical societal processes and their temporal constraints, and pay more attention to creating the spatial conditions for a diversity of temporal-spatial activity patterns, including the potentially desirable ones rather than those that have merely been identified empirically. To improve the way physical urban systems are geared to the social and economic processes that they 'carry', transformation design should give way to process-directed design. It is quite possible that this will lead to modifying the design legends in current use.

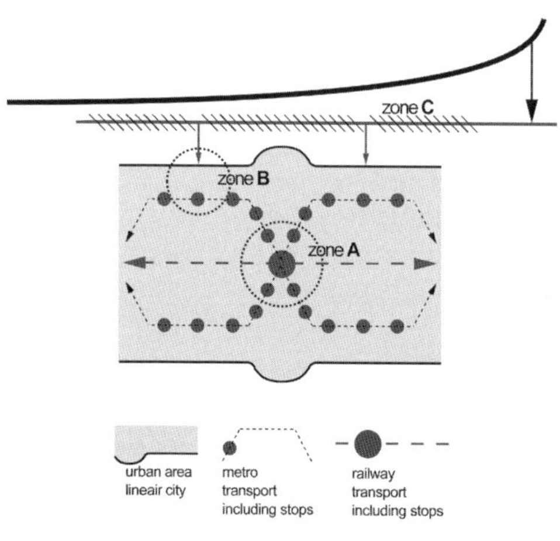


**Illustration 14.25:**
**Around the main station (zone A), the spatial conditions have been created for regional functions with a collective transport profile. Around one of the secondary stations (zone B), the spatial conditions have been created for regional functions that have a combined collective/private transport profile. In the zone along the secondary access route (zone C) – a motorway – functions that have a private (i.e car) transport profile can be located**

## References


Bailer-Jones, D.M., 1999, Tracing the Development of Models in the Philosophy of Science; In: Magnani, Lorenzo, Nancy Nersessian & Paul Thagard (eds), *Model-based Reasoning in Scientific Discovery*, Kluwer Academic/ Plenum Publishers, New York, pp. 23-40

Barnett, J., 1982, *An Introduction to Urban Design*, Harper & Row, New York

Broadbent, G., 1988, *Design in Architecture*, David Fulton Publishers, London

Dantzig, G.B. & Th.L. Saaty, 1973, *Compact City; a Plan for a Liveable Urban Environment*, W.H. Freeman and Company, San Francisco

De Boer, N.A., 2005, *De Stad van Niek de Boer*, Publikatieburo Bouwkunde, Faculty of Architecture, Technische Universiteit Delft

Dijst, M. J. 1995, *Het elliptische leven - Actieruimte als integrale maat voor bereik en mobiliteit*, PhD Thesis, Faculty of Architecture, Technische Universiteit Delft

Doxiadis, C.A., 1968, *Ekistics - An Introduction to the Science of Human Settlements*, Hutchinson & Co, London

Faludi, A., 1973, *Planning Theory*, Pergamon Press Ltd., Oxford

Giere, R., 1999, Using Models to Represent Reality; In: Magnani, Lorenzo, Nancy Nersessian & Paul Thagard (eds), *Model-based Reasoning in Scientific Discovery*, Kluwer Academic/Plenum Publishers, New York, pp. 41-58

Jacobs, M., 2000, *Multinodal Urban Structures, a Comparative Analysis and Strategies for Design*, PhD Thesis, Delft University Press, Delft

De Jong, T.M., 1992, *Kleine Methodologie voor Ontwerpend Onderzoek*, Boom, Meppel (Neth.)

Klaasen, I.T. (ed.), 1993, *Het Stromende Stadsgewest, Plananalyse Derde EO Wijers-prijsvraag*, Publikatieburo Bouwkunde, Faculty of Architecture, Technische Universiteit Delft

Klaasen, I.T., 2002, Modelling Reality; In: De Jong, T.M. & D.J.M. van der Voordt (eds), *Ways to Study and Research - Urban, Architectural and Technical Design*, Delft University Press, Delft, pp. 181-188

Klaasen, I.T., 2004, *Knowledge-based Design: Developing Urban & Regional Design into a Science*, Design/Science/ Planning Series, Delft University Press, Delft

Klaasen, I.T. & M. Jacobs, 1999, Relative Location Value Based on Accessibility: Application of a useful concept in designing urban regions, *Landscape and Urban Planning*, 45, pp. 21-35

Lawson, B., 1990, *How Designers Think - The Design Process Demystified*, Butterworth Architecture, Oxford

Lynch, K., (1960) 1992, *The Image of the City*, MIT Press, Cambridge Mass.

Mey, M.G., 1994, *Het stedelijk mozaiek* (The urban mosaic; preferences of households as guidelines for the design of urban settlements), PhD Thesis, Faculty of Architecture, Technische Universiteit Delft

Mey, M.G., 1996, *Tijd-ruimtelijke dynamiek en stedelijke inrichtingen*, Delft University Press, Delft

Mey, M.G. & H. ter Heide, 1997, Towards spatiotemporal planning: practicable analysis of day-to-day paths through space and time, *Environment and Planning B: Planning and Design*, pp. 709-723

Nio, I. & A. Reijndorp, 1997, *Groeten uit Zoetermeer- stedebouw in discussie*, NAi Uitgevers, Rotterdam

Parkes, D. & N. Thrift, 1980, *Times, Spaces and Places, John Wiley & Sons*, Brisbane (Aus)

Stadsdeel Amsterdam-Noord, 2002, *Voorontwerp Masterplan Noordelijke IJ-Oever*, Stadsdeel Amsterdam-Noord

Thüsh, M., 1993, *Almere, uitgeslapen stad*, Masters Thesis, Faculty of Architecture, Technische Universiteit Delft

Vidaković, V., 1988, *De Ruimte voor de Tijd*, Rijksplanologische Dienst/Gemeente Amsterdam, Den Haag/Amsterdam

Wijnen, D.M.P., 2002, *Bereikbaarheid beleven, Leefbaarheid bereiken - Structuur van de openbare ruimte Rotterdam*, Masters Thesis, Faculty of Architecture, Technische Universiteit Delft

# 15 Time in Urban Planning and Design in the ICT Age [1]

Paul Drewe

## Introduction

Time is crucial to network thinking in urbanism. Households and companies use technological systems selectively for their own specific purposes, as revealed by space–time budgets or logistic spaces. Urban planning and design can be seen as a search for spatial concepts to satisfy the needs of household and company networks. Therefore it is important to focus on the mutual interactions of time and space.

Spatial concepts often lack a temporal dimension, and also the impact of information and communication technology(ICT) (which supports the development of new temporal regimes) on space-time budgets. In the search for new planning and design instruments, it seems more promising to be guided by network thinking than by speculations about emerging, continuous cities.

By asking ourselves 'What about time in urban planning and design?', we are also asking 'What about people in urban planning and design?'. We are dealing, at the same time, with Lynch's time-honoured question 'What time is this place?'

## Time and the 'Network City'

The 'Network City' is a new way of thinking about cities (Drewe, 2003; http://www.networkcity.bk.tudelft.nl). It can be seen as a paradigmatic challenge for urban design and planning, as it translates into an integrated planning of land use and urban technology networks, including ICT. The 'Network City' is not a ready-made spatial concept, but rather a matter of network thinking in urbanism. Hence it is not to be confused with *'réseaux de villes'* (France), *'Städtenetze'* (Germany) or *'stedelijke netwerken'* (The Netherlands).

Illustration 15.1 has become the trademark of network thinking. It focuses on three interacting levels of network operators who (re)organise urban space (Dupuy, 1991:119):

- Level 1 involves the suppliers of technical networks such as water and sewerage, energy, transport and ICT. This level covers the infrastructure, the services offered and the operators (Graham & Marvin, 2001).

---

1 This chapter is a reworked and shortened version of Drewe, P. (2004), *What about time in urban planning & design in the ICT age?*, Design Studio 'The Network City', Faculty of Architecture, Delft University of Technology.

- On level 2, we find the functional networks of common-interest users centring on consumption, production, distribution and social contacts. Specific location factors apply to each of these networks.
- On level 3, both households and companies use technical networks and services selectively for their own specific purposes.

It is on level 3 that time comes into play. Household networks can best be measured in terms of space–time budgets or action spaces (Dijst & Vidaković, 1997). Company networks, on the other hand, are closely related to logistics, logistic chains and logistic space (Hesse, 2002).

The framework shown in Illustration 1 has also been applied to the Internet (Drewe, 2002). Level 3 refers to how the data actually flow on the Internet between points of origin and points of destination. Trace routes of the data flow have revealed an important time dimension of the Internet: the number of milliseconds that it takes a ping packet to move from the point of origin to each hop (from the node of origin to the node of destination) and back. The response time of accessing and downloading the home pages of websites is an important performance indicator of the Internet.

The interrelations between level 1 (the Internet infrastructure), level 2 (the Internet industry) and level 3 (traffic on the Internet) can be conceptualised as interrelations between supply, demand and performance, respectively (see Ill. 15.2).

## Time and space, mutual interactions

If the spatial concepts applied in urban planning and design lack a temporal dimension (Klaasen, 2004) then, first of all, this temporal dimension needs to be highlighted. Why is time important? Why has it become so important that it features – at least in some countries – on the political agenda? (Boulin & Mückenberger, 2002). Could it be that time has undergone significant changes that clash, with cities or space lagging behind? Has not the demand of citizens expanded and diversified, whereas the urban service supply, to a large extent, still functions according to traditional rhythms – as Gwiazdzinski (2002) sees it?

It does not suffice, however, to highlight just the time dimension. If the focus is on urban planning and design, one needs to introduce space as well: first, the spatial impact of significant changes in temporal patterns, changes which, in turn, are caused by societal changes and, then, the impact of changing spatial structures on temporal patterns.

Changes in spatial structure can result from urban design and planning interventions. This is where new spatial concepts come in, concepts that take into account the mutual interactions of time and space. Hence urban planning and design can be seen as a search for spatial concepts to satisfy the needs of household and company networks. To reveal the complexity is far from simple.

Illustration 15.3 includes some of the dichotomous terms used to describe time and space and their mutual interactions.

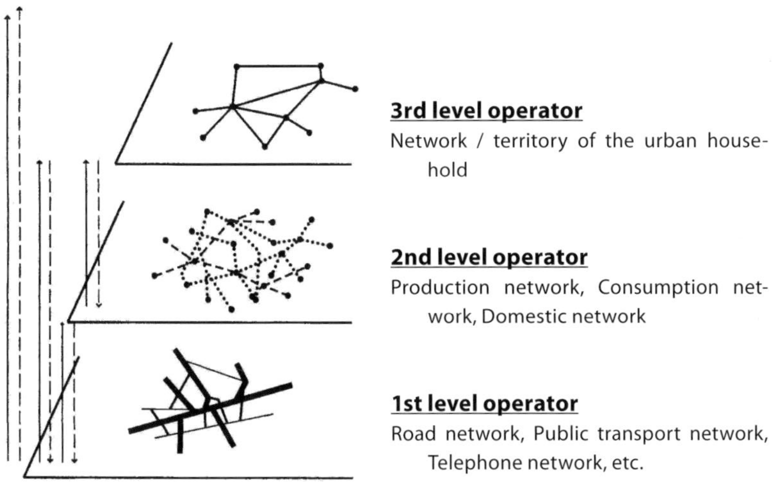


**Illustration 15.1: The three levels of operators who (re)organise urban space (Dupuy 1991:119)**

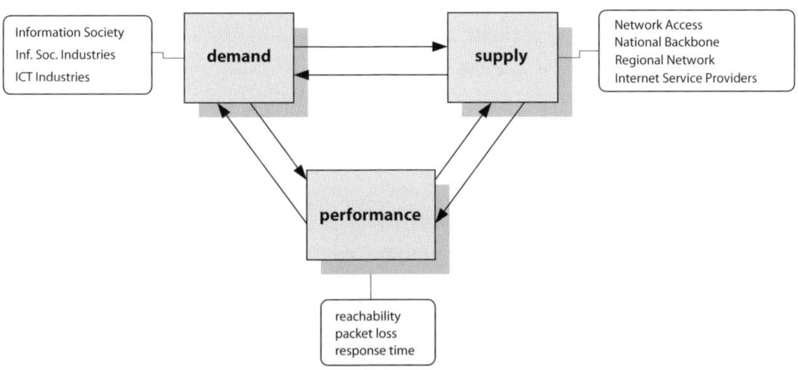


**Illustration 15.2: The Internet – interrelations between supply, demand and performance**

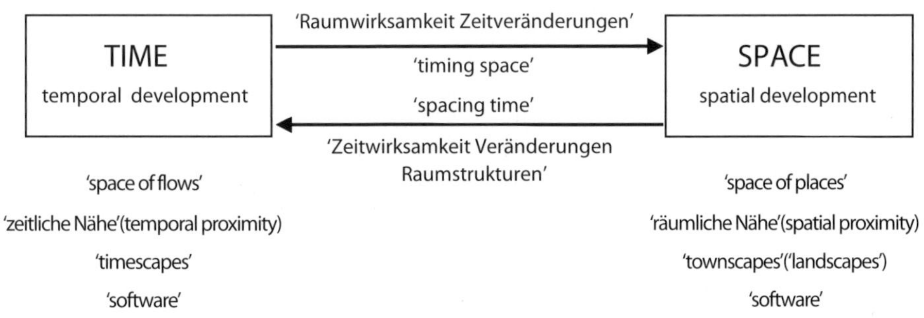


**Illustration 15.3: Time and space, mutual interactions**

## Temporal development – major trends

The root causes of major trends in temporal development are economic, social and technological in nature. There are six major trends according to DATAR (2001):

- A globalising economy puts pressure on companies to be operative around the clock, 'following the sun';
- Life styles are tending to become more individualistic, so daily life is becoming less synchronised;,
- The traditional ways of synchronising life through working hours no longer applies, because the working population is required to be more flexible;
- From the 60s onwards, women's participation in the employment process has increased significantly;
- New types of multiform mobility have emerged daily, weekly and annually;
- ICT has spread massively.

The overall effects on temporal development have been threefold: acceleration, expansion and flexibilisation. Take, for example, shorter product or branch life-cycles; faster transport systems, in particular air traffic; and, similarly, acceleration in ICT. Temporal expansion affects all kinds of supplies and individual options, whether they are related to consumption or to work (e.g. flexible working hours). Acceleration is also pushing expansion and flexibilisation, the latter causing instability in temporal patterns – see Henckel & Eberling (2002) for a wealth of details. To become more specific, the question of how to measure the use of time must be tackled.

## Time use: basic rhythms

The use of time is usually described in terms of basic rhythms: daily, weekly, yearly and life cycles. A time budget describes what is done, when and by whom. One usually distinguishes activities such as work, leisure, education, shopping, public and private services, and travel. The actors can be classified in terms of sex, stage of life, socio-economic position, ethnic origin, and the like. In the case of multiple-person households, different rhythms need to be coupled or synchronised. If one didn't distinguish different kinds of actors, one would gloss over the distributional side of time use, and the issue of distributional justice. With the increase in women's participation in the employment process, the issue is often related to the position of women (versus men). Another trend is the massive spread of ICT which, however, is far from evenly distributed among actors, as can be seen from the so-called digital divide (Drewe, Fernández-Maldonado & Hulsbergen, 2003). Once time budgets have been measured, it also becomes possible to specify the impact of the acceleration, expansion and flexibilisation of temporal development (Henckel, 2002).

So much for the temporal analysis. What about the 'where', the location of activities or activity spaces? Time budgets more often than not reflect spatial effects, in particular spatial constraints. In order to prevent speculations about temporal developments, some empirical illustrations will be presented. The degree to which acceleration, expansion and flexibilisation actually occur is an empirical question. And as to future developments, it is still uncertain. A French example for coping with major uncertainties is to use DATAR (2001) to construct scenarios of time use. Three possible futures can be sketched:

- *La fuite en avant*: time use dictated by market forces, in which continuous time (around the clock) and individual times dominate;

- *La fuite en arrière*: growing resistance to flexibilisation, and the call to return to traditional rhythms and collective times;
- A scenario combining individual and collective rhythms of time use.

Stiens (2002) also constructed three scenarios, but he focused more explicitly on the spatial impact of different 'timescapes'.

## Time use and space-time budgets

Time-use surveys are carried out in most countries. The European Communities' (2003) Eurostat contains results from 13 European countries, focusing on time use at different stages of life. The survey provides a cross-section of time use. Some of the highlights are:
- There (still) is a large difference between the genders in the time spent on gainful work and domestic work,
- Watching TV occupies about 40% of free time both for women and men, in most countries ,
- People who are employed sleep less and have less free time,
- There are greater differences in time use between women and men, in households with children.

A Dutch time-use survey covering the years 1975-2000 (Breedveld & Van den Broek, 2001) can serve as an illustration of changes in 'time use in an increasingly busy society'. The outcomes show that time pressure due to increased participation in labour (especially of women) has increased over the 25-year period. Hence there were more people who needed to combine domestic, care and work tasks during this period (although women still performed two thirds of the domestic and care tasks).

The structuring of time can be measured in terms of collective peak times for activities such as work, or in terms of individual routines. However, for shops, although there have been more opportunities for varying working hours and opening hours, time structures have shown little change. And, finally, the total free time available has declined at an accelerating pace, with one major exception: the time spent on electronic media. While the use of computers has grown, the 'digital divide', measured in terms of PC use, has widened.

These are interesting results; however, one has to take into account a number of contextual variables: October 2000 represented the peak of the economic cycle, which happened to coincide with a period of bad weather. Moreover, two laws have come into force since 1996 that allow shops greater variation in working hours and opening hours, including evening and Sunday openings. Mobility is also a factor in time use. The time spent on mobility stabilised in 2000 at the 1995 level, while, at the same time, car use increased within the mobility budget at the expense of other means of transport.

The above examples point to the importance of time-use surveys and hence of monitoring actual temporal developments. However, they do not show what changes may still lie ahead, nor do they make explicit the spatial dimension. Ever since Hägerstrand (1969), time and space uses have been combined. His research shows the differences in daily cycles for members of a household, and attempts have been made to model the individual action space (Dijst & Vidaković, 1997). This included constructing a typology of actual action spaces, and a simulation model. It is important to separate travel time from time spent on activities at destinations, because a household's time (and money) budget restricts its mobility behaviour.

In this respect, networking and the impact of ICT on space–time budgets is an essential question (http://www.networkcity.bk.tudelft.nl). With ICT, the individual action space may become much less constrained, provided of course one has access to ICT. A shift to e-activities is a shift to a higher speed; and a higher speed means a higher reach. A little time goes a long way in 'travelling along the electronic highway' (a longer way even than in air travel and travelling by high-speed trains). But does the digital reach really have a spatial dimension? Or are the innumerable websites and e-mail addresses located in the 'blue nowhere'?

As far as companies are concerned, it is the business logistics chain that provides the framework for 'inserting' locations into order cycles (24 hours, rush or stock orders). There are places of activity where material management is carried out as well as those centred on physical distribution management. Logistic chains from factories to points-of-sale tend, of necessity, to be tailor-made (Drewe & Janssen, 2002). The resulting space–time budgets can be analysed in terms of 'logistic zones' and (non-spatial) 'logistic regimes' (Hesse, 2002). To handle the information flows, and synchronise the different activities involved, ICT is indispensable. One notes a trend towards *infogistique* as the French call it (Irepp, 2000). E-commerce is generally considered to be one of the new driving forces behind logistic chains, i.e. business-to-business (B2B) e-commerce, and business-to-consumer (B2C) e-commerce. As space or physical distribution still matters, one needs to construct space–time budgets for companies, too.

## In search of new instruments of planning and design

In searching for new instruments of urban planning and design, the focus is on the spatial dimension. But to be able to deal with time, practitioners need to be aware of time use, basic rhythms and major trends, and the mutual interactions between time and space. From which urban concepts can these new instruments be derived?

Concerning time-oriented city concepts, Gwiazdzinski (2002) lists a number of continuous cities that are emerging (cities that function around the clock):

- *The global city* (Sassen, 1991) such as New York, London or Tokyo where the consumer can find (almost) everything every hour of the day, all year round;
- *The linear city* of international transport connections (motorways, railways) and their oases of continuous time (stations, motorway stops, seaports, airports, taxi stops);
- *The archipelago city* with their emerging 'bastions of continuous time', services and production still largely organised in a 24-hour sequence: industrial plants, restaurants, hospitals, hotels, police stations, fire brigade stations;
- *The festive city*, such as Ibiza or Las Vegas: specialised in the utopia of leisure and permanent fun;
- *The virtual city* of networks, the WWW and its electronic appendices (computers, telephones, television, etc.) which have colonised our homes and urban space.

These city concepts may inspire some designers, but to reassert the grounding powers of urbanism they, too, need to be translated into instruments. Moreover, the assumption of an emerging continuous city needs to be checked against known facts of time use and the uncertainties of future developments.

In 2003, the 'New Charter of Athens' was published (European Council of Town Planners, 2003). Though it contains many of the right 'sound bites' – such as sustainability, connectivity and the like – it still needs to be translated into new instruments of planning and design, if it is to become effective in reasserting the grounding powers of urbanism. Time, by the way, is not a topic in the New Charter.

## Emergent network-based concepts

Two examples of network-based concepts from the US which show how to design with ICT are: the 'Millennium City' (Page, Phillips & Siembab, 2003) and the 'Hudson County Cyberdistrict' (WRT, 2003). These are examples from planning and design practice, in which urban technology systems are integrated with land-use design. The winning entry for the Millennium City competition for Orange County tries to achieve smart sprawl in seven steps, by means of a network-oriented development, using a variety of complementary initiatives on various scales and with different actors. Illustration 15.4 shows how the existing urban pattern can be 'recovered' by means of a networked urban pattern.

As a follow-up to the Millennium City, the Hudson County Cyber Strategy requires three initiatives: the Network Neighbourhood Initiative, the Network Enterprise Initiative, and the Cyber Strategy Coordination Initiative. The first and second initiatives contain design elements. According to the Network

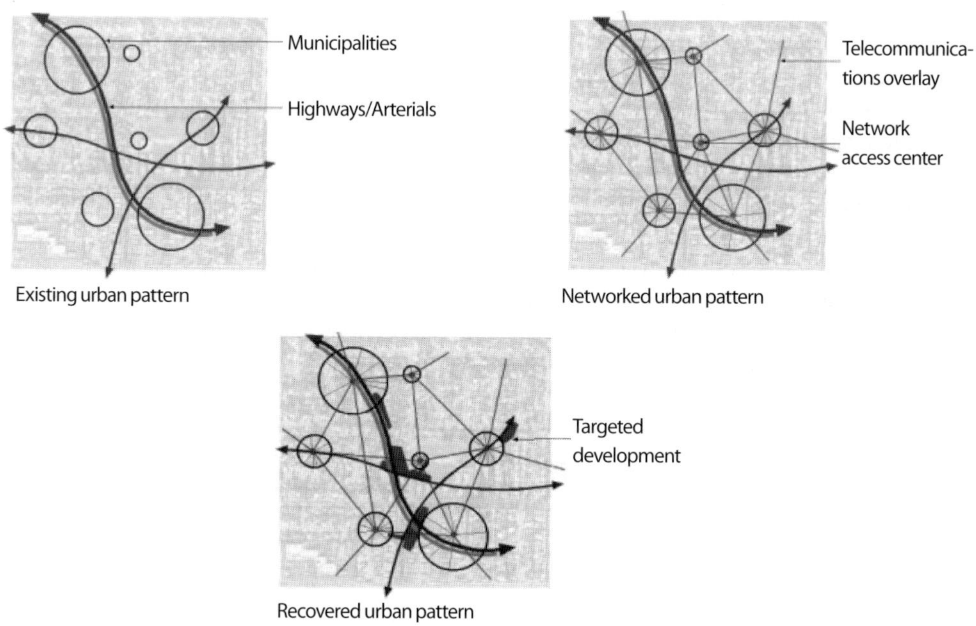


**Illustration 15.4: A network diagram (Page & Phillips, 2004)**

Neighbourhood Initiative, ICT is used to increase the pull of existing activity centres to the immediately adjacent residents and businesses, thus reducing demand for trips of over a mile or two. This initiative is also meant to reinforce a local sense of pride in that neighbourhood. The Network or Electronic Initiative is not an initiative in itself, but rather a complement to the Network Neighbourhood Initiative. The general idea is to facilitate the transition of the delivery activities of each enterprise (production and services) from 'bricks and mortar' to digital networks, and to create a demand for mixed function facilities. The proposed projects not only affect the space–time budgets of households, but also those of companies that play a pivotal ICT role.

## Rereading the classics

Today's investigations of time and space, and especially the space–time budget analysis, own a lot to the space-time model that the geographer Hägerstrand developed in the 1970s. "Over thirty years after it was first introduced, Hägerstrand's space–time model continues to provide new ways of understanding human activity in space, and promises novel solutions for solving difficult issues of transportation and access in modern society" (Corbett, 2003:3).

The 'Network City', in fact, is based on Dupuy's (1991) re-reading of classical network thinkers. The modern meaning of network is characterised by three principal criteria:

- The "*Topological Criterion*. The research into direct relations without intermediaries, and the desire for ubiquity, produces a very specific interest in the topology of a network.
- The *Kinetic Criterion*. Instantaneousness, homogeneity of speeds, the interest for rapid transfers without losses of time or interruptions give the network a tendency to move, thus defining the kinetic criterion.
- The *Adaptive Criterion*. As presently conceived, networks are based on a choice of space and time connections. The connections may necessitate a permanent support, a fixed infrastructure. On the other hand, ideally, the network has to be able to adapt constantly to the need for new connections, when requested and chosen by its users." (Dupuy, 2000:5).

The kinetic and adaptive criterions both relate to the time dimension.

As shown in Illustration 15.5, the classical network thinkers in urbanism have put different emphasis on the three criteria (Dupuy, 1991:105). Wright is one of the few who has dealt fully with all the dimensions. Wright also inspired Fishman (1988), who laid the foundation of the very framework of today's network urbanism. There may still be lessons to be learnt from Wright's 'Broadacre City' (Wright, 1940, 1943; Grabow, 1977). Cerdà's urbanistic propositions for Barcelona may also be a source of inspiration for new instruments of planning and design (Magrinyà, 1996).

Salingaros, with his 'theory of the urban web', in his book *Principles of Urban Structure* (2005), is a modern network thinker, using nodes, connections and hierarchy as principles. He, too, could be ranked among those who have produced emergent network-based concepts, even though his 'connecting the fractal city' – unlike e.g. the Hudson County Cyberdistrict – is primarily a conceptual contribution. Salingaros raises a number of fundamental questions: "(i) what these fractal properties are; (ii) the intricate complexity of the living urban fabric; (iii) methods of repairing urban space; (iv) an effective way to overlay pedestrian, automotive, and public transports; and (v) how to integrate physical connections with electronic connections".

| | TOPOLOGY | KINETICS | ADAPTATION | HISTORIC MARKS |
|---|---|---|---|---|
| Haussmann | O | - | - | |
| Paxton | O | O | - | |
| Cerda | X | X | - | Railroad |
| Henard | O | O | - | |
| Soria y Mata | X | X | X | |
| Wagner | X | X | X | Electricity |
| Insull | X | X | X | |
| Chambless | - | X | X | |
| Hart | O | - | - | |
| Pinchot | X | - | - | |
| Ford | X | O | O | Automobile |
| Wright | X | X | X | |
| Rouge | X | O | O | Telephone (in France) |
| Riboud | X | X | - | |
| De la Rochefoucauld | X | O | O | |
| Virilio | - | X | - | Telematics |

(X) Very large emphasis (O) Large emphasis ( - ) Middle or weak emphasis

**Illustration 15.5: Emphasis on topological, kinetic and adaptive dimensions of networks (adapted from Dupuy, 1991:105)**

The reason for referring to Salingaros, a mathematician and former nuclear physicist, in this section is because of his re-reading of Alexander, and in particular because of what he says about patterns (Alexander, Ishikawa, Silverstein, Jacobson, Fiksdahl-King & Angel, 1977). Alexander's patterns do not have an explicit time dimension, nor do the so-called functional properties that are linked by 'functional notes' to the original patterns (Alexander, 2005). Moreover, neither the patterns nor the properties take ICT into account. Salingaros has posed the question of how to integrate physical connections with electronic ones. Both Alexander and Salingaros are committed to the living city or living urban fabric (Salingaros, 2000), so combining their concepts, with those of urban time policies, would only strengthen their case. Similarly, Barabási (2002), a physicist who conducts research on complex networks, has demonstrated the multidisciplinary character of the 'new science of networks' awaiting to be applied to the complex network of the city. These 'outside' contributions are important stepping-stones to knowledge-based design, that are transforming urban design into a science (Klaasen, 2004).

Another classic worth re-reading is Jacobs. Apart from her plea, more than forty years ago, for organised complexity, her concept of city diversity is still valuable today. ICT only strengthens the case for organised complexity. The 'need for mixed primary uses', one of the conditions for city diversity, fits in with the objective of urban time policies: "The district, and indeed as many of its internal parts as possible, must serve more than one primary function; preferably more than two. These must ensure the presence of people who go outdoors on different schedules and are in the place for different purposes, but who are able to use many facilities in common" (Jacobs, 1992:152). Note also the emphasis on mixed-function districts, buildings and public facilities in the Millennium City as well as in the Hudson County Cyberdistrict.

Our last example is Lynch, perhaps best known for *The Image of the City* (Lynch, 1960). This work has inspired Roberts *et al.* (1999) to develop the network-based concept 'integrated metropolis', and also Page & Phillips (2003) in their urban design for Jersey City in relation to electronic space or electronic networks.

Practitioners do not usually work with spatio-temporal concepts, so they find it difficult to deal with time. Lynch – a true practitioner – was already an exception to this, in 1972, when he published *What Time is this Place?* He may even be considered as a forerunner to urban time policies (Lynch, 1972:76-77). According to Lynch, the following instruments or design methods are available to us: "One is the visible accretion of the signs of past events which makes apparent the depth of historical time [– the temporal collage –]. Another is the display of recurrent, opposed states which makes us aware of rhythmic time by contrasting the present state, with remembered and expected states [– episodic contrast –]. Still another is the direct display of environmental change, when, by transforming the scene, or shifting the viewpoint of the observer, the change can be made sufficiently palpable to be perceived in the experiential present [– the direct display of change and design for motion –]. Finally, there may be ways of symbolically speeding or slowing down otherwise imperceptible changes – changes too glacial or too feverish [too slow or too fast] to be seen [visible to the naked eye] – so as to bring them within our perceptual grasp [– patterning long-range change –]. All these modes have their own characters. None of them is practiced today [– in 1972 –] in any systematic way" (Lynch, 1972:168).

In addition to these four methods, Lynch speculates about extending "our perceptual reach by artificial means, in order to sense environmental changes" ('mutoscopes'). Using ICT as a visualisation tool, this is within reach today, and the question 'What time is this place?' is indeed an intriguing one.

## A design studio

The Network City is a specific approach to the search for new spatial concepts. It started as a co-operation between the Dutch Ministry of Housing, Spatial Planning and Environment, and the Faculty of Architecture at Delft University of Technology back in 1997. The studio can be visited at http://www.networkcity.nl. It involves staff, PhD researchers, contract researchers and undergraduate students in their final year. There are also (international) links to experts working in the same field. A new spatial concept similar to a new product of service, results from a complex interaction between three pillars: a societal or social, an economic-legal and a technological one (see Ill. 15.6).

Because design can demonstrate and visualise what could be, the studio approach is one of design-oriented research aiming at possible futures. Possible futures tend to be oriented towards the long term. In order to bridge the gap between a distant future and today's practice a number of location-specific test-beds have been chosen:

- The future urban agglomeration;

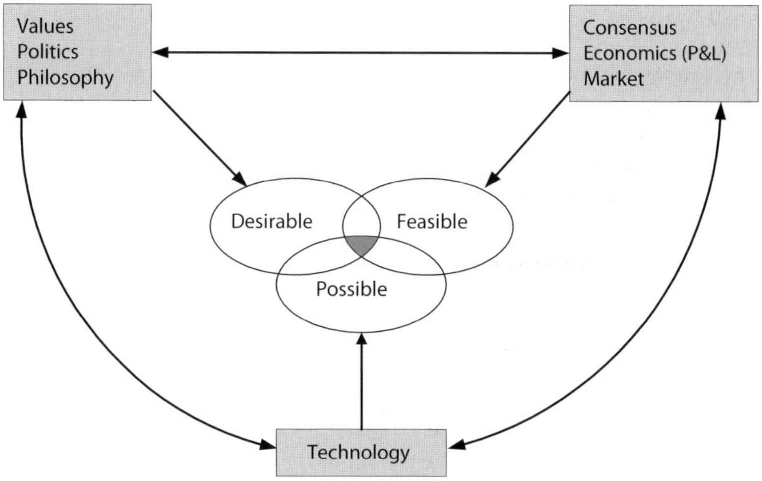


**Illustration 15.6: The three pillars of innovation (Van Mieghem, 1999:5)**

- The 'rest' of the Netherlands, beyond the 'periphery';
- The 'mainport' as a node in a logistic network;
- The 'euroregion plus', beyond the Dutch border.

The studio approach has focused on the opportunities offered by ICT, right from the beginning (Drewe, 1996). It was the rise of ICT in the 1990s that triggered network thinking in urbanism. In order to understand the new technology and its implications for urban form, there are lessons to be learnt from other urban technology systems such as water, energy and transport (their infrastructures, services and uses).
One of the projects was about mobility and the space–time behaviour of the elderly (the outdoor mobility of older people) (see chapter 1 by Mart Tacken). This was a European project, with partners in Germany, Finland, Italy and Hungary (Mollenkopf, Marcellini, Ruoppila & Tacken, 2003 & 2004). Another project dealt with so-called seamless multimodal passenger mobility, i.e. chain mobility and activity-travel behaviour. Logistics were studied in relation to the positioning of seaports and airports (Drewe & Janssen, 2002). The work of LABSAT at the Politecnico di Milano was an important source of inspiration. (LABSAT stands for the Laboratory for the Research and Design of Temporal Architectural Systems.) This brings us to urban time policies.

## Urban time policies

'Tempi della Città' or times of the city is the expression coined for urban time policies in Italy in the 1980s. This was imitated by other European countries later on. The foremost aims of these policies were to improve citizens' quality of life by organising and allocating living and working hours better, and by up-grading public spaces to encourage socialisation networks (Mareggi, 2001:4-5). To achieve these aims, interventions were implemented, such as services to improve public accessibility for those with less mobility than the norm (focusing on different age groups such as children and the elderly), integratedly designed public spaces, mobility agreements to desynchronise the start or end of activities in an attempt to improve traffic conditions and promote the use of public transport and sustainable mobility, making the opening hours of shops and school hours more flexible and changing the opening hours of museums. Similar examples can be found in Germany, Finland, the Netherlands and France (Boulin & Mückenberger, 2002). A set of techniques has been developed at the Politecnico di Milano to help identify, simulate and monitor the transformation processes of an area's spatio-temporal architecture, the key concept being the 'chronotope'. A chronotope shows how people use space in time. Searching for new instruments of urban planning and design, the so-called chronotopic maps are of primary importance; chronotopic maps may be synthetic, cinematographic or chronomatic systems for local mobility (Guez, 2002).

Urban time policies started before the rise of ICT. However, chronotopic analysis does not seem to incorporate ICT properly yet. To do so, should not be too complicated if one takes, for example, the design proposals for the Hudson County Cyberdistrict as a starting point. It is important to integrate ICT, because it supports the development of new temporal regimes.

Urban time policies, if they have a spatial dimension, are essentially geared to site planning. But urban planning and design also deal with the urban whole, or claim to do so. This requires links with urban technological networks (see Ill. 15.1) and mobility, which brings us to our final topic.

## The prospects for mobility

In the future, mobility will have to cope with new urban rhythms. Bailly and Heurgon (2001) from RATP provide directions for this future mobility. This includes ICT, not as a substitute for transport, but as an element of choice and convergence. European, and in particular Italian experiences, give a clear indication of the implications of mobility for organisations: peak hours will stay, but quiet hours will decrease. A continuous transport function is called for, with more tailor-made services. It will not be easy for existing public transport to adapt to these requirements. Nevertheless, the report gives information about various initiatives to adapt services to evolving urban rhythms. The use of ICT is part of these initiatives.

Future mobility requires new approaches to mobility research (models, mapping such as chronotopes and monitoring), innovations in mobility services that require the concerted actions of different players, and new regulations.

Density can no longer be used as an indicator of potential exchange, unless it is combined with accessibility (taking into account different speeds, from pedestrians to electrons, and different kinds of transport, i.e. passengers, goods and information). The continuity of the built environment that used to delimit political-administrative boundaries needs to be revisited, because passengers travel further than 30 km/h. Physical proximity or distance become less important. As the increase in the distance covered by trips correlates with increases in speed, access time becomes a better indicator than distance. A nearby place can be less accessible than a place far away, so geographical centrality within an agglomeration no longer guarantees accessibility. It can even be a handicap for private-car users. New centralities emerge at interconnection nodes, and at multimodal platforms that polarise various urban functions. Consequently, the notion of the inner city, or city centre is changing. Polarisation, functional specialisation and thematic concentration will be the more operational notions of the future. Hence, as soon as time is taken seriously in the ICT age, there will be a clear tendency for (future) mobility to clash with conventional urban planning and design. This will make the paradigm challenge even more urgent.

# Epilogue

In 1969, Hägerstrand, the founder of the space–time model, presented a paper to the European Congress of the Regional Science Association in Copenhagen entitled *What about people in regional science?*

"Historically, social scientists studying the effects of space on human behaviour have tended to treat time as an external factor, something that is relevant to understanding a given phenomenon, but not essential. Activity choices were viewed the context of distance alone, such as with the gravity model, and often these decisions were taken in an aggregate sense; individual decisions were viewed as minor variations of the decisions made by larger zonal-based groups" (Corbett, 2003:1).

Does not the question 'What about people?' also hold for conventional urban planning and design? Asking 'What about time?' brings people back into urban planning and design. This is most clearly proven by urban time policies. Chronotopic analysis reveals echoes of the past, present rhythms and the simulated presence of the future (Laousse, 2003). What better way of answering Lynch's time-honoured question 'What time is this place?'

## References


Alexander, C., 2002, *The Nature of Order, Book one, the Phenomenon of Life*, The Center for Environmental Structure, Berkeley, California

Alexander, C., S. Ishikawa, M. Silverstein, M. Jacobson, I. Fiksdahl-King & S. Angel, 1977, *A Pattern Language*, Oxford University Press, New York

Bailly, J.-P. & E. Heurgon, 2001, *Nouveaux rythmes urbains et organisation des transports*, Conseil National des Transports, Paris

Barabási, A.-L., 2002, *Linked, the New Science of Networks*, Perseus Publishing, Cambridge Mass.

Boulin, J.-Y. & U. Mückenberger, 2002, *La ville à mille temps*, éditions de l'aube datar, La Tour d'Aigues

Breedveld, K. & A. van den Broek (eds), 2001, *Trends in de tijd, een schets van recente ontwikkelingen in tijdsbesteding en tijdsordening*, Sociaal en Cultureel Planbureau, Den Haag

Corbett, J., 2003, *Torster Hägerstrand: Time Geography*, http://www.csiss.org/classics/content/29, 3 pages

DATAR, 2001, *Temps & territoire, prospective et experiméntations*, Groupe de prospective no. 6, Paris

Dijst, M. & V. Vidaković, 1997, Individual action space in the city; In: Ettema, D.F. & H.J.P. Timmermans (eds), *Activity-based approaches to activity analysis*, Pergamon Press, Oxford, pp. 73-88

Drewe, P., 1996, *De Netwerkstad VROM, bijdrage van informatietechnologieën aan nieuwe concepten van ruimtelijke planning*, Faculty of Architecture, Delft University of Technology

Drewe, P., 2002, Understanding the virtual space of the Internet, a network approach; In: Carmona, M. & M. Schoonraad (eds), *Globalization, Urban Form & Governance - globalization & urban transformation*, 6, Delft University Press Science, Delft, pp. 135-142

Drewe, P., 2003, 'The Network City', a paradigm challenge for urban design and planning; In: Carmona, M. et al (eds), *Globalization, Urban Form & Governance - Globalization and the Return of Big Plans*, 7, Delft University Press Science, Delft, pp. 27-40

Drewe, P., 2004, *What about time in urban planning and design in the ICT age?*, Design Studio, The Network City, Faculty of Architecture, Delft University of Technology

Drewe, P. & B. Janssen, 2002, Seaports and airports in Europe – what port for the future? In: *Critical Infrastructures, Delft 2001*, 5th International conference on technology, policy and innovation, Lemma, Utrecht (CD Rom)

Drewe, P., A.M. Fernández-Maldonado & E.D. Hulsbergen, 2003, Battling urban deprivation: ICT strategies in the Netherlands and Europe, *Journal of Urban Technology*, Vol. 10, nr. 1, pp. 23-37

Dupuy, G., 1991, *L'urbanisme des réseaux, théories et méthodes*, Armand Colin, Paris

Dupuy, G., 2000, A revised history of network urbanism, *Oase*, 53, pp. 3-29

European Communities, 2003, *Time use at different stages of life, results from 13 European countries*, Eurostat, Luxembourg

European Council of Town Planners, 2003, *The New Charter of Athens 2003, The European Council of Town Planners' Vision for Cities in the 21st Century*, Lisbon, http://www.ceu-ectp.org/e/athens

Fishman, R., 1988, The post-war American suburb: a new form, a new city; In: Schaffer, D. (ed.), *Two centuries of American planning*, Mansell, London

Grabow, S., 1977, Frank Lloyd Wright and the American City: the Broadacre Debate, *AIP Journal*, pp. 115-124

Graham, S. & Marvin, S., 2001, *Splintering Urbanism, Networked Infrastructures, Technological Mobilities and the Urban Condition*, Routledge, London and New York

Guez, A., 2002, La chronocartographie dans le développement d'un urbanisme des temps et de la mobilité: In Boulin, J.-Y. *et al.*, *La nouvelle aire du temps*, éditions de l'aube datar, La Tour d'Aigues, pp. 101-106

Gwiazdzinski, L., 2002, *La ville 24 heures sur 24*, éditions de l'aube datar, La Tour d'Aigues

Hägerstrand, T., 1969, What about people in regional science? *Papers of the Regional Science Association*, Vol. 24, pp. 7-21.

Henckel, D., 2002, Wer verteilt die Zeit? Oder: Zwei Hände voll Wind; In: Henckel, D. & M. Eberling (eds), *Raumzeitpolitik*, Lesket Budrich, Opladen, pp. 209-230

Hesse, M., 2002, Zeitkoordination im Rahmen der modernen Logistik – mehr als nur ein Impulsgeber für die räumliche Entwicklung; In Henckel, D. & M. Eberling (eds), *Raumzeitpolitik*, Lesket Budrich, Opladen, pp. 107-126

Irepp, 2000, Infogistique, nouvelle frontière du commerce sur Internet, *Médiation*, no. 23

Jacobs, J., 1992, *The Death and Life of Great American Cities*, Vintage Books, New York (first published in 1961)

Klaasen, I.T., 2004, *Knowledge-based Design: Developing Urban & Regional Design into a Science*, Delft University Press Science, Delft

Laousse, D., 2003, La ville en représentations: l'expérience de la RATP; In: Debarbieu, B. & S. Lardon, *Les figures du project territorial*, éditions de l'aube datar, La Tour d'Aigues, pp. 143-156

Lynch, K., 1960, *The Image of the City*, The MIT Press, Cambridge Mass.

Lynch, K., 1972, *What Time is this Place?*, The MIT Press, Cambridge Mass.

Magrinyà, F., 1996, Les propositions urbanistiques de Cerdà pour Barcelone: une pensée de l'urbanisme des réseaux, *Flux*, no. 23, pp. 5-20

Mareggi, M., 2001, *Multipartner table of co-design to manage city actions: Italian experiences in urban time policies*, Paper presented at the Danish Building and Urban Research / EURA 2001 Conference, May 17–19, Copenhagen

Mollenkopf, H., F. Marcellini, I. Ruoppila & M. Tacken, 2003, *The MOBILATE project, enhancing outdoor mobility in later life: personal coping, environmental resources, and technical support, final integrative report*, Heidelberg

Mollenkopf, H., F. Marcellini, I. Ruoppila & M. Tacken, 2004, *Ageing and outdoor mobility, a European study*, IOS Press, Amsterdam

Page, S., B. Phillips & W. Siembab (2003), The Millennium City: Making Sprawl Smart through Network-Oriented Development, *Journal of Urban Technology*, Vol. 10, nr. 3, pp. 63-84

Page, S. & B. Phillips, 2003, Telecommunications and Urban Design, representing Jersey City, *City*, Vol. 7, No. 1, pp. 73-94

Page, S. & B. Phillips, 2004, *Urban Interfaces: Designing In-Between*, Presentation at the Inaugural Conference Delft School of Design, June 2004.

Roberts, M. *et al.*, 1999, Place and Space in the Networked City: conceptualizing the integrated metropolis, *Journal of Urban Design*, 4, no. 1, pp. 51-66

Salingaros, N., 2005, *Principles of Urban Structure*, Design/Science/Planning Series, Techne Press, Amsterdam

Sassen, S., 1991, *The Global City*, Princeton University Press

Stiens, G., 2002, Wie unterschiedliche 'Timescapes' Staatsraum und Landschaften verändern können; In: Henckel, D. & M. Eberling (eds), *Raumzeitpolitik*, Lesket Budrich, Opladen, pp. 163-186

Van Mieghem, P., 1999, *Het draagvlak van de 21ste eeuw*, Inaugural Lecture, Faculty of Information Technology, Delft University of Technology

Wright, F.L., 1940, The New Frontier: Broadacre City, *Taliesin*, 1, October, entire issue

Wright, F.L., 1943, *Broadacre City*, Box six of an autobiography, Spring Green, Wisc.

WRT (Wallace Roberts & Todd, LLC), 2003, *Hudson County Cyberdistrict, feasibility study*, Philadelphia

# 16 ICT Infrastructure Networks as Supports for New Urban Processes

Ana María Fernández-Maldonado

## Introduction

In just one decade, the use of Internet and related digital technologies has affected almost every important sphere of urban life in modern societies. Internet and mobile telephony have become pervasive in developed countries and increasingly present in cities of the developing world, while each year, thousands of new digital devices and applications are brought onto the market and into the hands of millions of people. Digital technologies have changed the way we inform, entertain and educate ourselves, as well as our ways of working, carrying out research, doing business, and keeping in contact with each other. Not less important, the spatial reach of our daily activities and contacts has expanded from the local to the global scale. For people living in cities of the Western world, everyday life is no longer what it was twelve years ago, before the commercial introduction of the Internet to the public.

Digital technologies are, however, dependent upon conventional telecommunication infrastructures, even though the wire-less character of many of them makes people unaware of the territoriality of the supporting networks. These information and communication technology (ICT) networks undoubtedly have an urban character, since they are mainly deployed within and between cities. Surprisingly, the features and scope of these relatively new urban infrastructures are still an understudied field.

Graham & Marvin (1996) explain this neglect as being due to the invisibility and intangibility of ICT infrastructures, the conceptual challenge posed by the increasing space-time complexity inherent in digital technologies, and the conservatism of the urban and regional planning disciplines, which are still focused on the concepts and models of the industrial city. For Drewe (2000b), the failure to approach this topic is related to the high level of uncertainty regarding which direction telecommunications-related changes in cities will take, the indirect character of these changes, and the resistance of practitioners. Priemus (2004) claims that the lack of government involvement in the deployment of ICT networks, and the minimal awareness of the spatial implications of these networks, help to explain why government authorities and planners show so little interest in the topic.

There are certainly many reasons why ICT-related urban issues are difficult subjects to study, but because of their significance to cities, attention should be given to them, beginning by the material geography of ICT networks. The connections that are made through electronic networks are today as important as those made by means of transportation networks. Traffic networks have attracted academic attention, thus ICT networks should be brought into the picture of urban and spatial planning to a comparable degree (Priemus, 2004). The purpose of this chapter is to address three issues related to ICT networks in urban settings:

- The context of the telecommunications sector, which is currently in charge of deploying and operating ICT networks;
- The main features and global geography of the ICT networks; and
- Some aspects of the new urban trends and processes arising from ICT networks.

## Major transformations in the telecommunications sector

During an amazingly short period, the telecommunications sector has undergone huge transformations: it has developed from being an industry in itself, to its present position as a vital enabler of all other industrial sectors. Considering just the economy, the telecommunications sector is now both the core (the major economic activities are mostly information processing and transmitting, or activities critically dependent on them) and the infrastructure of the new information economy (World Bank, 2000). The main changes that produced this amazing transformation of the sector took place in its technical and political aspects.

### Technical aspects: From local and analogue to global and digital networks

Enormous technical advances have led to the convergence of informatics and telecommunications in the early 1990s. These concern four main technologies: data digitisation, data transmission, data communication and network technologies. These technologies have made it possible to convert text, images and sound into digital information to be transmitted through the networks.

In the case of text, each character is converted into an ASCI code number; in the case of images the picture is divided into a grid, in which the colour and intensity of each box has its own code number; for sound, the intensity of the sound wave is measured so that it can also be represented by a code number. These code numbers are then converted into binary digits, the so-called bits. Bits are able to travel between computers in flows of 1000-byte packets (one byte contains 8 bits). Powerful switches and routers located in hubs and ports then send and redirect the packet-flows through the networks to their final destination.

Packet flows can travel through four different types of lines: the basic copper pair of the traditional telephone networks; coaxial cables lines, such as those from cable TV networks; fibre-optic cables; and wire-less technologies. Because of its unmatched transmission capacity[4], fibre-optic is the material used for the 'backbones', the main networks, for transmitting large amounts of data over long distances. To give an idea of their amazing transmission capacity: a hair-thin optic-fiber cable allows the transmission of 60 thousand telephone calls at the same time (Graham & Marvin, 1996).

The rapid spread of digital technologies to the public has demanded the expansion and adaptation of the existing telecommunication infrastructures to the technical requirements of digitisation in a very short period. Transcontinental fibre-optic backbones have been deployed, while the traditional telephone networks have been transformed from analogue into digital, and from territorially-based local networks into networks with a global reach. The present Internet networks are still largely traditional telecommunication networks that were used for voice transmission. What has completely changed are the backbone circuits, switches and routers.

This rapid process of expansion, computerisation and modernisation of the networks has demanded high investments, and at the same time, it has made the business much more complex and diversified than before, greatly multiplying the number of services provided to end-customers.

Telecommunications services are traditionally delivered by networks of different types, generally a combination of (wire-less) waves and fixed-line networks. Before digitisation, most long-distance calls were delivered over wire-less lines (by satellite), but they started and finished in land lines. With the completion of powerful intercontinental fibre-optic submarine cables laid around the globe, the very logic of the infrastructure system has radically changed: "Radio is now being increasingly used to provide access networks, while wired networks provide the long-distance component" (International Telecommunications Union, 2002:4).

New technical advances may be bringing drastic new changes to the sector. There is still no clarity or consensus among experts about the advantages of completely converging voice and data transmission with digital TV, or about voice and data network specialisation, or about the consequences of introducing WiFi, WiMax or other new technologies. Investments have to be made in advance in this sector, and there are many uncertainties that have to be faced, because it is very difficult to predict consumer behaviour. For example, despite the great enthusiasm of large firms for full convergence between Internet and mobile telephony, users have been very cautious and have not fulfilled the investors' expectations. On the other hand, users have embraced other technologies such as text messaging, which experts originally did not consider interesting from an investment point of view.

## Political aspects: From an unimportant state-owned public utility to a highly profitable and highly competitive private business of strategic importance

The telecommunications sector has rapidly changed from a regulated public utility into a highly competitive business, privately run by large, mostly global, corporations. This means a great change of legal status and entrepreneurial logic. Before the privatisation reforms, the local telephony segment was providing a service to 60-75 % of the population in rich countries and to the elites of the developing countries. Its emphasis was on standardisation and social and geographical equalisation, that tended towards tariff equalisation, despite cost disparities (Graham & Marvin, 1996). However, the networks needed very high investments to expand and modernise themselves. This was difficult to acquire from State finances, especially in developing countries.

The new model for telecommunications, implemented by most countries of the world, has brought about a shift from a nation-wide telecommunications strategy with local networks to a model of large private and highly competitive telecommunications carriers. A few large US and European firms now dominate the telecommunication scene of entire regions of the world, expanding their worldwide networks according to their corporate strategies. In these private strategies, cities and locations are no longer the central object, but merely a point on the map of the world. As such, cities can become nodes of the network or they may be by-passed depending on the firms' objectives (which are linked to company profitability). Graham & Marvin have illustrated the consequences of introducing this model in cities and regions (see Ill. 16.1). Large firms with global reach now compete among themselves to provide telecommunication services and infrastructures to the most profitable financial, commercial and residential markets on local and global levels.

The introduction of the commercial Internet triggered massive investments and huge expansions among telecommunication firms, whose tactic was to expand by acquiring former monopolies and other local firms, the reasoning being that only large companies would survive in the new global context. This led to mergers and strategic alliances, with the aim of gaining a global foothold in all markets. The expansion of these firms was not only horizontal (in as many countries as possible), but also vertical (in as many different segments of the market and strategic sectors as possible).

As these telecommunication companies were the favourite assets during the 1996-2001 period, the money to finance this explosive expansion came from the world stock markets. An additional source for financing these expensive acquisitions and expansions were bank loans. These were easily granted, since, with estimated gross operation margins in the order of 40%, telecommunications was – and is – a very profitable business (Nellist & Gilbert, 1999). In 1996, the top 10 telecommunications firms earned more than the 25 largest banks in the world (Alaedini & Marcotulio, 2002). According to the International Telecommunications Union (ITU), the revenues of the total industry had reached $1.37 trillion by 2003, while over the same period, consumer spending on communications grew faster than any other household spending (Standage, 2003).

However, investors' confidence in the telecommunications business decreased greatly with the collapse of the stock market in the early 2001, and the resulting enormous job losses in the sector (International Telecommunications Union, 2002). At present, there is still a huge over-capacity in telecommunications networks, which has caused a significant drop in long-distance traffic prices. Only 14% of the trans-oceanic submarine cables was in use in 2003 (TeleGeography, 2003), the rest is still dark fibre, which is still not lit up to allow the digital packets to travel. Despite the massive revenues, the telecommunications sector has suffered great financial problems that have gradually increased the instability of the sector and finally produced the crisis in which it is currently immersed. The huge miscalculations, moves and risks taken by the telecommunication companies in their expansion fever to become the largest, led some large firms into bankruptcy. The main flops have been caused by massive network overbuilding, the European 'third generation' (3G) spectrum license auctions fiasco, and the US management and bankruptcies-related scandals (Shaw, 2002). One of these examples is WorldCom – the biggest ever bankruptcy in the history – the assets of which were 103.9 billion before liquidation (Standage, 2003).

Since the dotcom crisis, the telecommunications sector has been heavily affected. For example, the international bandwidth market has been caught in a deflationary spiral that has caused huge price reductions of international bandwidth. However, there are recent signs of recovery in the sector.

## The main features of the Internet as a network

The Internet is generally experienced as a seamless network serving and delivering digital flows from terminal to terminal. But when it comes down to its physicality, the Internet is composed of an enormous number of private networks of different sizes that interconnect at certain points, and which depend on a structured hierarchy and a different protocol to operate them (Gorman, 1998). The origin and circumstances of the creation of the Internet help to explain its current features.

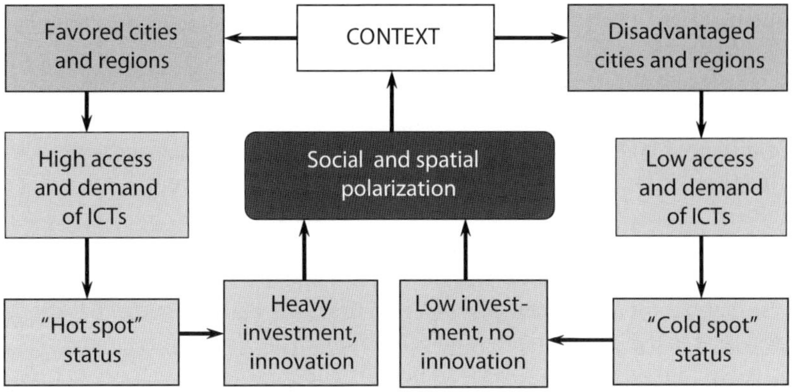


**Illustration 16.1: The trends in the telecommunication sector are contributing to an unequal development of cities and regions (adapted from: Graham & Marvin, 1996)**

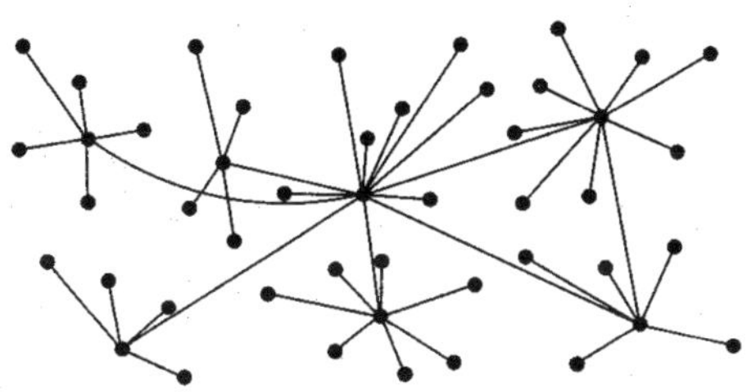

**Illustration 16.2: The configuration of a decentralised hub-and-spoke network structure**

ARPANET, the experimental network that preceded the Internet, became operational in 1975. ARPANET was established by the US Defense Department, although its goals were not military but research applications. The initial configuration of the ARPANET, however, responded to design principles of military origin: to minimise vulnerability to foreign attacks. These principles are still present in the Internet infrastructure in the form of a decentralised network structure, computing power distributed throughout the nodes, and links that can be made redundant (Castells, 2001:17). Illustration 16.2 shows the basic configuration of the backbone network of the Internet, a decentralised network that works rather like the routing system in the airline industry in which the most direct and fastest connections are given to a few hub cities.

When the NSF (National Science Foundation) privatised the first backbone network of the Internet, between 1993 and 1995, its original configuration changed. The new situation required an open network that would allow computer networks anywhere in the world to connect with it. The architecture then changed towards a configuration based on access nodes: NAPs (Network Access Points). These are the physical points where computer networks converge, and where the packet-switched flows are transferred between networks. The openness required for commercialising the Internet has transformed it into a self-organised network.

Because the Internet consists of thousands of different networks, direct interconnections among them would be unworkable. The thousands of networks are structured in a hierarchical way, which can be disaggregated into five different functional levels. NAPs constitute, then, the first level of the Internet infrastructure. The second level consists of the largest global backbone providers. Operating high-capacity, long-distance transmission facilities, these providers are interconnected with each other. The third level is made up of regional networks, the 'local backbones'. The fourth level is that of the ISPs (Internet Service Providers), which provide access to end-users, the lower level of which consists of the internal networks of institutional and residential users (Gorman, 1998). The system works by, for instance, a small network (ISP) sending data packets from one of its customers to the large network that the ISP uses for backbone services, which in turn sends the data to another backbone network, which then delivers it to the ISP serving the end-user to whom the data is addressed.

To understand how a network grows, it is necessary to be aware of the logic that guides the behaviour of its operators. The operators of the backbone networks function according to four main principles. Dupuy (2004) explains these under the headings: (a) market principles, (b) proximity, (c) connectivity and (d) reliability.

a. *Market principles*: The development of infrastructure is primarily driven by the demands of the market. New networks are deployed according to future market expectations. Population concentrations and the economic activities of certain cities or regions are considered to be good predictors of a future market.
b. *Proximity*: Although the Internet is generally equated with 'the death of distance', the deployment of fibre-optic networks is very expensive: between $ 100 thousand and $ 500 thousand per mile. Because of the extensive resources and time investment needed to connect two nodes on the network, designers prefer to connect to the closest node that has sufficient bandwidth. This clearly favours the shorter links (Barabási, 2002).
c. *Connectivity*: Good connectivity requires adequate bandwidth and a multiplicity of links, so operators need to connect to nodes with the required bandwidth and to deploy their network in such a way that they maximise the number of links between the nodes of their network and the nodes of external networks.
d. *Reliability*: To avoid congestion and ensure reliability, global operators tend to diversify and distribute the links and nodes of their network. That links can be made redundant is therefore good for both reliability and connectivity.

Barabási (2002) has shown that the Internet is not a random network but a 'scale-free' type of network, characterised by an uneven distribution of connectedness. Links on web pages follow a 'power law' degree distribution, which are "the patent signature of self-organisation of complex systems" (Barabási, 2002:77). This means that there are many nodes with few links, and a few nodes with a huge number of links. This is because, in a network that expands by adding new nodes to the existing network, as in the Internet, new nodes prefer to attach themselves to nodes that are already well-connected – a network behaviour that is called 'preferential attachment'. Tracings of data routes along the Internet networks made by Lucent Technologies and TeleGeography in 1999, as shown in Illustration 16.3, show how this hub-and-spoke configuration with short links operates.

The Internet backbone network, then, is a large, complex, global, self-organised, hierarchical and unevenly distributed system. Its complexity makes it possible for data packets to travel from one node to the other through countless possible routes. The unevenness of its distribution and capacity comes

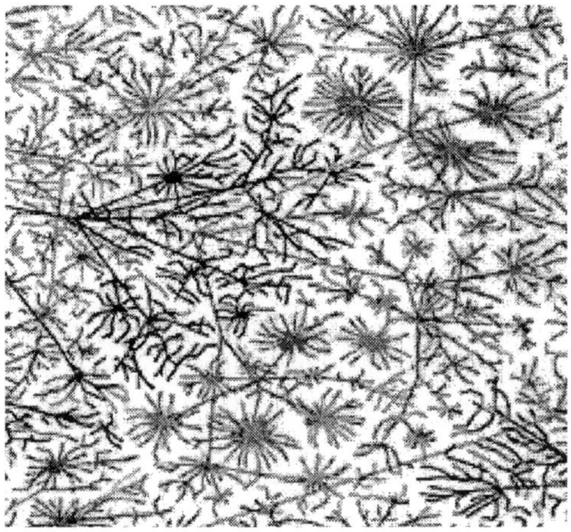

**Illustration 16.3: The hub-and-spoke configuration of the ICT network down to earth (source: TeleGeography and Lucent Technologies)**

from two principles that guide operators' logic: to apply market principles and the principle of connectivity, translated as preferential attachment behaviour.

Regarding the main components of the Internet infrastructure, Townsend (2003) made a useful analogy for urban planners, who are generally used to analyse city process under the paradigm of the industrial city. He compares the role of the components of the new system that supports the new informational economy with the important elements (such as ports, highways, warehouses and factories) of the system that supports the industrial economy (see Ill. 16.4).

## The global geography of the Internet networks

Tracing the infrastructure of the backbone of the Internet, and measuring its transmission capacity, is a very difficult endeavour. Since the NSF gave up the Internet in 1995, there is no central source of information about the Internet backbone and its capacity. The private character of the networks, the existence of thousands of networks and the continuous evolution of the Internet makes it difficult to gain an accurate image of it. In this highly competitive sector, telecommunication firms and ISPs do not easily share information about their networks. A US American consulting firm – TeleGeography – is one of the best sources of consolidated data on Internet networks. Its annual report on *Global Internet Geography* describes: "the number, capacity and complexity of these (backbone) networks… cataloguing where the international links can be found and which companies control them. But it also traces the market forces which effect the evolution of the Internet's architecture." (TeleGeography, 2005).

To analyse the Internet's network geography, the second level is the best level – the level of regional backbone providers – as Gorman (1998) claims. Backbone providers are the best indicators of the geographical location and agglomeration of the Internet. The main operators are large telecommunication firms of European or US origin who work on a global scale. The top 10 international carriers are: AT&T,

MCI, Deutsche Telecom, Sprint, France Telecom, BT, IDT Corporation, Cable and Wireless, Telecom Italia and Telefónica, in that order (TeleGeography, 2005).

Internet backbones began to expand from US territory outwards, and for that reason they are still dominated by US centrality. Internet's main nodes are located in the US and a large proportion of international Internet traffic still flows along the backbones that traverse US territory. As a result, every region and nearly every country has a direct Internet connection with the US, while direct connections between cities of different regions are more infrequent. The US functions as a central switching facility for inter-regional data traffic, and is used as a transit point for data packets travelling from one major region to another Although US-centeredness is diminishing, it is unlikely to disappear, since the US is well located, both geographically and culturally, between Europe, Asia, Australia and Latin America (Townsend, 2001).

Initially, Internet development was still mainly concentrated within and between developed countries and major urban agglomerations. However, following the huge deployment of trans-oceanic, satellite and terrestrial fibre-optic networks in the late 1990s, the expansion of the global Internet backbone has been remarkable, and with the emergence of new Internet exchange points (NAPs), this has led to a more diffused distribution of the Internet, one that is less centred on the US. Gradually, Western Europe has emerged as another important Internet hub: there is currently more bandwidth linking European cities with each other, than with the US. Intra-Asian network links are also growing at a very fast rate (TeleGeography, 2005). However, it is obvious from a simple look at the distribution of Internet backbones in 2004 that the ICT backbone network still passes over most parts of the developing world. Illustration 16.5 shows the position of the different world regions and cities.

Looking at the geography of the Internet, we observe that it mainly connects the largest agglomerations, which then become the hubs of this decentralised network structure. Anthony Townsend, whose dissertation has chronicled the development of the ICT networks during the 1990s and early 2000s, explains this: "Instead of trailblazing into the wilderness, opening a path to new settlements, digital networks have been built to reinforce existing connections between centres of power and influence in the world's greatest cities and metropolitan areas." (Townsend, 2003:16). Castells states shortly that the Internet is basically a "network of metropolitan nodes" (Castells, 2001:228).

Metropolises are at a clear advantage, since, due to preferential attachment law, they have high capacity and many more redundant links, whereas peripheral cities have lower capacity and therefore fewer redundant links. However, each region has different characteristics. The historic origins of the Internet have determined, once again, the geography of its core backbones, causing the US network topology to be different from the topologies of other world regions. When the NSF privatised the Internet networks in 1995, it was decided to place network access centres (NAPs) at the four nodes of the existing network that connected university computing centres: San Francisco, Washington, D.C., Chicago and New York. The Internet then developed outwards from these nodes. Three years later, there were already 11 major peer interconnecting points linking the most important cities in the US (Gorman, 1998).

| Industrial economy | Informational economy | Referring to: |
|---|---|---|
| Highways | Information *highways* | The backbones: those transcontinental and undersea fibre-optic- lines which move data from city to city, at the speed of light |
| Ports | Information *ports* | Network Access Points (NAPs): the sites where telecommunication carriers interconnect their systems into a single, global network |
| Warehouses | Information *warehouses* | Data centres: the secure structures which house rows of telecommunication equipment, such as Internet servers and switches. |
| Factories | Information *factories* | The wired businesses and homes that produce (and consume) information products |

**Illustration 16.4: The four main components of the global Internet infrastructure (Townsend, 2003)**

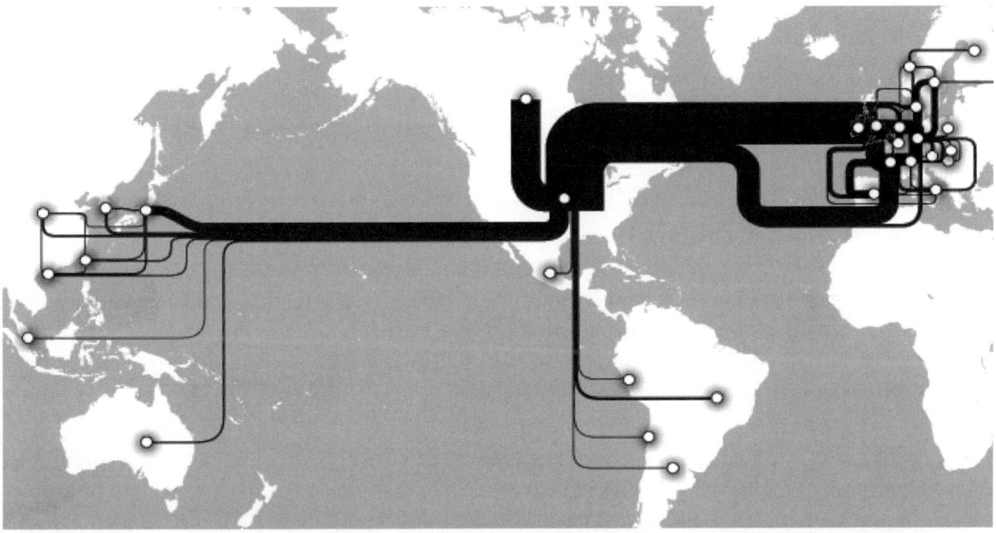

**Illustration 16.5: A map of the major international Internet routes in 2004 (TeleGeography, 2005)**

For his doctoral thesis, Sean Gorman continued to map in detail the physical infrastructure of the Internet in the US. However, after the events of 11 September, distributing the information he had collected became a security threat in the eyes of government officials and corporate executives, because it was easy to trace critical points and vulnerable areas from it. This information was also withheld from the public, for the same reason. On the other hand, Townsend's examination of the ICT backbones in 2003 has shown that Washington D.C., rather than New York, is the leading hub of backbone networks in the US, closely followed by San Francisco – a fact that defies what is generally considered to be the role and definition of global cities (Townsend, 2003).

Regarding the European backbone system, Paul Drewe elaborated a series of reports for the Ministry of Housing, Spatial Planning and Environment, that examine the ICT backbone networks in Western

Europe (Drewe, 1999a; 2000a) and the Netherlands (Drewe, 1999b), and the position of the cities within them (see http://www.networkcity.bk.tudelft.nl). His studies show that the European Net is multinodal and that the bandwidth 'ranking of cities' has been changing constantly in line with the evolution of the networks. Recent research (Rutherford, Gillespie & Richardson, 2004) has confirmed that, unlike the US American backbone system, the European backbone relies on "a minimum of 12-15 cities to deliver high bandwidth networks and services across Europe" (2004:29), even though a significant number of networks and bandwidth capacity are concentrated on the two largest metropolises of London and Paris. This suggests that the spatial distribution of the Internet networks in Europe is closely related to the features of its urban system.

The correlation between ICT backbones and the distribution of population world-wide can be more clearly seen in the developing world, where the networks are less diffuse. For example in the case of the Latin American backbones, the largest nodes correspond to the largest metropolises of the region: Mexico City, Sao Paulo, Buenos Aires, Lima and Santiago. Nevertheless, they are clearly dependent upon the US backbone networks as the main regional hub is not located inside Latin American territory but in Miami (Fernández-Maldonado, 2004).

## Some urban implications of the geography of ICT networks

The previous sections have described how, from a technical point of view, the expansion principles of ICT networks favour situations where "only the rich get richer" (Drewe, 2003), and so are obviously detrimental to the objectives of sustainable development. From an economic point of view, the trends towards urban concentration are twofold at important nodes, the one feeding the other. While, on the one hand, bandwidth deployment is located according to existing and potential markets, on the other hand, empirical studies have established that access to high-speed networks is an important location factor for knowledge intensive and information handling firms and institutions.

Generally speaking, the best served locations in terms of ICT networks are large metropolises that concentrate on educational, governmental and financial activities. At the same time, they attract the main offices of global firms and advanced producer services (APS) – those highly specialised and sophisticated industries that provide the tailor-made services to large, globally operating corporations and act as one of the 'main connectors' between global cities (Rocco, 2005). Therefore, if a city is to attract foreign firms and foreign investments, and offer a smooth link between urban economies and the global economy, then its position within the global backbones is relevant.

The combination of these factors produces the unevenness that characterises the distribution of ICT networks. The notion that global ICT networks would create a world of uniform connectivity, in which geography and location would not matter any more, has been proved wrong. Futurists and utopianists claimed that the challenges posed by ICTs to the forces of agglomeration would eventually lead to the dissolution of cities, but these deterministic predictions have also been disproved. ICTs are indeed powerful technologies for decentralising and dispersing activities in space, but the drive towards deconcentration that they can bring is challenged by ICT-related agglomeration trends that have both a technical and economic character.

While the spatial geography of ICT networks is still evolving, the highly dynamic and complex urban territorial processes that arise due to the expansion of ICT networks, and their corresponding use, still offer an unchartered world for research. Most of the literature has focused on the influence of telecommunications on the location of firms, and the approach that has been used stems from the logic of the industrial city, extrapolated into the informational city. There are, however, many new urban issues that have not been sufficiently explored, to which urban planners and designers should turn their attention; new urban issues that do not respond just to (macro) economic concerns. Zook, Dodge, Aoyama and Townsend (2004) have recently pointed out some of these important issues. They are related to space and time dimensions, and to methodological issues, respectively:

1. *To the spatial effects of ICT-related processes in commercial functions.* Electronic business, both B2B and B2C, has been steadily increasing and affecting most firms and business sectors, although in different depths, ways and scope, according to business sector. But while the adoption of e-commerce and e-retailing manifests itself differently, place remains as a critical factor. The spatiality of B2C e-commerce in Japan, for example, is linked to short-distance exchanges in convenience stores, while in strictly regulated West European countries it is related to traditional mail-order shops located outside the city centres. Informal and illegal businesses are also taking advantage of the ICT revolution. High-profit online gambling and pornography businesses are mainly 'located' in East Europe and the Caribbean countries.

2. *To the changes in 'city metabolism' brought about by the widespread use of wireless mobile technologies.* Anthony Townsend (2000, 2003) has focused his research on WiFi and mobile-phone use and the urban dynamics related to them, especially regarding changing patterns of mobility and social interaction. The power of mobile phones to coordinate daily activities indeed helps to increase the pace of personal or business-oriented exchanges and transactions, and, in this way, it also increases the metabolism of the city, providing more flexibility in travel time and supporting an increased mobility. "... the introduction of mobile connections further amplifies the dynamic complexities of contemporary digital geographies." (Zook *et al.*, 2004:157). WiFi and mobile phone use are also encouraging a new type of emergent behaviour in cities, in which the coordination of activities can have different outcomes, strong enough, even, to destabilise political regimes, as evidenced by the events that followed the Madrid train bomb of 11 March 2004. These new processes represent the convergence of ICT technology and social activism in urban settings. Howard Rheingold (2002) has studied these new digital phenomena, and the increasing importance of what he calls 'smart mobs', at global level.

3. *To the need for instruments for mapping and visualising new digitally enabled spatial processes.* In *Mapping Cyberspace*, Dodge & Kitchin (2001) have illustrated the many ways of representing cyberspace graphically. However, urban professionals are still unfamiliar with new methods such as these, which do not resemble topological data and Euclidean geography. Cartographic visualisations of the geographical distribution of ICT network infrastructures, ranging from global backbones to city districts and neighbourhoods, are useful for grasping and analysing the territoriality of the new telecommunications infrastructures. But these mappings show more of the 'where' than the 'how'. For example, they show the nature of traffic flows and what people actually do when connected to the networks.

4. *To the coupling of self-organised processes at micro and macro level.* "... digital communication technologies actually do their 'work' at the level of the individual, everyday performances of space."

(Zook *et al.*, 2004:158), but the coupling between these small-scale mobilities and larger urban processes, in order to measure and document ICT-related transformations at aggregate level, is very difficult methodologically. Techniques are evolving that might help in future to overcome this problem in modelling and tracking down the evolution of contemporary cities and regions. For example, tracking technologies such as GPS (Global Positioning Systems), have been used in recent research in attempts to aggregate multiple individual activity patterns at neighbourhood and city level. This may eventually lead to being able to visualise maps of dynamic human processes in real time.

These four topics provide a small glimpse of the new possibilities and challenges that the use of ICT has brought about in cities. ICTs are making contemporary cities more complex than ever before. In the current city, different spatial and temporal organisational principles, communication cultures, modes of economic organisation and ways of life co-exist, and all of them matter in this transitional period. Each city has a different mixture and measure of the different principles, but those of advanced economies have a larger proportion of the recently emerged ICT-related organisational principles. Attempts to analyse the city from one perspective only, from an 'objective' viewpoint, and with an emphasis on the industrial layer are too limited in their approach to interpret contemporary urban transformations adequately.

## Conclusions

The first section of this chapter describes the huge changes that have taken place within the telecommunication sector, showing how the new trends in the sector, based on market principles, are contributing to an unequal development of cities and regions. The second section, outlining the main technical features of ICT networks, has shown that the global Internet is a complex, self-organised, hierarchical and unevenly distributed system, with an uneven distribution and capacity caused by preferential attachment behaviour – a characteristic of self-organised complex systems. These two political and technical trends are favouring a situation in ICT connectivity whereby 'the rich-get-richer'. The third section has described the distribution of the ICT networks in different regions of the world, illustrating the new digital geography, with its concentration on the richest parts of the globe. The fourth section has briefly described some of the urban implications that arise from the spatial distribution of ICT networks. The aim here was to bring the range of new possibilities and challenges that ICT use has brought to cities into the picture. To understand urban places and urban phenomena, a good level of awareness of the features of these new ICT infrastructures and ICT-related trends and processes is indispensable.

## References


Alaedini, P. & P.J. Marcotulio, 2002, Urban Implications of Information Technology/New Electronics for Developing Countries, *Journal of Urban Technology*, Vol. 9, No. 3, pp.89-108

Barabási, A.L., 2002, *Link: The New Science of Networks*, Perseus Publishing, New York

Castells, M., 2001, *The Internet Galaxy. Reflections on the Internet*, Business and Society, Oxford University Press, Oxford

Dodge, M. & R. Kitchin, 2001, *Mapping Cyberspace*, Routledge, London

Drewe, P., 1999a, *The Internet - The Randstad and the rest of the Netherlands*, Report for the Ministerie van Volkshuisvesting, Ruimtelijke Ordening en Milieubeheer

Drewe, P., 1999b, *The Internet - beyond the hype. How to position the Randstad Holland?*, Report for the Ministerie van Volkshuisvesting, Ruimtelijke Ordening en Milieubeheer

Drewe, P., 2000a, *The Internet - How to position cities on the European net. Release 1.1*, Report for the Ministerie van Volkshuisvesting, Ruimtelijke Ordening en Milieubeheer

Drewe, P., 2000b, *ICT and Urban Form. Urban Planning and Design - Off the beaten track*, Report for the Ministerie van Volkshuisvesting, Ruimtelijke Ordening en Milieubeheer

Drewe, P., 2003, *Internet Industry and Internet Infrastructure. Matching Networks*, Report for the Ministerie van Volkshuisvesting, Ruimtelijke Ordening en Milieubeheer

Dupuy, G., 2004, Internet: une approche géographique a l'échelle mondiale, *Flux*, International Scientific Quarterly on Networks and Territories, No. 58, pp. 5 – 19

Fernández-Maldonado, A.M., 2004, *ICT-related transformations in Latin American metropolises*, PhD Thesis, Delft University Press, Delft.

Gorman, S., 1998, *The Death of Distance but not the End of Geography: The Internet as a Network*, Paper for the Conference of the Regional Science Association, Santa Fe, October

Graham, S. & S. Marvin, 1996, *Telecommunications and the City. Electronic Spaces and Urban Places*, Routledge, London and New York

International Telecommunications Union, 2002, *World Telecommunications Development Report 2002. Reinventing Telecoms*, ITU, Geneva, March

Nellist, J.G. & E.M. Gilbert, 1999, *Understanding Modern Telecommunications and the Information Superhighway*, Artech House, Boston, Mass

Priemus, H., 2004, *From a Layers Approach towards a Network Approach: A Dutch Contribution to Spatial Planning Methodology*, Planning, Practice & Research, Vol. 19, No. 3, pp. 267–283

Rheingold, H., 2002, *Smart Mobs: The Next Social Revolution*, Perseus Publishing, Cambridge

Rocco, R., 2005, *Foreign Direct Investment and Urban Change: A contrast between Randstad-Holland and São Paulo*, Paper for the II PhD Seminar, Barcelona 2005

Rutherford, J., A. Gillespie & R. Richardson, 2004, The Territoriality of Pan-European Telecommunications Backbone Networks, *Journal of Urban Technology*, Vol. 11, No. 3, pp.1-34

Shaw, R., 2002, *Regulation: what changes are needed?*, Keynote Speech II Rio Telecom, Rio de Janeiro, Brazil, July 30, 2002, http://people.itu.int/~shaw/docs/rio-30-jul-2002.html , November 2002

Standage, T., 2003, Beyond the bubble. A survey of telecoms, *The Economist. Special Report Information Technology*, October 11th 2003

TeleGeography, 2003, *Global Internet Geography Database and Report. Executive Summary*, http://www.telegeography.com , October

TeleGeography, 2005, *TeleGeography. A research Division of PriMetrica, Inc.*, http://www.telegeography.com , May

Townsend, A., 2000, Life in the Real-Time City: Mobile Telephones and Urban Metabolism, *The Journal of Urban Technologies*, Vol. 7 No. 2, pp. 85-104

Townsend, A., 2001, The Internet and the Rise of the New Network Cities, 1969-1999, *Environment and Planning B*, Vol. 28, No. 1, pp. 39-58

Townsend, A., 2003, *Wired / Unwired: The Urban Geography of Digital Networks*, Unpublished doctoral dissertation, Department of Urban Studies and Planning, MIT

World Bank, 2000, *Telecommunications Sector: Background and Bank Group Issues*, World Bank Board Seminar, http://www.worldbank.org/html/fpd/telecoms/boardseminar/telecom.html, November 2002
Zook, M., M. Dodge, Y. Aoyama, & A. Townsend, 2004, New digital geographies: Information, communication and place, Brunn, S.D. *et al.* (eds), *Geography and Technology*, Kluwer Academic Publishers, the Netherlands

# 17 Metamobility: [1] In Search of Connections within Mobility Networks

Ernesto Philibert Petit

## Introduction

Mobility is one of the most critical issues that spatial design and planning will have to tackle in the coming years. It is fundamentally important, therefore, to gain a better understanding of the complex implications of mobility, because this is an agent in configurating the, mostly urban, space in which we live. The reason why mobility is so relevant is that it is an integral function of space and *time*. This distinguishes it from other topics studied in the discipline. Time brings people into perspective in spatial design and planning, their patterns of organisation, and their ways of interacting with physical and virtual space, and mobility is integrally involved in all of them. We will also consider the concept of mobility in this chapter. We will add a third main component – information – to space and time, because it is information, as a concept, that contributes to making mobility a critical issue. First, however, I would like to contextualise this topic briefly by referring to four factors that give it form: the emerging network paradigm which is used as a tool for conceptualising mobility; the current fragmented reality of our urban space; the recent discovery of principles of urban structure; and the opportunities we can find for intervening in mobility spaces to build new connections to help sustain our civilisation. The perspectives from which these factors are observed are first explained, and then a framework is constructed for developing a theoretical mobility network with theoretical connections. The aim in constructing this meta-mobility is to arrive at a point at which we can consider how to approach mobility. By so doing, we will be in a better position to construct an agenda for the future in spatial planning and design.

## Context

Mobility is studied here within the context of four dominant trends, events, and possibilities:

- *The emerging paradigm for conceptualising the world is the study of networks as part of complexity theory.* Networks appear to be pervasive in both nature and human society (Buchanan 2002). In less than a decade, and a few years after Dupuy and Drewe had suggested the use of networks

---

1 Meta: A prefix used to denote a higher level of thought about the subject, e.g. metascience (the level on which we consider how to approach science), meta-ethics the level on which we consider how to define normative behaviour. Each level in a complex system can be considered as a meta-viewpoint on the previous level. Meta relates to a category or type of theory; a higher-order logic. www.calresco.org/glossary.htm

as a tool for researching new spatial planning and design concepts (Dupuy, 1991; Drewe, 1997), and after the publication of Castells' book *The Rise of the Network Society* (1996), researchers, in particular Barabási, Buchanan, Capra, Johnson, Strogatz, and Watts, made a number of discoveries regarding the pervasiveness of networks in different realms of our world. The organised complexity problem foreseen by Jacobs (1961) is with us and networks are here to help us understand how a city works.

- *The recurrent fragmentation in today's urban environment.* Viewed from economic, social and environmental perspectives, our city spaces are more disconnected nowadays than ever before. They have either been splintered by the superimposition of a number of technological infrastructures, especially mobility infrastructures (Graham & Marvin 2001), or they are becoming more and more segregated by global socio-economic phenomena (Castells 1996). In spite of the emergence of a range of new virtual global links, an opposite process has taken place whereby physical structures and patterns have been fragmented on the local level, in our cities. To deal with these changes, and sustain our urban civilisation, modified and new instruments of spatial planning and design are required.

- *Nikos Salingaros' recent discovery that urban structure is not haphazard but is based on scientific principles that can be measured* (2005). The importance of this discovery for the future study of urbanism, and our understanding of how cities gain form, is paramount. These principles have "rules [which] are derived from connective principles in complexity theory, pattern recognition, and artificial intelligence" (Salingaros, 2005:17). Salingaros has demonstrated that any urban setting can be reduced to human activity nodes and their interconnections, which can then be investigated as a mathematical problem. His theory relates the network approach to the study of urban complexity by treating spatial interventions with scientific rigueur.

- *The possibilities offered by urban mobility spaces, can be seen as opportunities for disconnecting urban tissues from their economic, social and environmental perspectives and for reconnecting them spatially.* In conceptualising mobility and the spaces it occupies within the city, new relationships can be found. Thus, a central issue in this work is connectivity. "A central component of the human intellect is the ability to establish connections. Urban design is most successful when it establishes a certain number of connections between activity nodes" (*ibid.*:17).

This chapter deals with a theoretical search for connectors to link the study of things and people who are subject to mobility (people, goods and information) with the spatial, economic, social and environmental realms that they impact on spatial scales ranging from the region to the house.

## Perspectives

In a previous article (Philibert 2004) , I approached this subject in the twofold sense of structures and patterns, in line with Anthony Giddens' definition of the relationship between social structures and human agency (Capra 2002:78): "the interaction between social structures and human agency is cyclical (...) social structures are both the precondition and the unintended outcome of people's agency". I stated that patterns of fragmentation had created structures of fragmentation, which in time, are causing even more fragmentation of the patterns we use in running our cities. If, within the flow of human conduct, these actions are considered to have the capacity to transform, then this fragmentation can present an opportunity or pose a threat. Fritjof Capra first represented this relationship and its process in the form of a three-node graph, the nodes being process, form (patterns) and matter (structure). Later, he postulated that "the systemic understanding of life can be extended to the social domain by adding the perspective of *meaning*..." (Capra 2002:80). I am incorporating this same hermeneutic dimension into the perspective used in carrying out this work. It is of central importance, since the human dimension is incorporated into the study of time and into considering information as an interrelated aspect of mobility. To be able to arrive at proposals for action, I will also adhere to the links identified by Jürgen Habermas between process and each of the other three perspectives: strategic action that links process to patterns (human relationships); instrumental action linking process with structure, and communicative action linking process with meaning.

## Metamobility

To gain a deeper understanding of the relationship between structure and patterns, in accordance with Giddens' and Habermas' ideas, an approach to the study of mobility is presented here. This begins with an empirical observation of reality, followed by interpretations (hermeneutics) of the observations, and concludes by trying to establish links in a theoretical network between observations, interpretations, and various combinations of them.

### Observation

Mobility was the theme of the 1st International Architecture Biennial of Rotterdam (2003). The architect Francine Houben, Professor of Mobility Aesthetics at the Technical University of Delft, and founding partner of the architectural firm Mecanoo, was appointed as curator and director. For the Biennial, she decided to invite academic institutions from many countries and to cover international research on different aspects of the relationship between motorways and cities. Twelve universities from widely different parts of the world were invited to take part, including China, Egypt, Indonesia, and Germany. To establish a problem statement, the Mexican team, which I had the privilege to lead, devised a survey on this theme, to try and establish the most important issues. Three general issues could be identified: safety, communication and fragmentation. Below is a list of the most frequently mentioned observations expressed in the survey questionnaires:

1. Motorways are associated with danger. The rates of fatal accidents in automobile spaces are relatively high;
2. Where automobile spaces meet those allotted to pedestrians, the interaction between pedestrians and automobiles has not been solved adequately;
3. When passenger and freight transportation are mixed, the roads are alarmingly unsafe;
4. Automobile spaces are confusing;
5. Motorway users are given insufficient information;
6. (Free) commercial information in the form of advertisements along the moving auditorium uses and abuses traffic infrastructure;
7. Traffic continues to grow relentlessly;
8. Congestion and pollution are degrading urban life;
9. Allocating city budgets to improving infrastructure for car traffic is a poor investment.

## Interpretation

According to Capra, understanding social reality is inextricably linked to reflective consciousness (Capra 2002:73). Reflections on the observations made by the team provided a series of interpretations about those observations. From the Mexican viewpoint, these interpretations formed a body of concepts regarding the city and the road systems that serve it. Weeks afterwards, in Rotterdam, at the seminar held during the International Biennial, the 12 universities presented and shared their observations, interpretations and proposals, and, by so doing, they constructed a global body of concepts on this theme. But now, to continue with our case, I will elaborate on the interpretations made by our team, and on what led us to consider fragmentation as a major issue. The numbers refer to the nine observations mentioned in the preceding paragraph.

1. Fatal accidents on Mexican federal highways reached a level of 75 in the month of July 2002. 2.5 fatal accidents a day is a relatively high rate for the number of inhabitants, compared with the other two countries in the NAFTA Treaty. This statistic, in addition to the perceived physical lack of safety, is why people view the motorway as dangerous. Although the perception of danger on urban motorways is not so acute, it is still significant, though difficult to generalise.

2. and 3. The general opinion of pedestrians who replied to the survey was that although streets and areas near motorways are unsafe because of cars, they are made more unsafe by buses and lorries that use the same route. Drivers' opinions were very similar, but their remarks were directed more towards the insecurity that arises when there is a mixture of freight and passenger transportation within the same road system. Cal y Mayor y Asociados, perhaps the most renowned traffic engineering firm in Mexico, states that this is the main problem for mobility spaces: the insecurity brought about by a mixture of different kinds (and scales) of traffic. This has much to do with the principle that states that inverse power scaling has been dangerously ruptured in the modern city (Salingaros, 2005). This phenomenon is more acute in spaces designed for large-scale transport, but where there is interaction with smaller scale forms of transport. This violation of power-scale distribution helps to explain the understandable and frequent sense of insecurity. From discussions during the seminar, it was clear that this is not a condition peculiar to Mexican mobility spaces; it is a frequently recurring issue worldwide.

4. Mobility spaces are confusing. This is another global issue. The pedestrians and drivers who responded to our survey agreed that the spaces reserved for four-wheeled transport (in the case of urban motorways) are so neutral and characterless that they often found them confusing. We interpret this observation very much in relation to the concept of the imaging in urban space (Lynch, 1984), whether it is present or not, and to the problem of not recognising patterns in modern spaces, as identified by Salingaros. The dimensions of the problem are much greater than initially imagined since "we don't appreciate how completely architectural patterns connect to social patterns; the former make up a significant part of the traditional culture in any society" (Salingaros, 2005: 204).

5. That the Mexican drivers in the survey mentioned lack of formal information for motorway users as being a problem, may be survey-specific, or perhaps it is more characteristic of developing countries. In the seminar, this problem was only expressed in the cases from

Indonesia and Mexico. Nevertheless, in developed countries, even though road signs and markings are more comprehensive, divergences and other temporary changes are often insufficiently indicated, causing stress, aggravation and accidents.

6. In Mexico, the use by the advertisers of commercial signs along the roadside can become an abuse for drivers, pedestrians and citizens in general. Along extensive sections of urban motorways, advertisements saturate the urban landscape and give it a chaotic appearance. Along the Periférico, the ring road in Mexico City, these enormous boards try to capture the attention of drivers and passengers every day. We found no similar situations in other cities represented in the seminar, with the exception of some parts of Beirut. There, as in Mexico City, there are extrovert facades, colours and signs of urban space. "It's amazing how Mexico is all about paint" Francine Houben observed, while on a visit to our university.

7. There is a well-founded perception that traffic is growing relentlessly, no matter how many public works are constructed to extend road surfaces. First of all, it is hard to compete with the increasing number of cars everywhere. The world production of cars reached 40 million in 2000. These use about 800 square kilometres of new parking space worldwide, and in some extreme cases, such as in the Phoenix development; they occupy more than 50% of the urban territory. Centrifugal forces have been (literally) winning territory from centripetal forces.
Urban sprawl is another common sight in cities today, and single function zoning. This has detracted from real traffic increases in cities. From statistics and prognoses, higher increases were expected.

8. There is a general perception that congestion and pollution degrade urban life. I said much the same in a book describing 19th century industrial cities in Britain, but recurrent as it is, congestion and pollution can reach alarming proportions in some large cities such as Mexico City or Beijing. Even though very low emission and hydrogen motors have recently been introduced to tackle the pollution issue, it is likely that congestion will persist and even increase.

9. The surveys registered a general feeling that it was ineffective to allocate money in budgets to construct more spaces for the automobile. We carried out these surveys at a time when a debate was going on in Mexico City about the mayor's desire to build a second level over the *Periférico* (see Ill. 17.1), the most important urban motorway in the metropolitan area. Two years later, with half of the second floor already built and operating, public opinion is at least the same, if not more negative, because of the many traffic jams and delays caused by construction activities.

## Connection

When observation and interpretation points to fragmentation as a preliminary conclusion. it is time to seriously consider *connecting* as a strategy for city planning and design. The fragmentation of city structures has contributed to fragmenting the organisational patterns of our society. Thus, if economic polarisation, social segregation and environmental disruption are the causes, they are also effects at the same time, since there is a cyclical process between structures and patterns. Connecting the spatially fragmented entities can help reduce fragmentation in social organisation patterns. New knowledge in the form of the emerging science of networks will help to orient these connection processes.

## What should be connected?

*Mobility networks.* The fragmentation observed and interpreted above suggests that the best line of action would be to connect mobility networks, starting with the smallest scale (Salingaros, 2005), and to increase the connectivity by adding to the quantity and quality of connections on that same small scale. At the same time, this system of mobility networks should be connected to one of a larger scale, which in turn will then develop more connectivity. By increasing connectivity, mobility will be improved, and most importantly, accessibility. These three phenomena are intertwined, as are spatial, social and economic phenomena, which again cannot be disassociated from a process such as this.

*The car and people.* Although the physical relation of the automobile with its user has been studied in industrial design, from the point of view of ergonomics, there have been few scientific studies about the relationships between cars and city users. This is a new, urban ergonomics field that will be introduced in the near future.

*Disciplines.* For centuries, our understanding of reality has been divided into a number of isolated compartments that sometimes relate to each other. This compartmentalisation of knowledge has contributed to the fragmentation of our structures. In recent times, new exciting connections between disciplines have appeared, that have shed new light on existing knowledge. Physicists are now exploring biology (Capra), mathematicians are building scientific theories of the built environment and emergence synchronisation (Salingaros, Strogatz), sociologists and industrial engineers are helping to build up a science of networks (Watts, Dupuy) and economists are using networks and IT as tools for designing the city of the possible future (Drewe). As Sue Roaf said in her closure speech of the International Conference "Teaching in Architecture", Oxford, July 2000: "it is so enriching to step one foot out of your discipline".

*Scales.* Disciplinary division has fragmented our knowledge even about scales. Sibling disciplines in three-dimensional design have become separated in that way. Industrial design, architecture, urban and regional design can learn a lot from each other, as one of the main aims of all these disciplines is to satisfy people's needs. Linking scales also means making our environment coherent again. We have observed and interpreted a fragmented inverse power-scaling in our cities, which has resulted in us thinking partly in separate scales about them. However, all these scales are interconnected and should be approached interconnectedly, within a dynamic framework of scales.

*Space and time.* Our way of looking at the world has also separated time from space. However, our spatial disciplines are finally taking time into account again, and this is bringing the real needs of people back into consideration (Drewe, 1997; Klaasen, 2004).

*Economic cycles.* In our globalised reality, these cycles are also fragmented, making the gap between the rich and poor ever wider. Network thinking and network science can help us to connect economic cycles, providing their proven feature of behaving in networks with emergent properties. This issue is

**Illustration 17.1: A general view of the Periférico, the ring-road infrastructure, in Mexico City, showing how commercial advertisements abuse the landscape**

related to the growth of our cities (or should I say built environment) and can be connected up to the mobility spaces where the investment sector currently focuses its activities.

*Social groups.* Our cities have also become fragmented in that social groups have become separated. This is by no means a new phenomenon; there has been social segregation in cities from the beginning of their history (Kostof, 1991). The difference is that the modern city has fewer and fewer spaces for social interaction. Whereas the public spaces of the pre-modern city functioned as effective mixers of social groups, those in the modern and post-modern city have lost this quality.

*The cultural and natural environment.* Another area of fragmentation is that artificial or culturally built environments have become separated from nature. With its heritage of linear, reductionist thinking, the modern built environment has been constructed as though natural resources are unlimited and as though wastes can be freely released into the biosphere. At the time when most of the urban built environment in the world was being constructed, this paradigm was unfortunately at its peak. Nowadays, we are starting to react to this by trying to reconnect nature and culture in our built environments.

*Physical and virtual structures.* Although spatial planners and designers can learn a lot from the Internet, only a few contemporary thinkers in spatial planning have proposed seriously considering this network as a means for studying the practice of urban design (Mitchell, Drewe). New scientific discoveries about the way the Internet is organised, and concepts such as growth, preferential attachment and fitness (Barabási, 2002) suggest new avenues here in urban design and planning, especially with respect to mobility spaces.

## Where should connections be made?

*Global and local processes* follow separate paths nowadays. While global capital is active in local scenarios, its performance as a generator of welfare is poor, as poor as the benefits local structures get from global investments. Due to economic and technological socio-fragmentation resulting from mobility and information divides all over the world, social segregation is also evident between global and local patterns of organisation. Global environmental processes are both the cause and outcome of local patterns; conceptual connection in this relationship is also network-oriented.

*The fragmented tissues* of our cities is another field of study where connections can be made. Their continuity has been broken by infrastructure that has become structure and sometimes even suprastructure, serving only part of the entire urban area. When an infrastructure is introduced to connect a given urban layer, the consequence is that another layer becomes disconnected (Graham and Marvin, 2001). New forms of, mainly mobility, infrastructure are needed to connect the entire area, which can be expected to be more physically impacting in the cities of the future.

Connections also need to be made where different forms of infrastructure have been superimposed, one above the other. Mobility spaces have been superimposed on different scales on top of existing urban structures, preventing the urban structure from evolving naturally. Superimpositions of structures for commercial communication (such as advertising) need to be reviewed, re-conceptualised and linked better to the urban whole. For example, superimposed infrastructures occur in the modern city where the needs of car users have been favoured rather than those of pedestrians; where global structures have been given priority, rather than local ones; where mobility infrastructures of people, goods, information, energy and resources have dissected existing tissues. In fact, there are opportunities everywhere for re-connecting urban societies.

## How should these connections be made?

As an architect, my first answer is "through urban projects", but in fact I think the answer lies in teaming up urban projects and urban planning, in using a combination of bottom-up and top-down intervention methods. The issue here is to find this "third way"; the method that Salingaros has suggested as providing a balanced answer. Urban projects are not always large-scale interventions; they emerge on every urban scale. Depending on their origin, they are the outcome of either top-down or bottom-up methods. In the same sense, planning can be top-down planning, as in the master planning tradition, or bottom-up, as in the strategic planning school of thought. In my opinion, an effective combination would be to combine the inductive approach of urban projects with the deductive approach of planning.

Connections can be fully integrated (Capra, 2002) by simultaneously linking up with actions, process and meaning (communicative actions), process and patterns (strategic actions), and process and structure (instrumental actions).

## Conclusion

Mobility spaces, while causing fragmentation in the urban environment, also provide an opportunity for applying an instrumental approach to connection. If connectivity can be increased in towns and cities, mobility will improve and, more importantly, accessibility, not only from a spatial perspective, but also economically, socially and environmentally. Increasing connectivity in the city is thus a step towards increasing the sustainability of urban conglomerations.

Mobility spaces act as a melting pot of disciplines, fields of study, interesting people, transport, and information. How we consider mobility is therefore relevant in that it can provide us with new concepts of what, where and how to connect.

One of the most important connections that needs to be made is to create a network to link people's mobility with the mobility of information, i.e. to connect people's contemporary mobility spaces with the rapidly emerging and developing information technologies.

The simultaneous emergence of information technologies, scientific principles by which urban structures can be analysed, and the discovery of the organised complexity of networks offer new and exciting perspectives in attempts to improve the connectivity, and thus accessibility, of our urban environment, and with that, the sustainability of our urban life.

## References


Barabási, A., 2002, *Linked*, Plume, New York

Buchanan, M., 2002, *Nexus: Small Worlds and the Groundbreaking Science of Networks*, W. W. Norton & Company, New York

Capra, F., 2002, *The Hidden Connections. A Science for Sustainable Living*, Anchor Books, New York

Castells, M., 1996, *The Rise of the Network Society*, Blackwell Publishers, London

Drewe, P., 1997, *In search of new spatial concepts, inspired by Information Technology*, Paper prepared for the conference 'Cities in the XXIst century, Cities and metropolis: breaking or bridging?', French ministry of housing, transportation and public works *et al.*, La Rochelle, 19–21 October 1998

Dupuy, G., 1991, *L'urbanisme des réseaux, théories et méthodes*, Armand Colin, Paris

Graham, S. & S. Marvin, 2001, *Splintering Urbanism*, Routledge, London

Jacobs, J., 1961, *The Death and Life of Great American Cities*, Random House, New York

Klaasen, I.T., 2004, *Knowledge-based Design: Developing Urban & Regional Design into a Science*, Design/Science/Planning Series, Delft University Press, Delft

Kostof, S., 1991, *The City Shaped*, Thames & Hudson, London

Lynch, K., 1984, *La imagen de la ciudad*, Gustavo Gili, Barcelona

Philibert, E., 2004, *Structures and Patterns of Fragmentation and Connectivity. A Preliminary Assessment of eight areas of the City of Querétaro*, paper prepared for the Urbanism International Seminar, Jiao Tong University, November 22–24, 2005, Shanghai

Salingaros, N., 2005, *Principles of Urban Structures*, Design/Science/Planning Series, Techne Press, Amsterdam

# 18 Small but Vital: The Influence of Small-Scale Mobility on Sustainable Urban Functioning

The paradox of transportation of the late twentieth and early twenty-first century is that while it became indeed possible to travel to the moon as a tourist, it also became impossible, in many cases, to just simply walk across the street.

Remon Rooij

## Introduction

### Small-scale connectivity

From all kinds of scientific research on the composition of complex networks (Salingaros, 2005:85-114) we know that the growth of complex interacting systems, such as the city, follows certain 'rules'. One of these rules is about hierarchy. It stipulates that the smaller scales have to be defined before the larger ones: their elements must be coupled in a stable manner before the higher-order groups can begin to form and interact. Applied in relation to how the urban environment functions, it can be hypothesised that a vital city depends on good connectivity within and between the lowest levels of scale.

It is the responsibility of urban planners and designers to incorporate in their designs and plans ideas that will help to sustain vital and successful cities, both now and in the future. The vitality and success of cities can be defined in many ways. Based on the European Council of Town Planners' (2003:12,14) New Charter of Athens, a city is considered successful (from a spatial, social, economic, and ecological point of view) if:

- There is a positive emotional connection between human beings and their environment;
- It capitalises upon its competitive advantages.

The importance of the smallest scales in cities is particulary evident with respect to the first of these gauges of success. The importance of the smallest scales also becomes clear if we look at human beings and their main forms of spatial mobility. The natural way for a human being to move around is to walk. Thermodynamically, humans on foot are more efficient than any motorised vehicle, and, on bicycles, where they can move three to four times faster, using five times less energy, they far surpass them (Tolley, 1997:3). What is more, walking and cycling have a negligible impact on the environment in terms of consuming energy, using space (directly and indirectly), and causing pollution, etc..

## How the city functions

Let us now focus more closely on the physical networks of the city and on how consumption and production take place along them. It should be the aim of every city planner and/or designer to generate a successful urban environment, by creating an efficient and livable, psychologically positive environment for human beings. The city should be used intensively, both in time and space! From ancient times onwards, the (physical) city has always been – and most certainly will be in the future – a place where people gather together and encounter one another.

The urban fabric or urban web is a complex structure, mainly present in the space between buildings. Each building encloses and shelters one or more human activity nodes, linked by coupling elements, such as infrastructure and public space. The role of the urban fabric should be to stimulate, facilitate, and accommodate human gatherings and encounters as much as possible. This is facilitated by components such as streets, shops, offices, houses, pedestrian zones, green spaces, plazas, parking lots, telephones, cable networks, etc., To be successful, the physical structures and surroundings of cities should also meet the physical and psychological needs of the human beings, in terms of efficiency, vitality, and livability.

To this end, it is helpful to be able to distinguish the most important elements of an urban pattern from the general fabric or 'tissue' of a city (Roberts, Lloyd-Jones, Erickson & Nice, 1999:51-66). These elements are likely to comprise the following: the most striking and frequently used streets, the main communication arteries, places of exchange and assembly, and the key buildings and complexes that serve social infrastructures (*ibid.*:51-66). These can be referred to as the armature of the city. The armature is not the network as a whole, but rather the key features of the various urban networks which, together, form the core movement, activity, and meaning. In identifying the armature, we are, at the same time, structuring and organising complexity. This leads us to the theoretical foundations of complex interacting systems.

Whereas a city without complexity can be said to be 'dead', too much complexity, with no organisation or structure, makes a city 'chaotic' and 'unlivable'. It seems that one of mankind's basic drives throughout the ages has been to raise the degree of organised complexity (Jacobs, 1968; Salingaros, 2005). One of the principal ideas behind the city complexity theory is that a city mimics human thought processes, in that both depend on establishing connections, patterns, and structure.

## Defining the problem and stating the objective

Urban planning and design is a practical science, entrusted with the task of designing the urban fabric. From both a scientific and practical point of view, to do this it needs knowledge-based design principles (Klaasen, 2004:5, 6). Based on the research into complex networks, we can conclude that an important problem for the field of urban design and planning is the lack of a scientific body of knowledge about small-scale design principles. The objective of this chapter is therefore to show how such a scientific body of knowledge can be constructed for these lower levels of scale. A M.Sc. project on 'The City Child' is used to illustrate how this body of knowledge can be acquired.

## Planning the chapter

The next section focuses in more detail on the role of the lowest levels of scale in complex networks. Then, in the third section, 'The City Child' project is presented. Based on this project, in the fourth and concluding section, an approach is suggested on how this body of knowledge about small-scale connectivity can be gained.

# The role of small-scale mobility in the urban network

## Small-scale mobility

Over short distances, slow methods of travelling tend to predominate (see Ill. 18.1). In the Netherlands, about one in three or four trips are made by bicycle or moped (Centraal Bureau voor de Statistiek, 2005: http://www.statline.cbs.nl). In 2003, the Dutch population cycled a total of 14.8 billion kilometres, which amounts to 925 kilometres per inhabitant per year. In that same year, they walked approximately 3.9 billion kilometres, or about 240 kilometres per inhabitant. It should be noted that accessing and/or leaving buildings/stations, etc. is not considered separately under the definition 'trip'. In the national statistics, trips such as walking-car-walking, or walking-bus-walking are referred to as 'unimodal'. The Dutch pedestrian union rightly claims many more walking kilometres in the national travel statistics, therefore, than those presented by the Central Bureau for Statistics.

## Slow means of transport mean proximity

In order to understand the position of slow methods of travel in Dutch society, we should always keep in mind the ratio of slow transport modes, to car transport, and to the number of kilometres travelled on public transport. There is a clear relation between travel distance and means of transport. In general, the longer the distance, the faster the means of transport. However, it is worth noting that the car

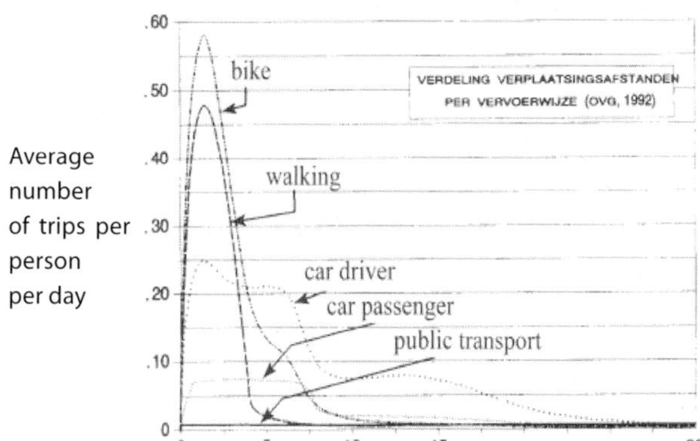


**Illustration 18.1: The distribution of travel distances by travel mode in the Netherlands (based on the National Travel Survey, 1992) (Bovy, 1995:5520-17)**

is used a lot for trips of all distances, even for those shorter than 2.5 km; and almost half of all the trips made are shorter than 2.5 km.

In research that focuses specifically on the use of the bicycle, the Project Bureau for Integral Traffic and Transport Studies (1998:5) concludes from their data that, in the Netherlands, for distances of up to 2.5 km, the bicycle is used twice as often as the car (see Ill. 18.2). For distances of 2.5-5.0 km, the bicycle and car are used to about the same extent. For distances between 5.0-7.5 km, the car is used twice as often as the bicycle. While for distances of over 7.5 km, the share of trips by bicycle becomes marginal. Above that, at distances over 7.5 km, public transport comes into the picture. Distances of between 5.0 and 7.5 km can thus be seen as a reasonable range for cycling. Considering that most trips are of distances shorter than 5.0 km, and that the car is used quite intensively for this distance, then there is still a large potential for the bicycle within present travel patterns.

The role of mobility on the lowest levels of scale becomes even more interesting when we view the city as a complex network, and consider the 'composition rules' that apply to complex networks and their hierarchy. We now present one of the main principles that explains how complex networks arise: the principle of hierarchy.

## Scale-free networks

The rules of hierarchy in complex networks have been studied by the physicist Barabási (2002). He shows that the complex networks that function well are, in fact, the so-called scale-free networks; those dominated by the power law (see Ill. 18.3). When the power law dominates, distribution will be characterised by patterns in which most nodes will only have a few links. These will be held together by a few highly connected nodes, the so-called hubs. Barabási explains that distribution under the power law forces us to abandon the idea of scale, or characteristic node. The largest (best connected) node is closely followed by two or three somewhat smaller hubs, followed by dozens of even smaller ones, and so on, arriving eventually at numerous tiny nodes. There is no intrinsic scale in these kinds of networks. This is why they are referred to as scale-free. Furthermore, Barabási argues that the scale-free topology is a natural consequence of the ever expanding nature of real networks. Starting with two connecting nodes (see Ill. 18.4), in each panel, a new node (shown as an empty circle) is added to the network. When deciding where to link, new nodes prefer to attach themselves to the more connected nodes (in accordance with the 'rich-get-richer' concept). In this way, a few highly connected hubs emerge.

But there is more. In a competitive environment, new nodes which have a high level of 'fitness' attract new links more easily than those lacking it. So, both fitness and connectivity are the driving forces behind preferential attachment. Where two nodes have the same number of links, it is the fittest one that will acquire links more quickly; and should two nodes have the same fitness, then the one that has been established longer will still have the advantage (in accordance with the 'fittest-get-richer' concept).

## Scale-free public-space networks

When we apply the principle of how scale-free networks arise to public-space networks in cities, it becomes clear that cities are in need of, for example, very many pavements and streets (these being at

| Travel distance | Car | Public transport | Bicycle | Walking | Total (in%) |
|---|---|---|---|---|---|
| 0.0 - 2.5 km | 22 | 1 | 40 | 37 | 45 |
| 2.5 - 5.0 km | 50 | 3 | 40 | 7 | 16 |
| 5.0 - 7.5 km | 64 | 5 | 26 | 5 | 11 |
| > 7.5 km | 75 | 11 | 7 | 0 | 28 |
| Total (in %) | 46 | 5 | 29 | 18 | 100 |

**Illustration 18.2: The number of trips per person per day (in %) for the main means of transport at different distance ranges (Projectbureau Integrale Verkeers- en Vervoerstudies, 1998:5)**

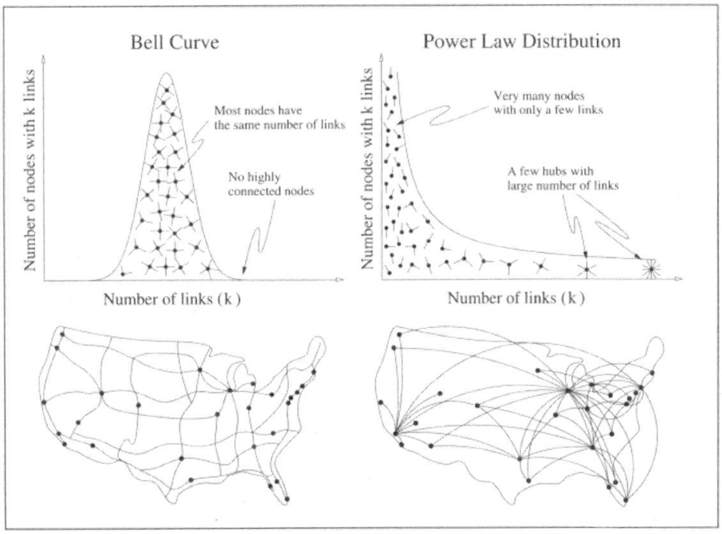


**Figure 6.1 Random and Scale-Free Networks.** *The degree distribution of a random network follows a bell curve, telling us that most nodes have the same number of links, and nodes with a very large number of links don't exist (top left). Thus a random network is similar to a national highway network, in which the nodes are the cities, and the links are the major highways connecting them. Indeed, most cities are served by roughly the same number of highways (bottom left). In contrast, the power law degree distribution of a scale-free network predicts that most nodes have only a few links, held together by a few highly connected hubs (top right). Visually this is very similar to the air traffic system, in which a large number of small airports are connected to each other via a few major hubs (bottom right).*

**Illustration 18.3: The distribution of scale-free networks, those dominated by the power law (Barabási, 2002:71)[1]**

1 Note that there is an important qualitative difference between the power law and a bell curve when it comes to the tail of the distribution. Bell curves have an exponentially decaying tail, which decreases much faster than the decrease seen under the power law. This exponential tail is responsible for the absence of hubs. In comparison, the power law causes a much slower rate of decay, which allows for 'rare events' such as hubs.

the lowest levels of scale of public space in the residential and district environments), but only a small number of (market) squares and motorways (these being public spaces on a higher level of scale).

In line with Bekkering (1997:179-186), we would like to point out here that this scale-free hierarchy of public space in the city is subject more to the meaning, readability and functionality of that specific public space than to (hierarchical) power, as some might reason: "...Coherence at a certain level of scale is conditional for recognising a unity at a higher level of scale..." (Bekkering, 1997:181). When elements of a certain level of scale are missing, the network becomes pathological and/or malfunctioning. CIAM city districts, where there are many dwellings with no or little low-level public space (no transition from private to public space), are often considered to be 'inhuman' environments, even though there might be a surplus of public space at the level of scale of the building block or neighbourhood.

## The lowest levels of scale, and specific user groups

Local connections are the strongest. Small-scale elements (and sub-networks) are necessary, not only to carry out the function for which they arose, but also because of their secondary role in linking other (higher order) elements. This clearly shows why it is so important to plan and design the slow-mode networks properly (e.g., the pedestrian zones and the cycling paths, etc.) within the total multimodal network of transport systems and the urban fabric as a whole. As mentioned above, the city will continue to be a place where people meet each other; it will continue to be a place where small-scale, personal interactions take place. It is components of the urban frabric, such as streets, shops, offices, houses, pedestrian zones, station buildings, green spaces, plazas, and parking lots, etc., that stimulate, facilitate, and accommodate these activities. Successful cities also meet the physical and psychological needs of the human beings who use them in terms of their physical structures and surroundings, if these add values such as efficiency, vitality, and livability. Successful cities provide spatial quality.

By definition, everyone is operative on the lowest levels of scale, which is why these are referred to as the human scale. However, we know from urban and mobility research and statistics, that some citizens use their direct living surroundings more intensively than other user groups whose action spaces cover a wider area. In particular, the activity and travel patterns of vulnerable user groups and the less privileged (those with low incomes, low levels of education, bad health, old age, etc.) take place mainly on the lowest levels of scale (Centraal Bureau voor de Statistiek, 2005).

One of the user groups whose action space is limited, and which can also be considered a vulnerable user group, is the group comprising children. So the next section focuses on the case study 'The City Child' project. To develop the field of spatial planning, it is useful to look at groups such as this for two important reasons.

- Today's city is a mobile city. Its growth has been influenced by the increase in individual action spaces, and the development of information, communication and transportation technologies. Today's network society offers many possibilities and chances, but it also embodies risks, especially for vulnerable groups such as older adults, poor people, the disabled, and children, etc. If 'children' are considered as a specific-user group, a number of possible risks are exposed. For example: increasing travel options, (could) imply increased traffic. Many vulnerable groups do not like (or are not allowed, in case of the children) to move around in busy traffic situations. Therefore parents may choose not to go out for traffic-safety reasons.

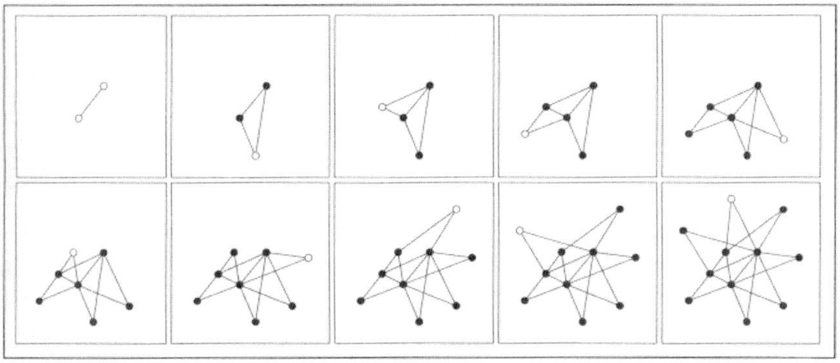

**Figure 7.1 The Birth of a Scale-Free Network.** *The scale-free topology is a natural consequence of the ever-expanding nature of real networks. Starting from two connected nodes (top left), in each panel a new node (shown as an empty circle) is added to the network. When deciding where to link, new nodes prefer to attach to the more connected nodes. Thanks to growth and preferential attachment, a few highly connected hubs emerge.*

**Illustration 18.4: The growth of networks (Barabási, 2002:87)**

- Designing (and planning) for vulnerable user groups very often means 'designing for all', i.e. planning and designing urban environments in such a way that all people, whether they are vulnerable or not, can interact freely within that environment. One of the basic ideas in this concept is that if a design can meet the demands of groups of vulnerable people, such as the disabled, or a parent trying to carry shopping bags and manage a number of young children at the same time, then it contributes to an environment that functions better for the non-vulnerable as well. For example: buses, trams, and metros with floors flush to the pavements or platforms are especially suitable for people whose mobility is impaired. But, they also enable non-vulnerable users, to board and step off them more easily and quickly. So the non-vulnerable benefit too. Thus travelling, in terms of time and comfort, is made easier for both groups of travellers.

# The case study: 'The City Child'

## The motivation for the study

The pressure on space in cities is increasing (see Ministerie van Volkshuisvesting, Ruimtelijke Ordening en Milieubeheer, 2000). The streets of old(er) city districts, which were developed in a car-free time period, have become increasingly cluttered with cars. Young people growing up in these urban districts are being deprived of qualitative and quantitative space. The negative effect that this is having on their development is in turn being felt by the societies in which they live, in the form of social problems

(Jansen & Schouten, 2001). If children's individual development is hindered because of overcrowding, the child ceases to function properly. It develops language problems, followed by learning problems. As it falls behind other children of the same age then social problems set in. The more isolated a child feels, then the more it will resort to vandalism, drugs, criminality and violence, which in the short or long term will also impair the lives of the inhabitants of the neighbourhood as a whole (Karsten, Kuiper & Reubsaet, 2001).

Most research that deals with the role of the child in the city is theoretical. In 1985, however, Joost van Andel (1985) started to investigate children's actual behaviour in congested urban districts. The MSc project 'The City Child' (Van Duijn, 2004a, 2004b), took Van Andel's work a step further by presenting what is known about children's behaviour in their living environment in pattern form. This research project offers urbanists clear reference material on how children use the city.[2] Furthermore, it gives an insight into design features that can improve children's outdoor playing spaces, to further, instead of impairing, their development (Ill. 18.5).

## The research approach

Illustration 18.5 shows the research approach that was used in this project. Firstly, an image was built up, based on literature, and especially on sociological, environmental, and psychological studies of children's developments and their needs for outdoor space for playing. These needs were translated into a pattern language of what children consider to be a child-friendly city design.

Secondly, the location chosen for the case study was the district of Delfshaven in Rotterdam. The shortage of space for young people is especially evident in Dutch pre-war city districts, i.e. those built before 1940. The design task was to make a child-friendly urban design for Delfshaven, Rotterdam. Children's needs for space and patterns based on a study of the literature were used as input for this design process, but new patterns were also developed from this design process, and these were also integrated into the pattern book.

## The pattern language and the pattern book

The design instrument 'pattern language' is an important development in environmental-psychological research (Alexander, Ishikawa, Silverstein, Jacobson, Fiksdahl-King & Angel, 1977). A pattern language translates non-physical-spatial starting points into physical-spatial ones. All patterns are connected through the structure of the pattern systems. This is known as the pattern field. Every pattern not only supports others, but is supported by them, simultaneously. In fact, this is how Alexander views spatial reality. He argues that objects should never be built in isolation. The world around the object to be designed must always be taken into account.

---

2 Sara van Duijn, a MSc candidate, did the actual research. The contributor, Remon Rooij, was involved in the project as one of her mentors. Machiel van Dorst from the Chair of Environmental Design, Department of Urbanism, and Maurice Harteveld from the Chair of Urban Design, Department of Urbanism, were also involved as mentors in this project.

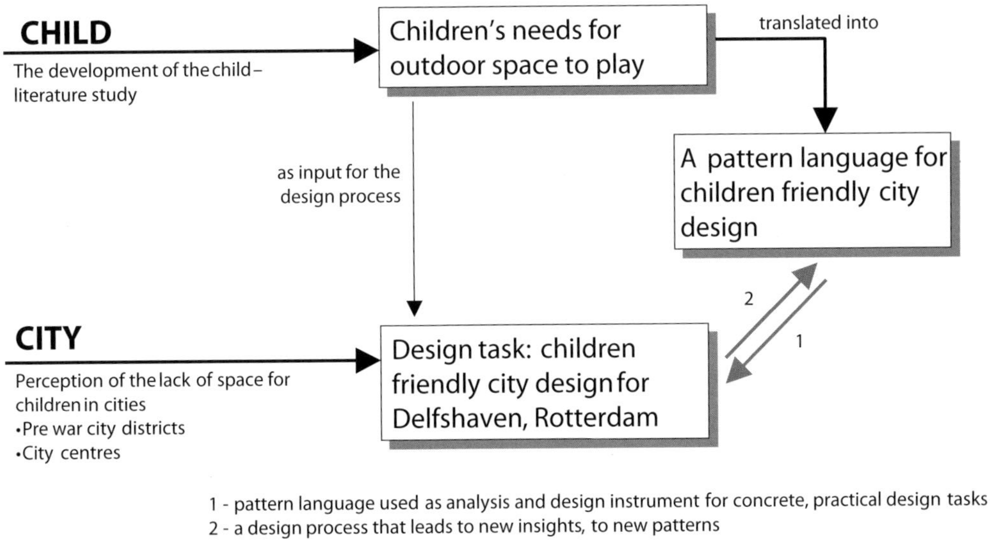


**Illustration 18.5: Research approach used in 'The City Child' project**

A pattern language is subject to changes, just as a normal language is. Changes in society cause new patterns to emerge, while existing ones are transformed or disappear. These patterns thus contribute to the spatial quality of buildings and cities, although this spatial quality is not the sum of a certain number of patterns. Furthermore, Alexander argues that we should talk about a pattern language, and not the pattern language, as there are many possible patterns, and he implicitly invites people to add patterns to the present body of knowledge.

A pattern language can also be culturally or climatologically bound. For example, Eskimos have a variety of words for the single word 'snow', in English, because snow is much more important to them. Likewise, the Dutch have many more words for drainage canals and canalised waters than the English, because draining land has been a much more important aspect of their history. Children playing in tropical countries experience different weather conditions from those in North-West Europe, so what is an optimal design principle for a children's playground in the tropics may be totally unsuitable in North-West Europe. So the generalised design principles presented in pattern books have their limitations when applied to specific projects. However, this does not mean that the patterns are useless, but it is important to bear in mind the context in which the pattern language is used.

Illustration 18.6 (in Dutch) shows the pattern field that was developed in 'The City Child' project. It is organised according to the four main factors that give children positive feelings about their living (and playing) environments (Van Duijn, 2004b:36):

- Affection for the environment;
- Behavioural opportunities;
- Knowledge about the environment;
- Presence of other people.

## An example of a child friendly pattern

In order to compare and increase the readability of all patterns, they are all structured in the same way, according to the structure that Alexander proposes.

**Title**
The title of each pattern can be seen as the shortest summary of that pattern.

**Statement**
The statement should give a (positive and catchy) spatial direction for a non-spatial starting point.

**Explanation**
Here, it should be argued why the statement is true and/or relevant to examples based on research results. If the statement is based on knowledge from experience, rather than scientific research, then that statement is known as a hypothesis.

**Solution**
Here, we can find actual design examples or guidelines. An example can both inspire and explicitly show specific qualities or characteristics.

**Image**
Images can underpin the theory, or represent the pattern visually.

**References**
Every pattern is related to other patterns. It is these links that cause a pattern field to grow.

**Literature**
If a pattern is based on literature, then we refer to those literature sources, or to some specific literature if it contains more information about the topic at hand.

Illustrations 18.7 and 18.8 show (in Dutch), the patterns for "Flowers in the neighbourhood" and "School routes that avoid conflicting confrontations". The statements that correspond with these patterns are: "Children appreciate a neighbourhood where there are many flowers in the spring and summer" (Ill. 18.7) and "Children use the route to school very frequently. During this trip, they tend to play, so if children are to be allowed to go to school independently, then it has to be a safe route" (Ill. 18.8).
The 'flower' pattern refers to several other patterns: (i) green pavements, (ii) living green structures, such as trees (iii) private gardens, (iv) tangible green, (v) public parks and gardens, (vi) a changeable environment. The 'school-route' pattern refers to the patters of (i) safe crossings, (ii) slow bicycle routes, (iii) a slow mobility street, (iv) pavement games, (v) children's networks, (vi) the alley as a playground, (vii) the spacing of small networks.

## Evaluating ' The City Child' Project

Due to the enormous numbers of (parked) cars and the high density urban environment, there is very little (outdoor) space for children in many pre-war city districts. This problem can be solved by redesigning the public space in such a way that it can function as both a traffic zone and a meeting place for

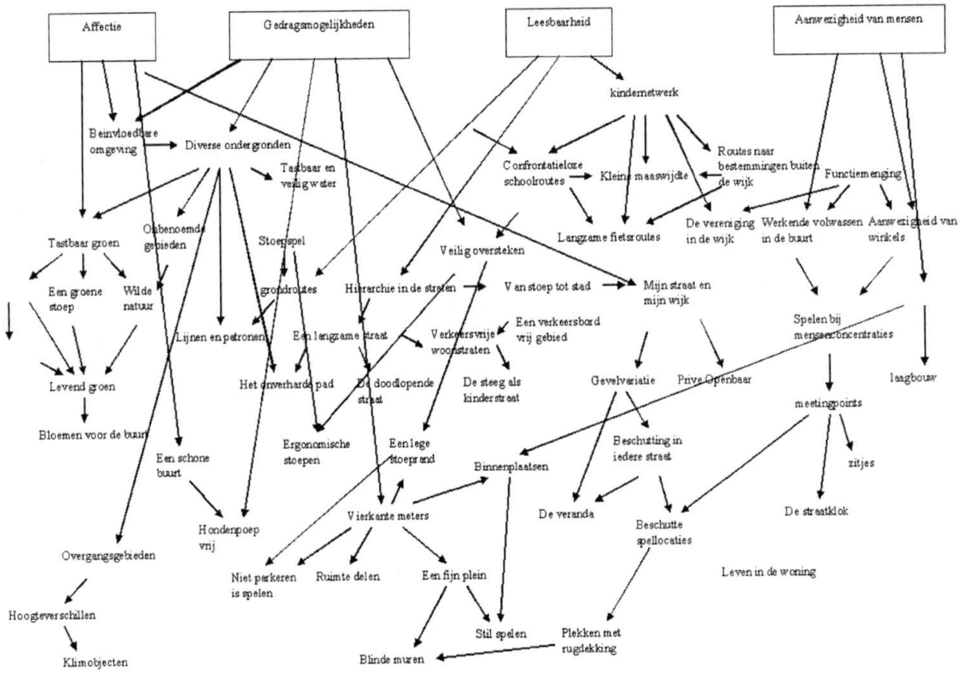


**Illustration 18.6: The pattern field of 'The City Child' project (Van Duijn, 2004a:15)**

those who live in the city, but also as a nice space where children can play. If children are given informal playing spaces within the public space, this often motivates them to play in a different way than they do in playgrounds specifically designated for children's use. Examples of informal playgrounds are: pavements, ditches, squares, streets, and bushes. If the public space is designed properly, ditches can become fishing ponds, and pavements can become cycling routes.

Sharing public space can lead to conflicts among those using that space. However, confrontation with other users is what makes public space the best playground, in the sense that children's games very often include the environment and its users. For adults, this could, however, be frustrating. They would find it even more difficult to park their cars, and driving would be even more hazardous due to playing/skating children and the danger of running over them. Such difficulties, however, are not so different from the annoyances experienced between adults. But for children, and their development, it is very important that they come into contact with other people, in a lively playground.

The patterns that contribute to a lively environment for children have been extracted out of what is known about their behaviour, development, and experiences. To a large extent, they are focused on the lowest levels of scale. What eventually emerged is that an environment that suits the needs and wishes of children, also appears to meet the conditions needed for a lively adult environment. It seems that many of the patterns created especially for children also generate general spatial conditions that create a livable environment for all.

## Towards a research agenda for small-scale patterns

Networks play, and will continue to play, a crucial role in the planning and design of 21st century cities (Rooij, 2005:246-251). A more fundamental knowledge of the principles of complex networks might help the planner and designer in their task of organising this complex interacting system, known as 'the city'. In fields other than urbanism (e.g. biology and computer science), knowledge and 'rules' have become available for explaining the composition of complex networks. Barabásiasi, for example, explains the 'rules' of distribution when dominated by the power law (see the second section of this chapter) and the growth of complex networks into so-called scale-free networks.

An important thing that we learn from the research on complex networks is the role of the lowest levels of scale. Salingaros formulates it as follows:

"... To achieve geometrical coherence in any system, a tightly-knit and complex whole is generated via general rules. Geometrical coherence is an identifiable quality that ties the city together through form, and is an essential prerequisite for the vitality of the urban fabric. The underlying idea is very simple: a city is a network of paths, which are topologically deformable. Coherent city form must also be plastic; i.e., able to follow the bending, extension, and compression of paths without tearing. In order to do this, the urban fabric must be strongly connected on the smallest scale, and loosely connected on the largest scale. Connectivity on all scales thus leads to urban coherence. " (2005:86)

Urbanists should therefore focus their future research on discovering knowledge-based design principles for these lowest levels of scale; for the level of the pedestrian, for example – a user of the urban space who was largely ignored in post-war architecture and urban design.

Analysing groups of users who intensively utilise the lowest levels of scale of the city – the residential environment and the neighbourhood – could give an accurate indication of what that small-scale environment should look like, and how it should be organised, both spatially and functionally. Examples of these user groups are: pedestrians, cyclists, pensioners, disabled people, children, intra-urban public transport and/or car travellers, etc. A lot of scientific or practical knowledge is already available about the needs, wishes, and behaviour of most user groups regarding the urban environment. However, hardly any of that knowledge is available for urban designers and architects through design tools such as a pattern language.

From 'The City Child' project, we concluded that many patterns not only indicate environments that function well for children, but also for everyone living in those urban environments. It can be hypothesised that the common denominator of all small-scale-user-group mobility patterns equates with the spatial conditions needed for a vital, sustainable city. The research aim for the future, therefore, is to identify these knowledge-based patterns of sustainability.

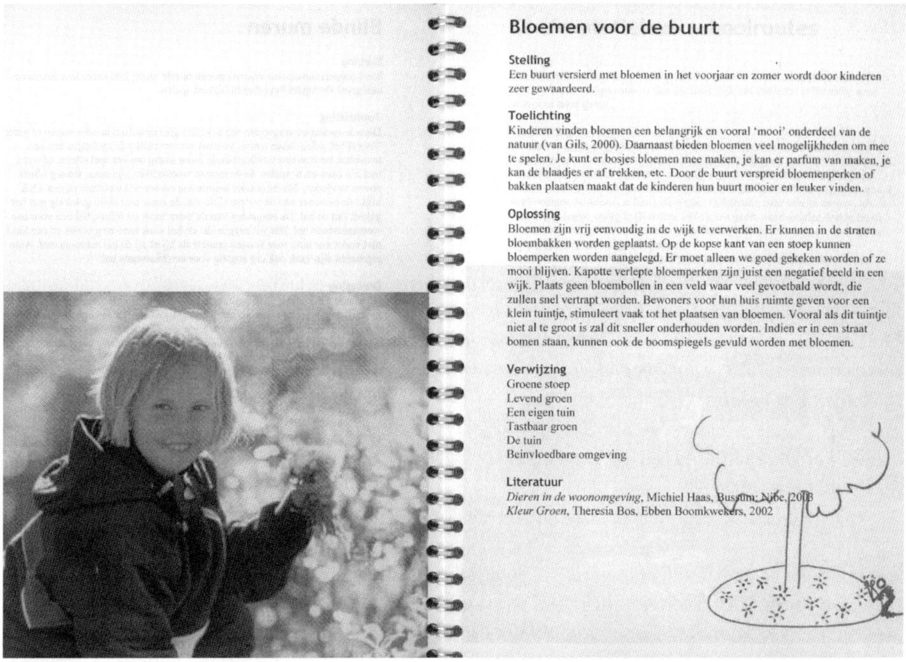

## Bloemen voor de buurt

**Stelling**
Een buurt versierd met bloemen in het voorjaar en zomer wordt door kinderen zeer gewaardeerd.

**Toelichting**
Kinderen vinden bloemen een belangrijk en vooral 'mooi' onderdeel van de natuur (van Gils, 2000). Daarnaast bieden bloemen veel mogelijkheden om mee te spelen. Je kunt er bosjes bloemen mee maken, je kan er parfum van maken, je kan de blaadjes er af trekken, etc. Door de buurt verspreid bloemenperken of bakken plaatsen maakt dat de kinderen hun buurt mooier en leuker vinden.

**Oplossing**
Bloemen zijn vrij eenvoudig in de wijk te verwerken. Er kunnen in de straten bloembakken worden geplaatst. Op de kopse kant van een stoep kunnen bloemperken worden aangelegd. Er moet alleen we goed gekeken worden of ze mooi blijven. Kapotte verlepte bloemperken zijn juist een negatief beeld in een wijk. Plaats geen bloembollen in een veld waar veel gevoetbald wordt, die zullen snel vertrapt worden. Bewoners voor hun huis ruimte geven voor een klein tuintje, stimuleert vaak tot het plaatsen van bloemen. Vooral als dit tuintje niet al te groot is zal dit sneller onderhouden worden. Indien er in een straat bomen staan, kunnen ook de boomspiegels gevuld worden met bloemen.

**Verwijzing**
Groene stoep
Levend groen
Een eigen tuin
Tastbaar groen
De tuin
Beinvloedbare omgeving

**Literatuur**
*Dieren in de woonomgeving*, Michiel Haas, Bussum: Nibe, 2003
*Kleur Groen*, Theresia Bos, Ebben Boomkwekers, 2002

**Illustration 18.7: 'Flowers in the neighbourhood': an example of a pattern (in Dutch) (Van Duijn, 2004a:30, 31)**

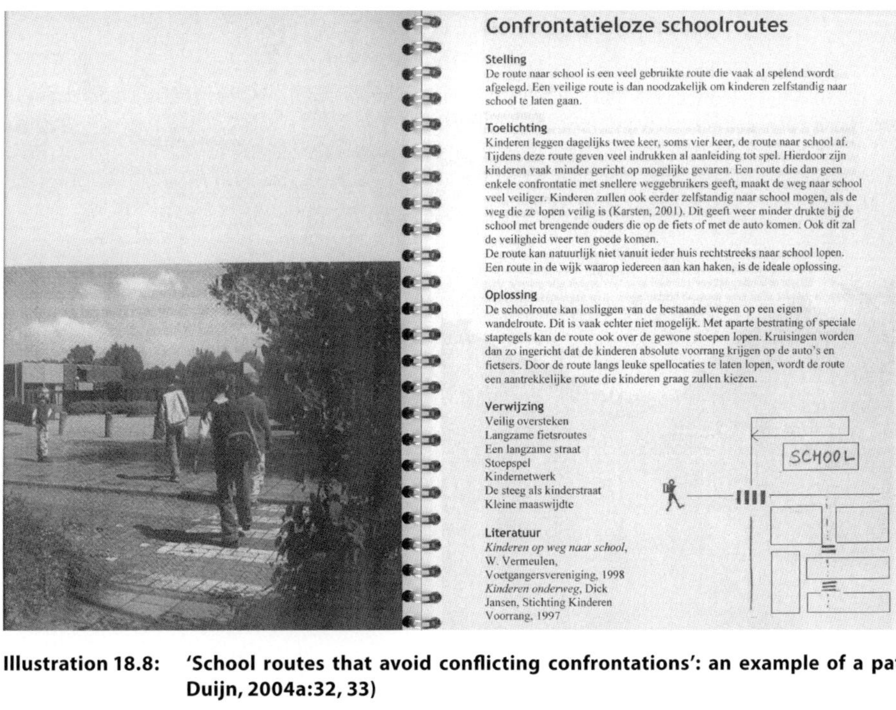

## Confrontatieloze schoolroutes

**Stelling**
De route naar school is een veel gebruikte route die vaak al spelend wordt afgelegd. Een veilige route is dan noodzakelijk om kinderen zelfstandig naar school te laten gaan.

**Toelichting**
Kinderen leggen dagelijks twee keer, soms vier keer, de route naar school af. Tijdens deze route geven veel indrukken al aanleiding tot spel. Hierdoor zijn kinderen vaak minder gericht op mogelijke gevaren. Een route die dan geen enkele confrontatie met snellere weggebruikers geeft, maakt de weg naar school veel veiliger. Kinderen zullen ook eerder zelfstandig naar school mogen, als de weg die ze lopen veilig is (Karsten, 2001). Dit geeft weer minder drukte bij de school met brengende ouders die op de fiets of met de auto komen. Ook dit zal de veiligheid weer ten goede komen.
De route kan natuurlijk niet vanuit ieder huis rechtstreeks naar school lopen. Een route in de wijk waarop iedereen aan kan haken, is de ideale oplossing.

**Oplossing**
De schoolroute kan losliggen van de bestaande wegen op een eigen wandelroute. Dit is vaak echter niet mogelijk. Met aparte bestrating of speciale staptegels kan de route ook over de gewone stoepen lopen. Kruisingen worden dan zo ingericht dat de kinderen absolute voorrang krijgen op de auto's en fietsers. Door de route langs leuke spellocaties te laten lopen, wordt de route een aantrekkelijke route die kinderen graag zullen kiezen.

**Verwijzing**
Veilig oversteken
Langzame fietsroutes
Een langzame straat
Stoepspel
Kindernetwerk
De steeg als kinderstraat
Kleine maaswijdte

**Literatuur**
*Kinderen op weg naar school*, W. Vermeulen, Voetgangersvereniging, 1998
*Kinderen onderweg*, Dick Jansen, Stichting Kinderen Voorrang, 1997



**Illustration 18.8: 'School routes that avoid conflicting confrontations': an example of a pattern (Van Duijn, 2004a:32, 33)**

## References


Alexander, C., S. Ishikawa, M. Silverstein, M. Jacobson, I. Fiksdahl-King & S. Angel, 1977, *A Pattern Language*, Oxford University Press, New York

Barabási, L., 2002, *Linked. A New Science of Networks*, Perseus Publishing, Cambridge Mass.

Bekkering, H., 1997, Betekenis en traditie, In: Bekkering *et al.* (eds), *Stedelijke transformaties*, Delft University Press, Delft, pp. 179-186

Bovy, P.H.L., 1995, Personenvervoer; In: *Planologische Kengetallen*, pp. 5520-1 – 5520-53, Kluwer Publishers

Centraal Bureau voor de Statistiek, 2005, *De mobiliteit van de Nederlandse bevolking*, http://www.statline.cbs.nl

European Council of Town Planners, 2003, *The New Charter of Athens 2003, The European Council of Town Planners' Vision for Cities in the 21st Century*, Lisbon, http://www.ceu-ectp.org/e/athens

Jacobs, J., 1968, *The Death and Life of Great American Cities*, Jonathan Cape, London

Jansen, D. & M. Schouten, 2001, *Ruimte voor de jeugd*, network document, Amsterdam: Platform for the youth

Karsten, L., E. Kuiper & H. Reubsaet, 2001, *Van de straat? De relatie jeugd en openbare ruimte verkend*, Van Gorcum, Assen

Klaasen, I.T., 2004, Knowledge-based Design: Developing Urban & Regional Design into a Science, Design/Science/ Planning Series, Delft University Press, Delft

Ministerie van Volkshuisvesting, Ruimtelijke Ordening en Milieubeheer, 2000, *Balans ruimtelijke kwaliteit 2000*, Den Haag

Projectbureau Integrale Verkeers- en Vervoerstudies, 1998, *Mobilopolis, De actieve fietsstad*

Roberts, M., T. Lloyd-Jones, B. Erickson & S. Nice, 1999, Place and Space in the Networked City: Conceptualising the Integrated Metropolis, *Journal of Urban Design*, Vol. 4, No. 1., pp. 51-66

Rooij, R.M., 2005, *The Mobile City. The Planning and Design of the Network City from a Mobility Point of View*, PhD Thesis, T2005/1, February 2005, TRAIL Thesis Series, The Netherlands

Salingaros, N.A., 2005, *Principles of Urban Structure*, Design/Science/Planning Series, Techne Press, Amsterdam

Tolley, R. (ed.), 1997, *The Greening of Urban Transport. Planning for Walking and Cycling in Western Cities*, Edition II, John Wiley & Sons Ltd., Chichester (UK)

Van Andel, J., 1985, *Woonomgeving en kinderen*, PhD Thesis, Technische Universiteit Eindhoven

Van Duijn, S., 2004a, *De woonomgeving, ook voor kinderen. Patronenboek*, Masters Thesis, Department of Urbanism, Faculty of Architecture, Technische Universiteit Delft

Van Duijn, S., 2004b, *De stedelijke woonomgeving als speelruimte voor kinderen*, Masters Thesis, Department of Urbanism, Faculty of Architecture, Technische Universiteit Delft

# 19 Integrating the Social and Spatial Aspects of the Urban System
## Comparing the Models of Heeling, Dupuy, Castells and Lefebvre

Jeroen van Schaick

## Introduction

The use of a layered approach to deal with spatial questions has become increasingly popular during the last 10 years. To carry out primary analysis using a layered approach, the different subsystems of the urban system are extracted as separate layers, which often correspond with separate public sectors, such as, infrastructure or housing. This has led to an ongoing alienation between sectors, and specifically between the social and the spatial aspects of urban issues. Nevertheless, this approach has been adopted in the *Vijfde Nota over de Ruimtelijke Ordening* [Fifth Memorandum on Spatial Planning] in which urban systems have been divided into three layers: 'substratum', 'networks' and 'land use' (Ill.19.1). A specific layered approach has also been adopted for educational and research purposes in the Department of Urbanism of the Faculty of Architecture in Delft. This last-mentioned model developed for the *De Kern van de Stedebouw* [The Core of Urbanism] project (Heeling, Meyer & Westrik, 2002), is also used as reference in this chapter for an in-depth analysis of layered models of the urban system as a whole.

In a recent critique of the 'Fifth Memorandum', Priemus (2004) – apparently following the Network City approach of Drewe (2004) – argued for a network approach. This article develops the argumentation further by a more general repositioning and critique of layered approaches. The main problem developed here is that the dynamic social subsystem and the physical-spatial subsystem cannot be meaningfully separated. This chapter shows that, as the urban system is both a social and physical system at

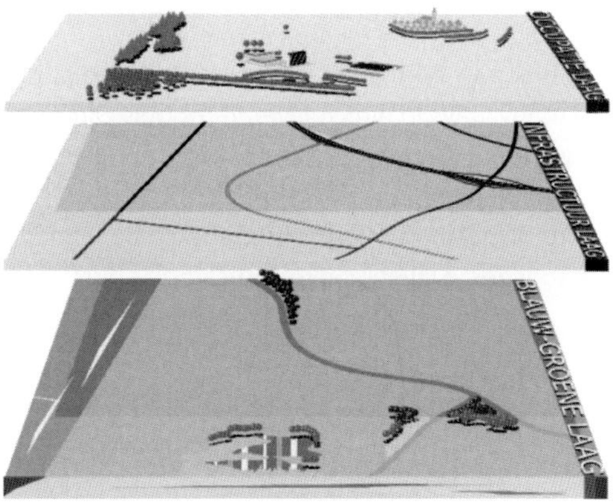

**Illustration 19.1:**
**The Layered Approach as adopted in the Fifth Memorandum on Spatial Planning, the Regional Plan for Noord-Holland South and the Structural Plan Amsterdam. From bottom to top the following layers have been identified: Blue-green layer (or substratum), Infrastructure (networks), and Land use (source: Streekplan Noord-Holland Zuid 2003)**

the same time, it is too complex to handle using a layered approach, because it is impossible to maintain a distinction between separate layers, especially those representing subsystems or sectors.

The urban-system models used in this chapter all give – more or less explicitly, but in different ways – shape to the relation between social and physical-spatial subsystems. By comparing two verbal, and two visual urban-spatial models in their original forms, a better understanding of the integrality of the urban system can be gained. In its place, a new network-oriented approach to urban-design problems is suggested, as a first step in bridging the gap between the social and physical sectors in both science and urban politics.

In the first part of this chapter, the four models are presented in their own right. The first two models, that used for the *De Kern van de Stedebouw* project (Heeling, Meyer & Westrik, 2002) and Gabriel Dupuy's 'Network City' model – are approaches to spatial planning and design. The last two models – Manuel Castells' ' Space of Flows' and Henri Lefebvre's 'Social Space' – have been developed by the author as visual representation of the original verbal models based on approaches used in social science. Each model will be analysed to establish the relation between the model and reality, and, specifically, to find out how it deals with the layered approach.

In the second part of the chapter, the four models are related to each other, each of them visualised in a new way to serve the comparison made in this chapter. Similarities and distinctions are visualised by analysing complementary and overlapping elements to gain some idea of what ingredients are needed for an integral approach.

The chapter concludes with the observation that the differences between the models cannot be traced back to the differences between social and technical sciences. The models differ according to the modelling method adopted: this is a strict reading of the layered approach, on the one hand, and an open network approach, on the other. Instead of layers, the latter adopts simultaneous, though varying, perspectives on the urban system. The network approach indeed appears to create better possibilities for integrating the social and physical-technical aspects of urban issues, though it needs to be seen in a broader light than Priemus (2004) sees it. This wider view is, however, offered by Dupuy, in his concept of the Network City, because he uses networks to integrate the social and spatial aspects of complex spatial questions.

## Four stratified approaches

This first part sketches a variety of stratified approaches. The many different sorts of models illustrate the many ways in which layered models can be used for urban questions, but, as will be shown here, they are not equally effective in facilitating an integral understanding of the urban system.

All the visual models presented should be read from bottom to top. In both Dupuy's and Heeling *et al.*'s approaches, a layer or level indicated as 'first' is the bottom layer.

## Superimposable layers: Heeling, Meyer and Westrik

In recent years, layers have become standard equipment for many an urban designer. In *De Kern van de Stedebouw* the socio-spatial model of urbanism is put forward as having five layers, of which the second from the bottom – the urban ground-plan – is viewed as being the focus of the urban design: the "mediating layer between substratum and possibilities for use" (Ill. 19.2). Heeling, Meyer and Westrik (2002:18-20) define the three middle layers together as the "physical-spatial manifestation of the urbanist work" – in brief: *urban work*. "The two layers above the urban ground plan, can be seen as elaborations of the urban ground-plan." This stratified model for practicing urban designers implies, according to Klaasen (2004), that there are divisions within an urban system, those represented by the layers shown in the illustration. In this model we recognise the Trancik's 'Figure Ground Theory' (1986, cited in Klaasen, 2004:80) which uses, as starting point, the composition of building mass and open space. How the top layer should be used is ill-defined. In the original model, in *De Kern van de Stedebouw*, this layer is devoted to programmatic, designated functional aspects of built-up areas, as if it were a land-use plan, while in the sketch of the same model, in the discussion issue of the magazine *De Blauwe Kamer*, it is the circulation aspects that are emphasised (Harsema, 2000). This social component is not considered part of the urban structure, and so it is set apart from the model (as emphasised in Klaasen, 2004:81). The relation between the different layers is also dependent upon which fields of practice are involved in the development, while these layers are restricted limited to a technical spectrum: architecture, landscape architecture, spatial planning, civil engineering (Heeling *et al.*, 2002:20). It can be concluded, therefore, that this model is more a physical-spatial model than a socio-spatial one.

In this model, which in theory can consist of any number of layers, the layers are superimposed without being directly related to each other, except for a spatial 'marker'. The different layers can be viewed as conditional strata in which – according to the layered approach – the speed of change on one layer conditions the possibilities on the other layers. Thus the substratum, as the most static layer, conditions the urban work, which in turn conditions the use to which the higher layers are put.

In itself, superimposition is not a bad way of relating the components of a single model, but the main objection to the way it is done here is that there is hardly any indication of how the

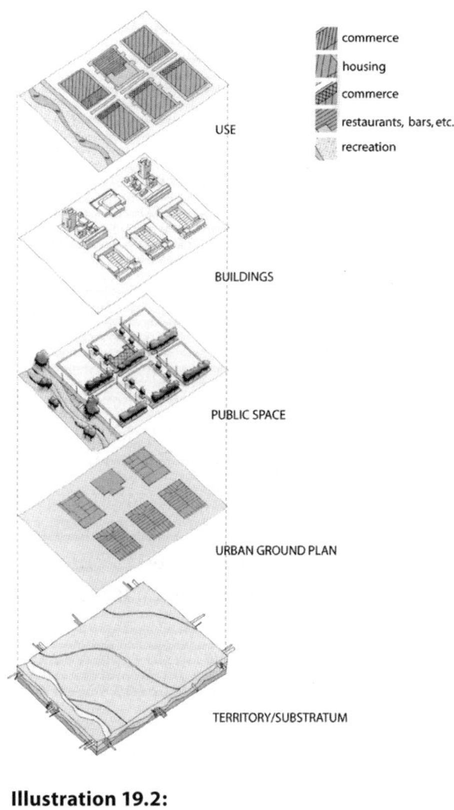


**Illustration 19.2:**
**The original model of the layer approach as suggested by Heeling *et al.* (2002:19)**

1 Scale, measured according to telescopic scale principles, is based on different sizes, measured in distance, i.e. a 'plot' is part of and included in a 'building block', is part of and included in a 'neighbourhood', is part of and included in a 'city', etc..

layers influence one another. All that is indicated is where one layer should be imposed on the other. Scale is understood in this model as a (continuous) telescopic principle.[1] However, as this is a static understanding of the concept of 'scale', the layers are also linked statically: only one scale at a time can be addressed through a set of layers. The other link between layers is by the middle layer acting as mediator between the other two. This means, however, that within the model a differentiation is created between 'passive' and 'active' layers, which, in my view, misrepresents not only those systems in which 'territory' and 'use' are significant features, but also the mutual interdependence of all the layers in the urban system. This model is limited, therefore, to a technical, physical-spatial understanding of the urban system as a whole, and thus its internal coherence is weak.

## Layered perspectives: Dupuy

The concept of the Network City, as described by Gabriel Dupuy in *L'Urbanisme de Réseaux*, represents a different approach to urban design. It falls under what Trancik calls 'Linkage Theory' (cited in Klaasen, 2004: 80). The layered character of this model must be understood as a set of levels, read from bottom to top, in which each layer focuses on the same item, but viewed from a different perspective, depending on what Dupuy calls the 'operator'. Thus Dupuy's model, the Network City, is actually a set of perspectives used to gain an understanding of the *workings* of the system as a whole. An important difference between this model, and that put forward by Heeling *et al.*, is that each layer is explicitly dependent on the other, and the different types of dependency are indicated by arrows (dotted and continuous, virtual-imaginative and real relations).

"Level One involves the suppliers of technical utility networks such as water and sewerage, energy, transport and ICT systems; Level One covers the infrastructure, the services offered and the operators. (...) On Level Two we find the functional networks of common-interest users centering on consumption, production, distribution and social contacts; specific location factors apply to each of these networks. It is on Level Three that the users of functional networks make actual, selective use of technical networks and services for their special purposes" (Drewe, 2003:29-30) (see also chapter 15 by Paul Drewe).

This socio-spatial model is thus internally more coherent than Heeling *et al.*'s model in two ways. Firstly, it is already apparent in the representation. The 'layers' – Dupuy calls them 'levels' – are visually connected by arrows (vectors) which imply more relationship than the superimposition markers used in Heeling *et al.*'s model (Ill.19.2, dotted lines connecting the corners of the layers). In *l'Urbanisme des Réseaux*, Dupuy focuses mainly on the relations between individual and different types of collectivity, as developed in specific administrative or political settings. This is complemented by Dupuy's view that these relations link the virtual/imaginative aspect and the real/technical aspect of the networks that form direct interdependences between the levels (Dupuy, 1991:108-115). Secondly the internal coherence becomes apparent in the content of the model. By taking perspectives, instead of different systems, as the layers, the model explicitly states that it deals with one system, but viewed in different ways. So each perspective, in its own way, relates different subsystems of the urban system, and thus, implicitly, each layer *does* relate the social to the physical subsystem.

At the same time, however, this model also presents some problems of internal coherence. One of these is that the levels are on different scales, another is that each level has a specific, but different, meaning for urban design and planning. This one model tries to relate spatial scales to time scales, but these are of such different orders that coherence becomes a problem. Drewe states that the 3rd level (network/

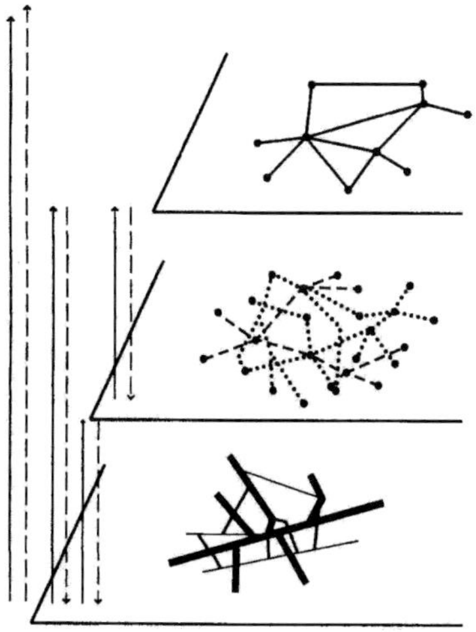

**3rd level operator**
Network / territory of the urban household

**2nd level operator**
Production network, Consumption network, Domestic network

**1st level operator**
Road network, Public transport network, Telephone network, etc.

**Illustration 19.3: The three levels of operators who (re)organise urban space (Dupuy 1991:119)**

territory of the urban household), in particular, offers a big challenge to urban design and planning in the form of incorporating space-time budgets, action spaces (Dijst & Vidaković, 1997; in Drewe, 2004:2) or logistics, logistic chains and logistic space (Hesse, 2002; in Drewe, 2004:2). Another, related, problem is that it becomes difficult, or impossible, to superimpose the layers visually, because of the differences and discontinuity of scale between the layers. The relation – explicitly made by Dupuy – between the layers tends to become meaningless if there is no knowledge in urban design and planning of how to apply these relations. Could this be because the model is flawed, or our representation, or our urban design and planning methods are flawed? So, although this model explicitly incorporates social aspects into the technological systems, in its present form, it fails to give a completely comprehensive and usable model that can be applied to the urban system as a whole. As this is one of the more appropriate models available in urban design at the present time, the question then is how can it be adapted to make it more useful for the urban designer.

Before I make some suggestions, I would first like to include two more models in this discussion, because these approach the urban system from a social/societal standpoint and therefore offer other ways of integrating social and spatial knowledge.

## Layered dominant power structures: Castells

At first sight the next model – Castells' 'Space of Flows' – appears to contradict Heeling *et al.*'s model. I say this mainly because in his 'Space of Flows' Castells rejects the idea of the city as an entity or a spatial power (and thus the 'urban ground-plan' as a mediator) and puts in its place a socio-spatial model based on the almost fluid organisation of powerful socio-spatial entities. However, the Space of Flows model is ambiguous about what it models: about how social aspects, and their power structure, dominate spatial organisation and/or the specific socio-spatial structure ('manifestation') of the model itself.

Castells distinguishes three layers (Ill. 19.4). In my view, these represent the strategic elements of the entire socio-spatial system which together comprise the main global power forces. In his own words "a new spatial logic, (...), this new spatial process, (...) the space of flows (...) is becoming the dominant spatial manifestation of power and function in our societies" (Castells, 1996:409). The space of flows is defined as "the material organisation of time-sharing social practices that work through flows" (Castells, 1996:442).

From bottom to top the Space of flows consists of:
"The first layer, the first material support of the space of flows, is actually constituted by a circuit of electronic exchanges (...) that, together, form the material basis for the processes we have observed being strategically crucial in the network of society. The second layer of the space of flows is constituted by its nodes and hubs.[2] The space of flows is not placeless, although its structural logic is. It is based on the electronic network, but this network links up specific places, with well-defined social, cultural, physical and functional characteristics. The third important layer of the space of flows refers to the spatial organization of the dominant managerial elites (...) that exercise the directional functions around which such space is articulated." (Castells, 1996:442 ff.)

Rather than representing perspectives in levels, as Dupuy does, Castells uses the term 'layers', but his divisions do not represent autonomous subsystems, as in Heeling *et al.*'s model. This socio-spatial model deals with three related types of structure, constituted by flows which are "the expression of processes *dominating* our economic, political, and symbolic life. If such is the case (flows being dominating), the material support of the dominant processes in our societies will be the ensemble of elements supporting such flows, and making their articulation materially possible in simultaneous time" (Castells, 1996: 442). The model is thus a representation of a part, albeit a dominant part, of the urban system.

Castells has never visualised his model, so the relationship between the layers can only be extracted from his texts. The internal coherence is organised asymmetrically, as Castells assumes that societies are "asymmetrically organised around the dominant interests specific to each social structure" (Castells, 1996:443). My interpretation is that Castells' model is highly hierarchical – in terms of dominance and conditions – but also quite complex – in terms of space and structure – as are the relations within the model. This complexity in the degree of coherence leads to considerable questions about how to handle conditional aspects and the nature of power structures in urban questions. How can we know what questions to ask about the relation between soci(et)al and physical-spatial issues? Maybe another model can provide us with ideas about this.

## A stratified approach to social space: Lefebvre

The last model presented here was drawn up by Henri Lefebvre, again not as a visual image, but in text. The visualisation is my interpretation of the model he describes in *The Production of Space* (1974/1991) (Ill.19.5).

2 The reader should be aware of the specific definition Castells uses for the terms nodes and hubs. Where in transportation and mobility these terms are exclusively used as transfer points in infrastructure networks, a kind of 'non-places', Castells refers to hubs and nodes in a broader sense as "specific places, with well-defined social, cultural, physical, and functional characteristics. Some places are exchangers, communication hubs (...). Other places are the nodes of the network; that is the location of strategically important functions (...)." (Castells 1996:442)

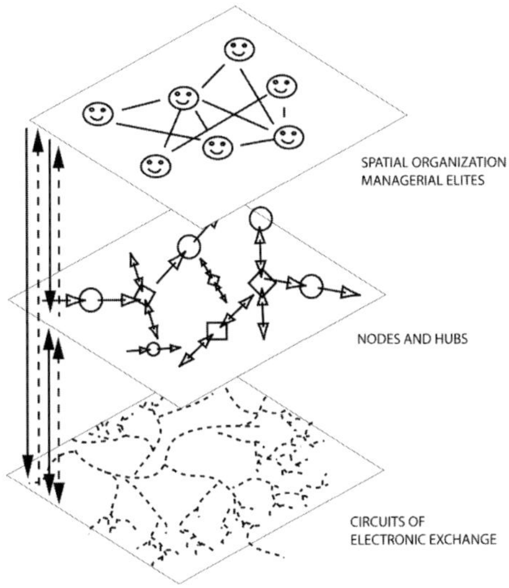


**Illustration 19.4:**
**Visual interpretation by the author of the three layers comprising Space of Flows. (Castells 1996:442)**

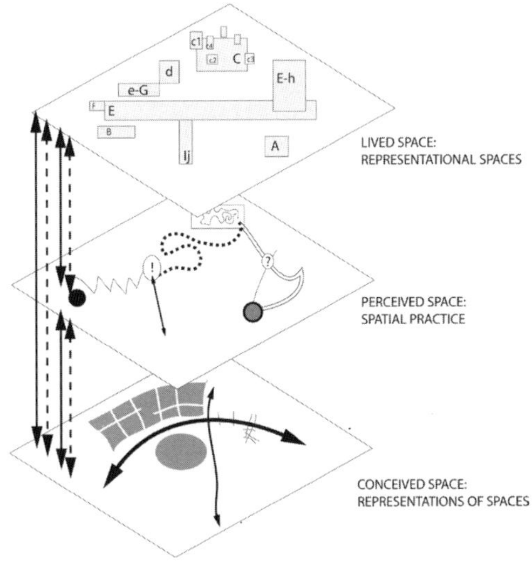


**Illustration 19.5:**
**Visual interpretation by the author of the three layers comprising Social Space. (Lefebvre 1974:38-40)**

Lefebvre approaches the problem of the socio-spatial model by introducing a new term: 'Social Space'.
"(Social) space is not a thing amongst other things, nor a product among other products: rather, it subsumes things produced, and encompasses their interrelationships in their co-existence and simultaneity – their (relative) order and/or (relative) disorder. It is the outcome of a sequence and set of operations, and thus cannot be reduced to the rank of a simple object. (….) Itself the outcome of past actions, social space is what permits fresh actions to occur, while suggesting others and prohibiting yet others" (Lefebvre, 1991:73).

Lefebvre defines social space by using an interrelated set of three spaces: conceived space, i.e., representations of space; perceived space, i.e., spatial practice (the relation between empirically measurable daily and urban reality); and lived space, i.e. representational spaces (directly lived through images and symbols) (Lefebvre, 1991,1974:38-40).

This socio-spatial model should be seen in relation to Lefebvre's understanding of space as consisting of a set of three interacting and interwoven levels of space: the public (or general) level, the private and the 'mediating and intermediary' level (Lefebvre, 1991:155-158). Furthermore, in his thinking about the idea of (spatial) scale, Lefebvre makes a distinction between simply defined micro, medium and macro levels (Lefebvre, 1991:388). This whole set of ideas about space counteracts the 'abuse of Reductionism', a situation in which "reduced models are constructed – models of society, of the city, of institutions, of the family, and so forth – and things are left at that" (Lefebvre, 1991:106).

Because Lefebvre positions his socio-spatial model as dealing with *interrelationships* and relativity, one expects to find, and will find, a strong internal coherence in the model. The problems of scale and superimposition found in the Network City model pose a possible threat here too. More important is the realisation that the layers within the model are equal, and, because of that, can be interpreted as a set of layers in which the relationship between whole and part can be understood, not by putting them in different layers, but by using the layers to illustrate the different ways of how we, as humans, attach meaning to this relationship (e.g., Lefebvre, 1991:366). You can say, therefore, that this model represents our body of knowledge about space. Moreover, it represents space as a process: as production.

Understood in this way the internal coherence of this model is very strong. However, in the specific context of urban and regional design and planning, it is not without problems. The philosophical and abstract nature of modelling social space makes it difficult to relate it directly to the notions of space normally used by the urban designer and planner. This is an important reason why it is difficult to position this model in urban practice. This problem is dealt with in the next part of this essay by comparing and contrasting all four presented models.

## Focus on networks

The models will be compared in three different ways. Firstly, the scene is set by analysing the representation of the models and how these relate to the ideas behind each of them. Secondly, the amount of overlap between the models is analysed by looking at parallels among specific parts of the models. Lastly, two sets of thoughts are extracted from the comparison of the models: themes that can be used by practitioners and thoughts on which to base integral urban-system models.

### Representation compared

Representing concepts in visualisations is an important part of designing. The difference in the degree of abstraction, and thus in the difficulty of representing these concepts is apparent when comparing the spatially-oriented models (of Dupuy and Heeling *et al.*) with the socially-oriented ones (of Castells and Lefebvre). This is strongly linked to how directly applicable they are in urban and regional design. However, Castells and Lefebvre do not ignore the issue of urban and regional design. They use a wider interpretation of the person of the designer. For Castells traditional roles in an urban design process such as client, designer, or political decision-maker are collapsed into one: the dominant actors in the space of flows located on his 'third layer' – the managerial elites. For Lefebvre, social space, as lived space, implies that every person is a designer of his or her own space. For applying socially-oriented models, however, the level of abstraction remains a major problem.

If these four models are viewed in one frame (Ill. 19.6), it is clear that they are all trying to integrate social and spatial aspects, although, as shown at the beginning of this chapter, the degree of integration differs. Dupuy and Castells' models show that social and spatial structures cannot be treated as two different, and fully separate, systems; not as conditional strata in the way that Heeling *et al.* perceive and separate them. Thus, to represent the interrelated co-existence of social and physical subsystems, network-thinking in general might offer a more coherent framework.

The concept of Social Space supports this on an abstract level: "(Social) space (...) subsumes things produced, and encompasses their interrelationships in their co-existence and simultaneity" (Lefebvre,

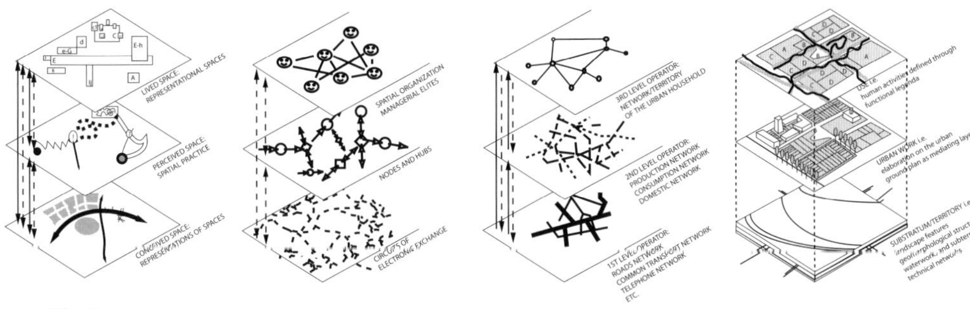


Social space
(Lefebvre)

Space of Flows
(Castells)

Network city
(Dupuy)

Urban design
(Heeling, Meyer and
Westrik)

**Illustration 19.6: Setting the scene: four stratified models in one frame made comparable as stratified models in three layers: Social Space (Lefebvre), Space of flows (Castells), Network City (Dupuy) and the layers approach by Heeling derived from the urban ground plan as mediating layer between substratum/territory and use/land use**

1991:73). A danger lurks behind the temptation to achieve clarity, although "at the outset (…) every scientific undertaking must proceed reductively. One of the misfortunes of the specialist is that he makes this methodological moment into a permanent niche for himself where he can curl up happily in the warm" (Lefebvre, 1991: 107). Using layers to represent a complex model is problematic. Layering has proved treacherous in the sense that it produces an artificial, inadequate demarcation of (design) research subjects. It is also static in character, and it is both risky and difficult to relate the layers to each other.

## Comparing angles

The models do not follow any specific line of demarcation with respect to scale. Where Heeling *et al.*, with their spatially oriented model, depict one scale level at a time, focusing on the neighbourhood and city scales, Dupuy is concerned with different scale levels in the same model. Castells, on the other hand, is specifically concerned with macro scales, which he links to lower scales by means of the concepts 'hubs' and 'nodes' (places). Lefebvre also uses a scale system for micro, medium, and macro social space, on three 'interacting and interwoven levels of space': the public (or general), the private and the 'mediating and intermediary' level (Lefebvre, 1991:106), thus creating a complex, though non-visualised, depiction of how different scales co-exist in Social Space.

Both Dupuy and Lefebvre favour the word 'level' rather than 'layer', but whereas these levels could also be regarded as perspectives in Dupuy's model, Castells uses very different levels of abstraction for his 'layers', which leads one to assume that these are not layers at all, but are again different perspectives on the system he is trying to model.

The models are represented here through the medium of the image, as is often the case in urban design and planning. This limits our capabilities, and therefore often our analysis and design, to a two-dimensional model, and although this can be extended to a three-dimensional model, this still omits an even more – or at least equally – important dimension of the urban system presented here: the dimension of time (as proposed by Klaasen, 2002) (see also chapter 14 by Ina Klaasen).

How they deal with time is an important feature that brings to light the differences and similarities among the four models. Klaasen makes a distinction between urban-design and planning approaches oriented towards patterns and those oriented towards processes. Temporal aspects hardly feature in the former, or are reduced to historical aspects and the linear transformation process of spatial patterns. In the latter, a wider understanding of temporal aspects is used in that a variety of temporal 'grains' (a notion explained in chapter 14, p. 184), including space and time-use issues, are addressed by combining cyclical and linear processes. Heeling *et al.*'s approach is an example of the former, and the Network City of the latter (Klaasen, 2004).

## Parallel thoughts

In addition to comparing the ideas behind the models, we can also distil a set of comparable elements out of a visual comparison of the models. We can immediately distinguish building blocks, or ingredients, that can be regarded as equivalent to each other in a certain way. The letters refer to those in Illustration 19.7.

A Castells' model is primarily one of power structure, showing how dependent aspects of networks are influenced. At the same time, this model puts forward the idea that flows are the dominant features in the spatial organisation/production of networks.

B The Space of flows model, however, is ambiguous about what it models. It also models the specific structure ('manifestation') of the space of flows itself, and this can be directly related to the Network City model. In Dupuy's opinion, Castells' electronic network is just one of the relevant physical networks, while Castells' middle layer (nodes and hubs) is a generalised form of Dupuy's middle level. The 'spatial organisation of managerial elites' in the Space of flows is just one specific form of a second-level network in the Network City model – albeit a dominant one according to Castells.

C Dupuy's idea of the Network City – or of networks in general – as a way of thinking about models is a specific way of 'representing space'. For Lefebvre, it is one of the concepts that lead to the production of Social Space.

D Underground networks in Heeling *et al.*'s territory layer are also accounted for in Dupuy's model by the technical-network level.

E Urban construction and, more specifically, the urban ground-plan is a form of representation belonging to conceived space.

F The definition of 'use' in Heeling *et al.*'s model is extremely general and seems to include both individual as well as collective use. These ideas are also represented in both Dupuy's and Lefebvre's models. However, as far as the model is concerned, this poses more questions than it answers.

This set of parallel thoughts shows that network thinking, and specifically the idea of the Network City, appears to integrate most of the fragments of social and physical-spatial aspects that can be found in other models.

## Integrating themes

In addition to the links between the models summarised above, there are a number of themes that are common to all four models. This means that, together, the models show a very interesting set of building blocks for the urban system (Ill. 19.8). Two or more of the models have four themes in common, and three of them offer an extra element that cannot be found in the other models. I regard the four shared

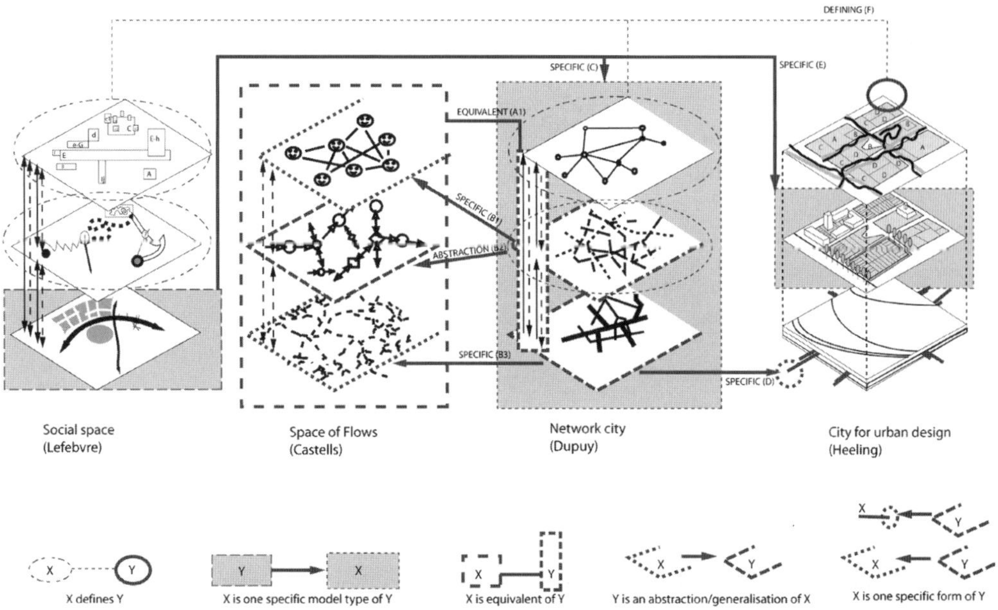


**Illustration 19.7:** **Comparing the four models for parallel thoughts. Dupuy' s model of the Network City is clearly prominent in integrating different issues from different models. The capital letters refer to the list in the paragraph "Parallel thoughts"**

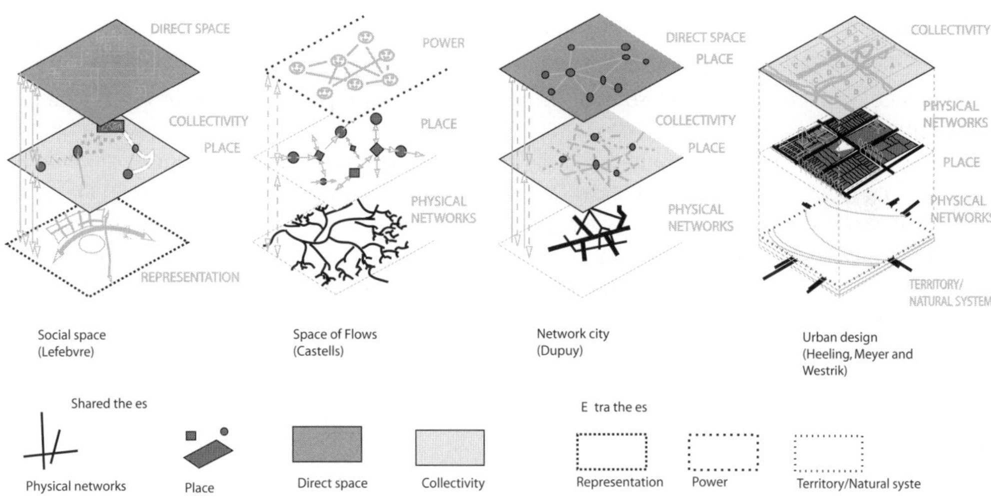


**Illustration 19.8:** **A comparison of the four models extracting seven lines of thought that can be regarded as a central set of building blocks for the socio-spatial system**

elements as the core of the model for the urban designer, but meanwhile, by viewing these models in one frame, a number of crucial problem areas come to light.

1. *Physical networks*: one of the most interesting problem areas of this type of network is developed in *Splintering Urbanism* (Graham & Marvin, 2001).
2. *Places (nodes and hubs)*: since Augé's provocative *Non-place* (Augé, 1995) the issue of the meaning of 'place' has been set firmly on the urban-design agenda.
3. *Collectivity*: in *Cities – reimagining the urban* (Amin & Thrift, 2002), a shift in types of collectivity has been sketched; a challenge for urban design and planning.
4. *Lived space*: this is a difficult issue that mainly gains attention in philosophical circles. Edward Soja investigated it further in *Third Space* (Soja, 2000), especially with regard to the position of urban design. This focuses attention on new, especially small-grained, temporal dimensions in the field of urban design and planning (see also Drewe, 2004).

### New issues raised

The different models above not only share a number of general principles but they also raise crucial issues in the discussion about how to integrate social and physical-spatial principles:.

i. The Space of Flows model introduces the issue of 'powers' or 'forces' that organise space and time.
ii. The concept of Social Space has introduced the importance of being conscious of how we represent ideas. All the shared ideas can be incorporated by using the network concept (see above).
iii. For urban and regional design, it is important to have a deep understanding of the territorial/natural system, especially in positioning the issue of sustainability.

The above comparison brings a range of issues to the fore that comprises some of the most challenging tasks for the urban designer and planner at the present time. The comparison also reveals that the Network City model is able to integrate these issues to a remarkable degree.

## Conclusions

In the introduction, the assumption was made that the layered approach, as it is widely used in the Netherlands at the present time – and especially in the way it is incorporated into courses given at the Faculty of Architecture in Delft – cannot deal sufficiently with the complex relation between the social and the physical subsystems of the city. In comparing the layered approach with other urban-system models, a network approach appears to offer better possibilities for integrating social and physical issues in complex urban systems. However, this chapter has shown the problem to be wider than this. The Network City model needs to be supplemented with other, related, concepts, such as those of 'lived space', 'place' and 'collectivity'.

As for the representation of models such as those studied in this chapter, two conclusions can de drawn. Firstly, explicit relations should be developed between levels or layers of urban systems. To be able to deal with the complexity of urban issues, urban designers and planners must acknowledge the importance of varied, co-existing perspectives on the same topic. Secondly, a deeper understanding should be developed of the dynamics of the urban system, in particular, with regard to representation and language, scale, complexity and (dis)continuity, relations and new perspectives (instead of sectored thinking), choice and probability.

As for the content of urban models in spatial planning and urban design, it can be concluded that the sets of related issues raised by both Dupuy and Lefebvre form the core around which the social subsystem and the related physical subsystem can be modelled. A new research agenda for urban design and spatial planning should focus on network (or relational) thinking, on the interrelation between the socio-cultural and the physical-technical systems and on how to integrate both in urban and regional designs and plans.

## References


Amin, A. & N. Thrift, 2002, *Cities – Reimagining the Urban*, Blackwell Publishers, Cambridge

Augé, M., 1995, *Non-places: Introduction to an Anthropology of Supermodernity*, Verso Books, London

Castells, M., 1996, *The Rise of the Network Society*, 2nd edition 2000, Blackwell Publishers, Oxford

Dijst, M. & V. Vidaković, 1997, Individual Action Space in the City; In: Ettema, D.F. & H.J.P. Timmermans (eds), *Activity-based Approaches to Activity Analysis*, Pergamon Press, Oxford, pp. 73-88

Drewe, P., 2003, *ICT and Urban Form – Old Dogma, New Tricks*, University of Technology Delft, Delft (available at www.networkcity.bk.tudelft.nl)

Drewe, P., 2004, *What about Time in Urban Planning and Design in the ICT Age?*, University of Technology Delft, Delft (available at www.networkcity.bk.tudelft.nl)

Dupuy, G., 1991, *L'Urbanisme des Reseaux – Théories et Méthodes*, Armand Colin Éditeurs, Paris

Graham, S. & S. Marvin, 2001, *Splintering Urbanism: Networked Infrastructures, Technological Mobilities and the Urban Condition*, Routledge, London

Harsema, H., 2000, Tussen functionele ordening en formeel ontwerp; In: Maarten Ettema *et al.* (eds), *De Kern van de Stedebouw* in het perspectief van de 21ste eeuw, *Blauwe Kamer*, Katern 5

Heeling, J., V.J. Meyer & J. Westrik, 2002, *De Kern van de Stedebouw in het perspectief van de 21ste eeuw, Deel 1, Het ontwerp van de stadsplattegrond*, SUN, Amsterdam

Hesse, M., 2002, Zeitkoordination im Rahmen der modernen Logistik – mehr als nur ein Impulsgeber für die räumliche Entwicklung; In: Henckel, D. & M. Eberling (eds) *Raumzeitpolitik*, Leske + Budrich, Opladen, pp. 107-126

Klaasen, I.T., 2002, Modelling Reality; In: De Jong, T.M. & D.J.M. van der Voordt, *Ways to Study and Research Urban, Architectural and Technical Design*, Delft University Press, Delft, pp. 181-188

Klaasen, I.T., 2004, *Knowledge-based Design: Developing Urban & Regional Design into a Science*, Delft University Press, Delft

Lefebvre, H., 1974, translation 1991, *The Production of Space*, Blackwell Publishing, Oxford

Priemus, H., 2004, From a Layers Approach towards a Network Approach: A Dutch Contribution to Spatial Planning Methodology, *Planning, Practice & Research*, 19/3, pp. 267-283

Soja, E.W., 2000, *Third Space: Journeys to Los Angeles and Other Real-and-imagined Places*, Blackwell Publishers, Oxford

# 20 Towards a New Urban Philosophy
## The Case of Athens [1]

Nikos A. Salingaros

## Introduction

This chapter presents some ideas on how to fix the disasters in European urban planning and design – how to repair Europe's damaged urban fabric. Governments have made tremendous efforts to implement solutions to problems that were obvious to everyone. Unfortunately, these solutions only exacerbate the situation, for reasons I will discuss. Urbanist ideas have been applied since the 1930s that contribute to the deplorable state of urban life in many European cities. Time and again, politicians are seduced into constructing showcase projects that boast an alien, 'contemporary' look. I also address the link between bad urban planning and ecological disaster. Finally, I focus on the case of Athens. No European country suffered more from misapplied architecture and urbanism in the postwar period than Greece did.

## The New Charter of Athens

I am extremely proud to be a contributor to the *New Charter of Athens, 2003* (European Council of Town Planners, 2003), which is shamefully unknown to most government planners in Europe. They continue to work on the basis of the discredited 1933 Charter of Athens written by Le Corbusier. The European Council of Town Planners decided in 1995 that the effects of applying the 1933 Athens charter were so disastrous for European cities that it had to prepare a new one. A draft was approved in Athens in 1998, and, after more revisions, the new Athens charter of 2003 was presented in Lisbon (not in Athens, because the Greek government then in power did not give its full support – and this at a time when it was funding 'fashionable' architectural and urban projects).

The new Athens charter presents an enlightened urbanism for the new millennium: one that accommodates people's needs and social forces; understands connective networks; promotes the principle of mixed use; respects irreplaceable elements of the past; and tries to integrate the built and natural elements of the environment. This vision considers spatial urban form as complementary to urban connections and movement, and gives priority to understanding their interdependence. It also emphasises monitoring dynamic changes in a living city so as to catch potential problems before they become entrenched.

The erroneous and untested ideas presented in the 1933 charter were primarily responsible for ruining cities around the world (Salingaros, 2005). The 1933 charter's main purpose was to erase pedestrian

1 Earlier version published online as "City of Chaos" in Greekworks.com (May & June 2004).

urban life as defined on vibrant city streets in prewar European capitals. Its ideas are an expression of megalomania and disdain for the individual. Everyone knows the seductive images of skyscrapers sitting in vast parks that come from the 1933 charter (along with the strict segregation of uses). I should mention that several generations of urbanists have been taught the principles of the 1933 charter as religious dogma, which is the reason they continue to apply them. At this moment, the Far East is fast destroying itself by following that *poleoctonic* (urbanicidal) model.

What we are dealing with here is a universal notion of isolation that extends over all scales. Anti-urbanist interventions cut human connections. High-tech architectural fantasies cut people off physically and emotionally from surfaces, and from the built environment in general. We cannot solve the present crisis until we acknowledge that the architecture and urbanism of the twentieth century had as its principal goal the isolation of people, from buildings and from each other. That admission necessitates the even more difficult acknowledgment that the idols of modernism were false gods, and that several generations of planners and politicians were deceived into destroying our cities by applying inappropriate urban principles.

## Practical suggestions for solutions

We now face an urban Europe partially destroyed, perhaps more so than after the Second World War, because its population is so much greater today. Then, we were confronted with an expanse of ruins; today we look upon inhuman city cores surrounded by a suburban cancer eating into the countryside. The cities (and countryside) require a radical reorientation if they are to survive in urban terms. I do not advocate radical top-down intrusions, since these work only in select circumstances. The best way to save European urbanism is to promote a correct urban philosophy, and to help people save their own cities with the government's encouragement and backing. This solution is independent of political orientation; I see no obstacle to its being embraced by all political parties. We cannot move forward unless we recognise, and get out from under, the ideology responsible for the destruction of urban landscapes.

What I offer is merely an outline rather than a full analysis. It needs to be filled in with considerably more detail and specific examples, which I leave to others. It is incredible that, for the most part, many European urbanists either do not know or choose to ignore the works of Christopher Alexander (Alexander, Ishikawa, Silverstein, Jacobson, Fiksdahl-King & Angel, 1977; Alexander, 2004) and Léon Krier (Krier, 1998), today's leading urban theorists. Let me outline some elements of this new approach: the following list can be used as a rough guide for developing more specific urban rules that better adapt to context.

### 28 elements of a new urban philosophy

1. Urban components should follow the universal distribution of sizes: many small buildings, structures, streets, sidewalks, and parks; a medium number of intermediate size; and only a few of large size (Ill. 20.1).

**Illustration 21.1:** **The correct distribution of urban elements. Very many small components; many of intermediate size; and only a few of large size**

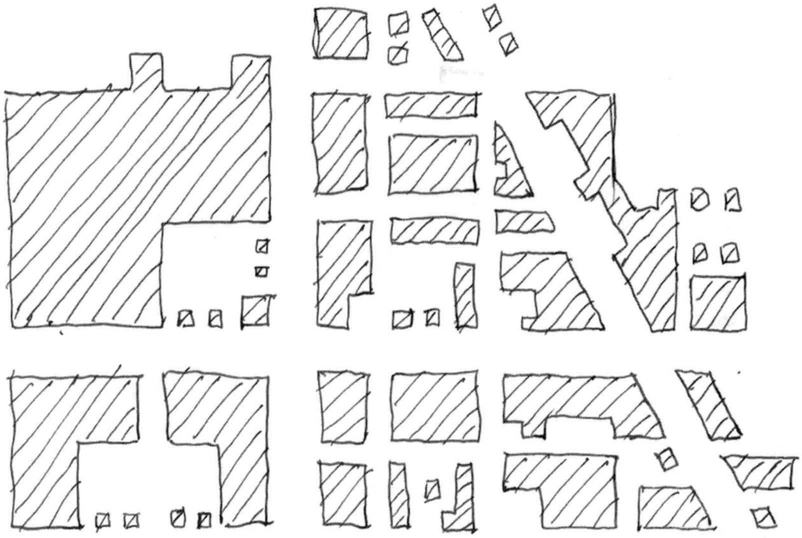

**Illustration 21.2:** **Urban components result from the combination and superposition of elements of different sizes. Here buildings are shown, yet the rule also applies to connective elements**

2. Since the smallest urban components commensurate to the size of a human being are the most vulnerable, they must be rigorously protected from encroachment by the larger urban elements (Ill. 20.2).

3. The majority of buildings ought to be mixed-use, combining different functions. This could be implemented by legislation or promoted by tax subsidies.

4. A 'neighborhood' is defined by its geography as a compact region where each point is no more than a 15-minute walk from any other point. Major impediments to pedestrians, such as highways, giant parking lots, or impassable barriers, have to be situated on the periphery (or otherwise raised or buried).

5. Zoning regulations should encourage every neighborhood to be mixed-use. I am now talking of an area with buildings of different uses, in rather close proximity (distinct from, and in addition to, mixed use in a single building).

6. City areas that are vacant at night will be populated during that time by marginal elements of the population and by the underclass. This is a natural phenomenon, in which an urban void is filled by available people.

7. Urban life occurs on the surface (sidewalk) level. This domain contains pedestrian activities, and has to be protected from stronger urban elements. It is also where links to other forms of transport must originate.

8. Building fronts must act as connecting interfaces between private and public space, not as barriers. The more permeable the interface, the more intense the street life it can support (Ill. 20.3).

9. Walls that are not perforated should instead be folded like a curtain, to provide a greater surface area for pedestrian nodes and interactions. Smooth, flat walls are essentially anti-urban (see Ill. 20.3).

10. Built elements provide the boundaries of urban space. The goal is to define a semi-enclosed outdoor space by arranging the buildings, and avoid buildings that stand apart. Vast, open spaces are not urban spaces (see Ill. 20.2).

11. If two distinct, vertically separated levels of pedestrian activity exist, either one will kill off the other, or both will be weakened.

12. When competing urban functions must be separated vertically because of density or danger, the pedestrian function has to occupy the ground level.

13. There is no sense in having strictly pedestrian areas larger than about 50 meters. It is essential, however, to protect primarily pedestrian areas from adjoining traffic by using physical structures such as high sidewalks, low walls, and bollards (Ill. 20.4).

14. A city, like the human body, works through network flow. Connect points within every neighborhood by alternative means of transport: pedestrian, private car, taxi, tram (if existent), and local buses (privately run jitneys or minibuses as well as public buses). Transport has to integrate into

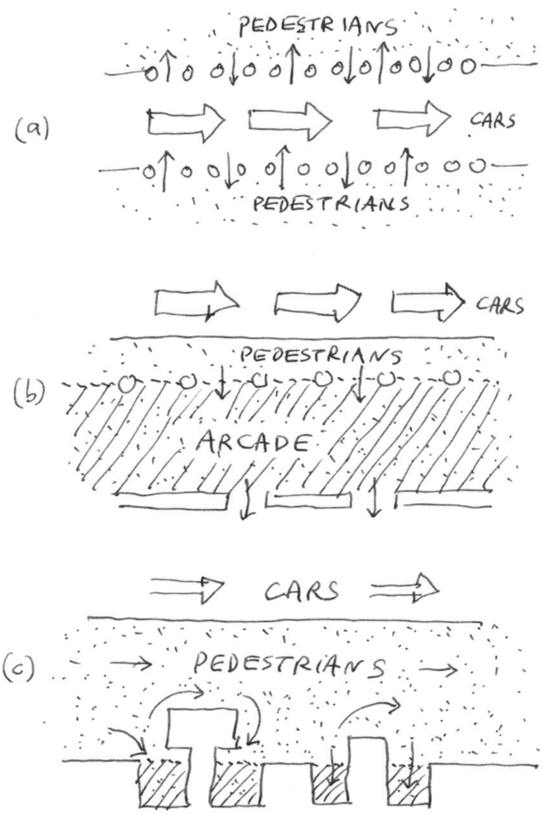


**Illustration 21.3:** **Permeability and folding of urban interfaces. (a) A pedestrian region has boundaries permeable to pedestrians but impermeable to vehicles. (b) An arcade provides a permeable transition between open public space, enclosed public space, and enclosed private space. (c) A folded or convoluted urban surface helps to generate street life by defining virtual nodes**

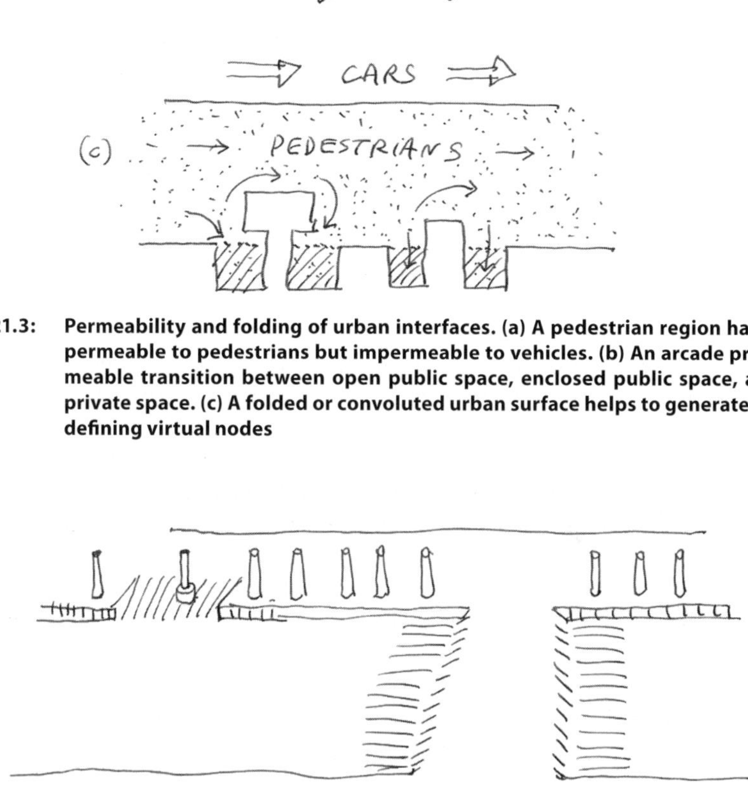

**Illustration 21.4:** **Schematic street showing several solutions. Bollards protect pedestrians psychologically from medium to heavy automotive traffic; ramp with removable bollard provides vehicular access to pedestrian realm; raised pedestrian crossing is a continuation of the pedestrian fabric**

a linked set of networks, each working on a distinct scale and speed and requiring different infrastructures (Ill. 20.5).

15. The city consists of interconnected modes of transport, made possible by permeable interfaces that allow one type of traffic to flow across while blocking another type (see Ill. 20.3).

16. Physically incompatible forms of transport, such as highways, the subway, and trains, should be located on a neighborhood's periphery or be vertically separated from pedestrians, trams, and small local buses – which is necessarily expensive.

17. However, it is infinitely more expensive (because it destroys society and culture) to sacrifice the ground-floor pedestrian urban realm to automobile circulation, parking, and other transport. Cars and trucks, if allowed, will occupy every available surface space. Pedestrians must be PHYSICALLY and PSYCHOLOGICALLY protected while closely interfacing with moving and parked vehicles (see Ill. 20.4).

18. Neighborhoods within the metropolis have to repave local roads so as to REDUCE traffic speed, thus making it possible to extend human life onto the street. Excellent solutions have been given by the Dutch in their *woonerven*, which are vehicular streets accommodating both pedestrians and cars.

19. Where transportation paths cross, the weaker link must be protected against the stronger. This necessitates defining pedestrian paths across a street, giving a visual cue and also physically slowing down vehicular traffic (see Ill. 20.4).

20. Primarily pedestrian areas (such as sidewalks lined with stores and apartments) have to be fed by transport such as cars and buses; otherwise, they will die off. That requires slowing traffic and making sufficient parking available nearby. The pedestrian urban element must be accessible to all transport networks (Ill. 20.6).

21. Parking in the dense urban core can only be accommodated by underground garages or vertical stacking, so that it doesn't encroach onto the ground-floor pedestrian realm. Multilevel parking garages ought to devote their ground floors to commercial use.

22. Neighborhoods need to be connected to each other by multiple transport, including cars, long-range buses, trams, subways, and trains. While the priority here is on non-pedestrian connections because of the larger scale, there must be at least one protected pedestrian connection between any two neighborhoods (Ill. 20.7).

23. The government has to invest in creating crossover points between different transport types to make all these competing transport possible, and to ensure their seamless interconnection (Ill. 20.8).

24. The city naturally divides into the car web surrounding and feeding pedestrian sidewalks and squares. The enclosed areas give priority to pedestrians, while being crisscrossed by cars constrained to specific paths. Cars are intentionally slowed down within a primarily pedestrian area, but are not excluded. Occasional access to all points in pedestrian areas for delivery and emer-

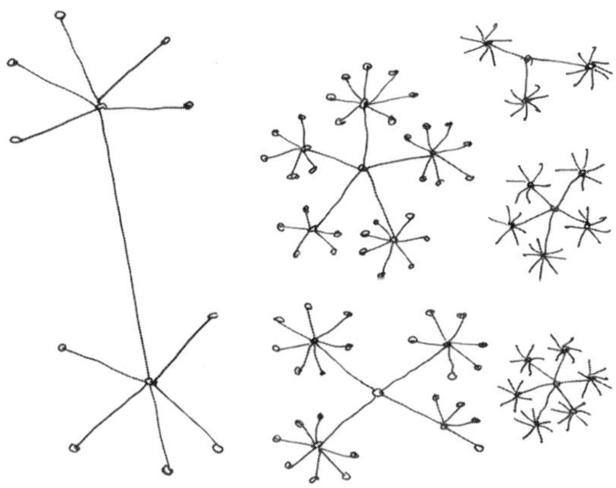

**Illustration 21.5:** **Distinct networks, each one working on a different scale, will combine to form the city's transportation system. They are here shown separated only for clarity.**

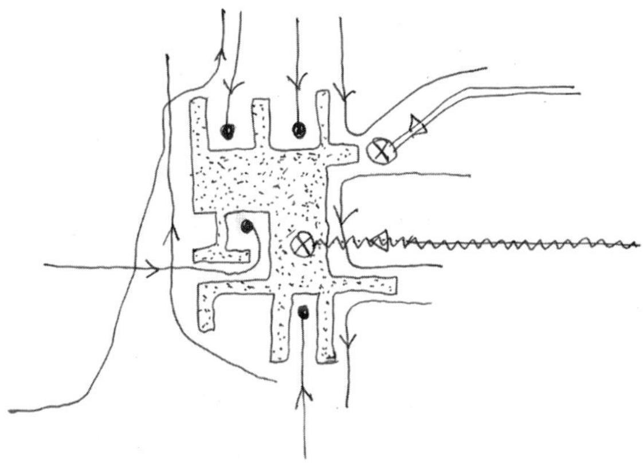

**Illustration 21.6:** **Pedestrian urban realm (dotted) is fed by several other forms of transport**

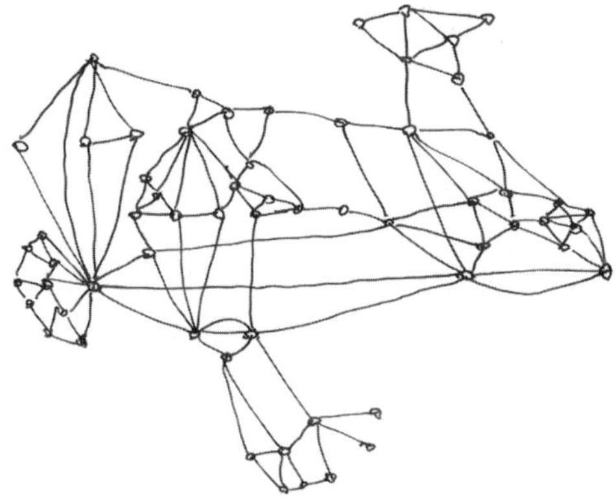

**Illustration 21.7:** **Networks of distinct type and working on distinct scales overlap and link to create an organized transportation system, in which every node has multiple connectivity**

gency vehicles must be guaranteed not by a wide road, but by a road surface that gives priority to pedestrians: vehicles should be allowed ACCESS, not SPEED (Ill. 20.9).

25. The car web contains all those functions that optimise fast automobile traffic, but are essentially hostile to human beings, including wide roads that connect non-pedestrian nodes such as heavy industry, military installations, warehouses, giant parking lots, car dealerships, garages and gasoline stations, among others.

26. The present trend to locate office buildings as isolated nodes in the car web must be reversed by tax incentives, so that offices can relocate within the pedestrian urban element. Isolating nodes that contain many people makes sense only if their activities conflict with residential and other uses, for they create a dangerous dependence on cars.

27. Using tax subsidies, light industry must be encouraged to relocate within mixed-use regions. Only heavy industry should be isolated from the city.

28. Skyscrapers (buildings higher than 10 stories) are not cost-effective, and they burden a city's infrastructure and transportation resources in a wide region around themselves. A city can only afford to support a very small number of skyscrapers for vanity purposes (Ill. 20.10).

The above propositions come from the work of Christopher Alexander (Alexander *et al.*, 1977; Alexander, 2004) and Léon Krier (Krier, 1998), as well as from my own studies (Salingaros, 2005). The three of us, drawing on work by others, are putting together a picture of the living city that can be used as a model for all future urban development. I have tried to orient this chapter toward the problems of the European city; yet most of these urban principles are, in fact, universally applicable. Joel Crawford (2000), Andrés Duany, Elizabeth Plater-Zyberk and Jeff Speck (2000) and David Sucher (2003) from the United States, Josep Oliva from Barcelona (Oliva i Casas, 2001), and Jan Gehl (Gehl, 1996) from Denmark have all published books of sensible advice on how to reconnect the urban fabric.

## The ecological dimension

A radically new urban philosophy can emerge from these suggestions. It is but a small step in the direction I am proposing to bring the natural environment into the picture. This way of looking at the built environment gives priority to human beings and small-scale structures. It represents a drastic reversal of twentieth-century urbanism, which emphasised the large scale and ignored the individual. An urbanism that destroys the small scale and treats human beings as expendable objects will never respect the natural world. On the contrary, it is an expression of human arrogance regarding nature. A new urbanism, which respects our sensibilities in the built environment, would also appreciate our natural environment (Roaf, 2005).

Once we begin to salvage the old, and now mostly lost, regions of our cities, we can also begin to appreciate the living elements within those cities. A tree grows naturally next to a low, crooked wall, and within a courtyard. A wide, uneven sidewalk has space to accommodate trees. An archeologically open space provides a habitat for some urban (if only avian) wildlife. This is more a philosophy of nature and of the earth than a conscious approach to urbanism. In the event, and as I said, an urbanism that is modest and respects human sensibilities will also respect the natural environment; it goes hand

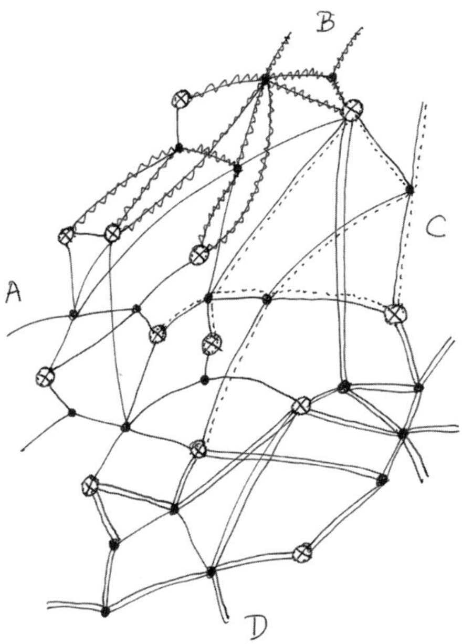


**Illustration 20.8:**
**Four distinct networks, labelled A, B, C, and D connect via interchanges. The crossover points are just as important as the paths**

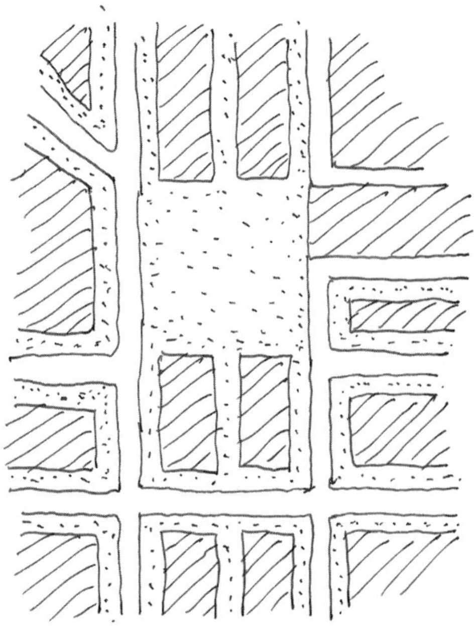

**Illustration 20.9:**
**Buildings (shaded), pedestrian realm (dotted), and car realm (blank) co-operate to define an interconnected urban fabric**

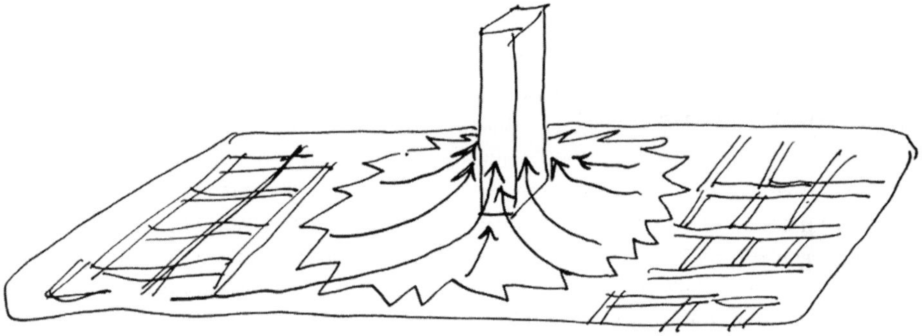

**llustration 20.10:** **A skyscraper draws enormous resources from a large region of the city, which weakens the surrounding urban fabric**

in hand with a modest architecture of human proportions and textures. The alien look of polished metal, glass façades, and smooth, windowless walls breeds an intolerance for living things precisely because it represents the opposite properties.

I am looking to the future, when we will use scientific knowledge about complex systems and their interactions to better plan our cities. Critics of such ideas dismiss them as nostalgic, belonging to the past. That is not accurate. What I propose has a striking commonality with some aspects of traditional urbanism, which accommodates human beings and not machines or abstract geometric forms. Those critics are stuck in an obscurantist mindset of inherited urbanist dogma. To them, any revolutionary proposal for progress threatens their own false promise of a 'progress' possible only through modernist principles. Those principles are the same failed ideas of the 1920s, recycled over and again. Each time, cities and nations are promised that they will work now, and that their previous applications were sabotaged by factors 'beyond' their planners' control.

Like a pathogen, modernist urbanism is easily recognizable once one knows what signs to look for. Some of its principal characteristics are: monolithic buildings and vast open spaces; geometrical alignment to arbitrary rectangular axes; elimination of the intermediate and smaller scales; insistence on industrial materials; insistence on the 'purity' of form and surfaces. This goes hand in hand with an intolerance of whatever helps to reinforce the urban fabric, such as pedestrian spaces, semi-enclosed urban spaces, permeable interfaces, folded urban boundaries, remnants of the past, modestly-sized structures, street furniture, and anything that 'clutters' an empty minimalist geometry.

Most telling is a static mindset that deceives anyone considering modernist solutions that look neatly regular on paper. A dynamic city constantly evolves because of urban forces, much like any ecosystem. Only those persons who are supremely arrogant assume that they can impose static geometric solutions, and that people will follow them exactly without eventual change. The same foolish assumption is made about materials: architects erect smooth, flat walls and complain that they stain and weather badly. They have never understood how materials age, nor how urban structure evolves in time.

Modernist prescriptions destroy cities by reversing hierarchies of connectivity. They remove organised structure and differentiation from the human scale. At the same time, they eliminate connective paths within human reach. The end result displays an artificial, mechanical movement as services have to be forced into over-concentrated downtown office nodes. Human beings need both structures and paths on the human scale – an obvious biological fact that has escaped modernist planners. Further, as in an ecological system, when certain levels of life are missing, they are replaced by organisms from nearby strata. This has led to many downtown areas being occupied, after hours, by homeless persons and/or petty criminals. It is not their fault; there are just no socially healthier elements willing to occupy that hostile niche in the urban ecosystem.

## A new generation of urbanists

A separate but subtler danger comes from postmodernist architects who appreciate correct urbanist principles, but misuse them to promote their own alien buildings. These people (some of whom occupy positions of great power and influence in the architectural community) are promoting good cities with faulty pieces. What they would have us build is similar to some northern European 'new towns', where all the right urban connections are present, yet the towns are still dead because the architecture is alienating on a human scale. We have here high-tech parasites of the living urban fabric.

To add insult to injury, some postmodernist urbanists have appropriated the terms 'ecological' and 'sustainable' to denote energy-saving buildings that are entirely alien to humanity. Up close, such buildings resemble a space station – as far removed from nature as can be imagined. Nevertheless, municipalities and national governments have been duped into spending money to build high-tech monstrosities out of extremely expensive materials simply because their architects add some solar panels to them or use elementary concepts of recycling and insulation. It is a mark of public gullibility that buildings have ignored such fundamental ecological concerns for so long that they can now be promoted as 'innovative' (Roaf, 2005). Just like their modernist predecessors, these architects deceive us with flashy and seductive images of industrial materials.

Some of these prominent architects are now destroying China's centuries-old sustainable urban fabric, replacing it with an unsustainable nightmare of concrete, glass, and steel. This will guarantee gasoline dependence and urban congestion for generations. One would think that these planners are employed by transnational oil companies to stake out profits for at least a century, but no. They have been invited, and are paid, by the Chinese government to 'renew' its cities. The damage they are doing, however, far surpasses that of the combined Mongolian invasions. I mention this only as a warning to other governments, which has been jealously looking to bring these same fashionable people to wreak havoc in their own countries.

The new generation of urbanists encompasses several diverse groups: those who wrote the *New Charter of Athens 2003*; the neo-traditionalists inspired by Prince Charles (Prince of Wales, 1989); those who cling stubbornly to the old modernist dogmas; and promoters of the network city (which includes me and others who propose radical solutions based on technology). Some of these persons understand how a city works, while others only think they do. Some have a good understanding of urban processes on a particular scale, but grasp neither other scales nor their need to integrate into one another. Others are impostors, plain and simple. Urbanism is an easy field in which to make wild new proposals without having to prove their effectiveness.

## Sociologists keep the old dogmas alive

One is hard-pressed to explain postwar urban destruction on such a massive scale, implemented by governments using all their power and resources. This anti-urban movement (disguised as progress) continues unabated, while the few voices proposing a humane urbanism are ignored by the entrenched circles of power. Sustainable urban models exist today, promoted by architects and thinkers who truly understand urban and social forces. We have adaptive solutions in our hands. It only remains to convince industry and governments to adopt them.

That is not happening, and the reasons are principally ideological. Even though new urbanist visions are cheaper and more effective in the long run, the madness now being pursued is supported by a fundamental belief system. Certain authors on the political left continue to cling to the dogma that technology can 'liberate' human beings from their own humanity; and that the gifts of the future are denied to those who connect to nature. The solutions offered are the same unworkable utopian dreams that in the past led to totalitarian interventions. Supposedly, the State knows all, whereas individuals are ready to betray progress so as to satisfy a sentimental desire for comfort.

A successful deception has been maintained for several decades: claiming that urbanism lies primarily in the domain of sociology. Not only are the geometrical foundations of urbanism ignored; but people don't even consider the possibility that urban structure has its own mathematical basis. When politicians seek advice on urban issues, they invariably turn to sociologist-urbanists, who offer the solutions we see implemented today. Why are sociologists complicit in this destructive act? I believe the answer lies in the original fascination the political left had with technology as a way of shaping society. Both Soviet Marxists and National Socialists were mesmerised by visions of skyscrapers, freeways, and centrally-planned industrial cities; i.e., all the paraphernalia of modernist urbanism.

Today's politicians, regardless of their political orientation, continue to be beholden to the same 'experts', who promise them quick technological fixes. The old images of shiny industrial objects have not lost their seductive qualities. International conferences planning the future of cities, on globalisation, economic advantages, etc. still invite people to present the latest untried utopian conjectures, in a direct parallel to the old modernist dogma from the 1933 Athens Charter. Its propagandists have now cleverly repackaged the original message using new buzzwords such as 'hypergrowth strategy', 'megaprojects', 'diversity', 'networks', and 'sustainability'. These words are unfortunately misused to promote an anti-urban agenda of skyscrapers and a dreary industrialised landscape.

Truly innovative urbanists have been pushed to the sidelines by the false accusation that urban solutions based on real human needs are backward-looking, and are furthermore politically conservative. New urbanists are condemned, along with their human-scaled proposals, as being outdated and reactionary. New urbanists who happen to be on the left, on the other hand, are dismissed as 'anarchists'. This brilliant propaganda ploy has preserved the power and income of a handful of wily individuals. As a result, no politician dares to risk his or her public image as a 'progressive' by sponsoring any traditional-looking urban project (that is, resembling pre-1933 prototypes). Modernist ideologues have even succeeded so far in their efforts to marginalise the 2003 Athens Charter.

Christopher Alexander demonstrates beyond any shadow of a doubt that urbanism consists of social processes that depend critically upon a geometric rubric (Alexander *et al.*, 1977; Alexander, 2004). Many social patterns simply cannot take place without the appropriate geometrical framework. This result (experimentally verified) invalidates claims by eminent sociologists that 20th Century society has freed itself from the constraints of the built environment. I can understand why sociologist-urbanists continue to ignore the geometry of the urban fabric, as a policy necessary to preserve their dogmas. According to Alexander, geometry is the core of all urbanism, and a geometrical understanding does not allow anyone to get away with specious proposals.

## Urbanism and ideology in Athens

The remainder of this chapter focusses on the case of Athens, as an example of a European city that destroyed its earlier, more human urban fabric. I argue that this was the result of misapplied urbanist principles. It is only by studying why things went wrong that we can ever hope to reverse the urban damage. The case of Athens is applicable, with only minor modifications, to many other European cities.

At the time of Greek independence in the early 19th Century, Athens was a fairly small town, ideal for the new government to begin erecting imposing new buildings and planning its urban structure for several decades. For the most part, Athens by the 1920s still followed the model of vibrant local neighborhoods partially connected by an electric subway (and soon to be even better connected by electric trams running on rails). Unfortunately, this balance between connective links and the built environment was shattered by both the tremendous influx of immigrants from the Asia Minor disaster in 1922 (an aftermath of World War I) and the onset of global economic depression at the end of the Twenties. These factors led to the overcrowding of Athens and to its future definition as the overflowing container of most of Greece's population.

Athens's urban collapse, coupled with the complementary collapse of villages that emptied their population into the capital city, generated social and political forces that are still unresolved today, and gave rise to a strongly ideological – and utopian – 'solutions' divorced from reality. For example, the former prime minister, Konstantinos Karamanlis (uncle and namesake of the current prime minister), eagerly dismantled the tram lines, as he obviously identified them with Athens's past and wanted to bring the country into the future. The world's top postwar urbanists recommended this step in order to speed up automobile traffic. But it was a mistake, as is now apparent by the reintroduction of the old tram (albeit in a technologically updated form) to restore this vital transport link to Athens.

While the developmental model chosen to deal with the devastation wrought by the German occupation and Greece's Civil War was wrongheaded in the extreme (based as it was, to a great extent, on construction alone), it was an obvious choice at a time of severely limited options. To be fair to the elder Karamanlis, he chose the fast – if illusionary – track to rapid economic growth in the 1950s. At the same time, however, this choice inevitably contributed to urban and social degradation. Now we recognize it as an economic choice that proved disastrous in the long run.

Worse of all was the ideology of 'progress', which could only be realised by rending the previous urban fabric. Certain essential elements of Greek urban culture – old Athenian homes with courtyards, small sidestreets, small pedestrian squares, kiosks, mixed-use four-story buildings – were condemned as useless and fit only for replacement. What was chosen to replace them were models imported from industrialised Europe and the United States (the latest in urban and architectural 'progress') that consisted of isolated villas, monofunctional high-rises connected by expressways, buildings closed to the pedestrian street, and even the total elimination of the pedestrian street. Along with these changes came an architecture that deliberately disdained life, and wore an alien face of polished metal, plate glass, sheer stone, or brutal concrete.

Greeks accepted this new 'look' as symbolic of architectural (and, by implication, social and economic) progress. The political left saw this utopian urbanism as a rejection of the 'old', traditional urbanism, which symbolised the right's power base, and as a necessary part of the socialist revolution that would guarantee the country a bright new future. The right, on the other hand, was equally willing to co-opt immigrant settlements that housed left-wing voters by razing existing structures and replacing them with 'modern' workers' housing. As for the right's upper-class constituencies, they wanted new highways through the city so that they could enjoy their cars. Wealthy residents eagerly embraced an isolating urbanism within their neighborhoods, since it offered protection from 'crime' (real or imaginary) and a chance to avoid mixing with those less well-off. In the end, a succession of governments, advised by respected urbanists, implemented policies that destroyed the functioning urban environment in place at the end of the Forties.

While more recent history easily confirms that the left has been responsible for its own share of urban disasters, the blame for the depredations of the Fifties, Sixties, and Seventies in Greece lies squarely with the political right. During a 40-year-long postwar monopoly on political and social power, right-wing governments consistently chose to apply anti-urbanist policies. Smaller cities survived better, simply because of neglect, as Athens concentrated most of the country's resources. Many provincial centers weathered postwar urban blight much better than Athens. It would be heartening to point to local civic pride as having tempered the worst of the urban assault, but this is not the case. Whenever they had the funds, cities beyond Athens immediately did the same damage to themselves, destroying what was most valuable in their urban environments.

## What went wrong in Greek urbanism

In the mad rush to 'equality' with the US and northern Europe, automobile ownership in Greece has skyrocketed. In rural areas, this is understandable, since cars have provided an efficient connectivity, perhaps for the first time in history. In cities, however, severe problems have resulted since no thought was given to how all these cars were going to get around; as a result, roads are now choked. In parts of Athens, even the smallest side street is crowded day and night with traffic. There was also no thought given to parking all these vehicles, either for the night, or once they reached their destination.

The present state of commuting in Athens is the result of a monumental misunderstanding. Cities are transportation networks connecting pedestrian nodes (Salingaros, 2005). The quality and density of connections within and around those nodes, and connections among spatially separated nodes, are what enable a city to function. How efficiently a city works depends on the degree to which distinct transportation (including pedestrian) networks are integrated. Government planners, however, have visualised cities as buildings fitted into an abstract geometry, allowing them to cover every available space. Their idea of connectivity is to build a highway out to a group of isolated villas or high-rises. This philosophy naively supposes that the urban fabric will magically reproduce and expand by itself.

The old courtyard house fronted by a calm side street provided Athenian urban life with a ground-floor pedestrian realm connecting internal, private space to external, public space that was available to children and the elderly. This was replaced by the four-storey apartment building, with shops on the ground floor. The street, consequently, carried a high traffic load, thus leaving only a narrow sidewalk for the urban realm. The lost pedestrian space was shifted to a number of small neighborhood parks, which represented a workable but primarily CAR-FREE solution, since the main transport in this pattern functioned through buses, trams, and subway. The increased pressure from cars promptly made it unworkable, although we were left with semi-functional pieces throughout Greece. What killed this model were greed and a total lack of government oversight.

The typical phenomenon of the high-rise (four- to six-storey) apartment house was both forced upon and eagerly adopted by the Greek public, for two reasons. First, it was propelled by the huge population (that is, internal-migration) pressure, which led to vertical stacking. Second, it was itself a driving force behind the construction boom that heated up the Greek economy in the Fifties and Sixties. For many citizens, the speculative building of apartment buildings became a gold mine, an employment opportunity, a route to a higher standard of living, or even all of the above. Politicians were unwilling, therefore, to criticise the postwar urban model in any way. The available ground space was far more useful (and valuable) for erecting more buildings.

It may indeed be possible to return to the four-storey, mixed-use apartment model. Today, however, one has to provide for underground (as well as limited surface) parking. Sidewalks have to be much wider, and urban space better defined, to enclose protected portions of the pedestrian realm (Alexander, 2004). A lot more green space needs to be made available. Finally, balconies have to be at least two meters deep (roughly); otherwise they don't work as raised living spaces (Alexander *et al.*, 1977). These improvements would not, in themselves, solve the circulation and parking problems in the city, however. Although the destruction of any vestige of urban green space amounts to criminal negligence, the consequences of ignoring the parking and circulation problems are just as serious over the long term.

I cannot overemphasise that urban society forms in the pedestrian realm, which itself has to be nurtured at street level. But the postwar residential urban model evolved into new and unsustainable typologies. The height of a typical apartment building in Athens has now increased beyond four stories, which surpasses the critical limit of density capable of sustaining urban life. Above four stories, there is no visual or spoken exchange with the street level (Alexander *et al.*, 1977). Children and the elderly are virtually imprisoned in their apartments, thus disconnecting society.

Even more serious is the elimination of mixed use. The parking garage has replaced the traditional commercial ground floor. A cheap solution – easing parking problems at home by putting cars under apartments (but not underground) – has dealt the final blow. Today, monofunctional apartment high-rises sit on stilts, with the ground floor entirely taken over by parking (following the 1922 *Citrohan* model of the hysterically antisocial Le Corbusier). This disconnects inhabitants from urban life, reconnecting them only through their cars. It is the same disconnection seen in North American suburbs, with their well-documented social alienation (Duany *et al.*, 2000). We have vertical isolation in Greece, as opposed to the horizontal isolation of the United States. Since there is no longer any connection with the ground, sidewalks have begun to shrink, and apartment buildings stand apart from each other, thus failing to define any urban space.

The pedestrian realm has been totally sacrificed to the needs of the automobile. A facile parking solution is to accommodate vehicles underneath the new, freestanding apartment buildings, but this is an illusion. Those cars start off in the morning to jam the streets and fight for nonexistent parking spaces at their destination. Some residents pretend that they need their cars only to take their families out of the concrete hell of Athens, so that they can live a 'normal' life in the countryside for a few days. It never occurs to them that it is possible to live a more connected life in the city itself – with the correct geometry.

## Conclusion

Modernist urbanism – based on the power to impose technology over nature – is essentially destructive of what already exists. It is also profoundly arrogant in its assertion of a brutal power over something it doesn't understand, and which it disdains. Europe's urban, social, and environmental devastation during the last two generations is due to interlinked causes. Today, progress requires a major change in worldview. The pairing of technological progress with an urbanism of alien forms is a great lie, but one fanatically believed in by many 'modern' Europeans. Technology can indeed help in the reconstruction, when applied intelligently. Science is essential to help urban residents live like human beings once again and regenerate their environment.

Misguided urbanists applying wrong ideas have done (and continue to do) so much damage that it is impossible to know where to begin a critique. Let me touch only on the topic of automobiles. Cars will not go away; everyone wants to own at least one and many persons own two. Automobiles are a tremendously useful, if very expensive, mode of transport, but they must be accommodated without destroying the pedestrian urban fabric. Since this has already been destroyed in most places, it must be rewoven. A living city connects its cars to people in a non-threatening way: automobiles should not take over a city. Planners have to understand how to interweave the car web to the primarily pedestrian urban fabric. Then they can work out how to optimise that web without destroying the rest of the city. These principles are very simple to understand.

Europeans used to have a well-balanced respect for the environment (at least in the mythical days of yesteryear, when people lacked the technological power to destroy it). But this respect was replaced by a new philosophy of intolerance. Old Europe stood in the way of grand urban and architectural projects deemed necessary to 'progress'. We sacrificed much to this progress, and it is now choking us. City-building consists of a series of compromises and accommodations. This, however, is not the same as sacrificing elements of our heritage and environment to antiquated visions of the future. I would like to see the countries in Europe repair their urban environment so that their citizens and their citizens' children can enjoy a better quality of life. I wish to save those pieces of uniquely European urban fabric from the senseless destruction to which it has been condemned.

The solution to Europe's urban problems is not merely contained in the points outlined above. It also lies in the adoption of a new philosophy of humankind's relationship to nature and the environment. It is contained in the serious, scientific study of what specific rules actually generate living cities. It lies in recognizing that the ideological urbanism of the postwar years has been discredited in practice. It lies in rejecting as toxic the high-tech 'look' of contemporary architecture (tolerable only in minute quantities). Once those philosophically linked steps are taken – accepting the full humility of human beings vis-à-vis their environment, their fellow beings, their historical past, and their urban heritage – then everything else will follow.

## References


Alexander, C., S. Ishikawa, M. Silverstein, M. Jacobson, I. Fiksdahl-King & S. Angel, 1977, *A Pattern Language*, Oxford University Press, New York

Alexander, C., 2004, *The Nature of Order, Books 1-4*, The Center for Environmental Structure, Berkeley, California

Prince of Wales, 1989, *A Vision of Britain: A Personal View of Architecture*, Doubleday, New York

Crawford, J.H., 2000, *Carfree Cities, International Books*, Utrecht (Neth.)

Duany, A., E. Plater-Zyberk & J. Speck, 2000, *Suburban Nation*, North Point Press, New York

European Council of Town Planners, 2003, *The New Charter of Athens 2003, The European Council of Town Planners' Vision for Cities in the 21st Century*, Lisbon, http://www.ceu-ectp.org/e/athens

Gehl, J., 1996, *Life Between Buildings: Using Public Space*, Arkitektens Forlag, Copenhagen

Krier, L., 1998, *Architecture: Choice or Fate*, Andreas Papadakis, Windsor, England

Oliva i Casas, J., 2001, *La Confusió de l'Urbanisme: Ciutat Pública Versus Ciutat Domèstica*, Pòrtic ECSA, Barcelona

Roaf, S., 2005, *Adapting Buildings and Cities for Climate Change*, Architectural Press, Oxford

Salingaros, N., 2005, *Principles of Urban Structure*, Techne Press, Amsterdam

Sucher, D., 2003, *City Comforts: How to Build an Urban Village*, City Comforts Inc., Seattle, Washington

# 21 Finding a New Meaning for Public Spaces in Postmodernity
## The Raval District in Barcelona

Francesc Magrinyà

## Introduction: On designing urban spaces – from modernity to a postmodern scenario

The agora and the market have been the landmarks of city public spaces in modern times. These spaces, built around the image of the Greek *polis*, based their appropriation on proximity relationships and everyday life. The city square was the mandatory meeting point where social control and community ritual took place. This scenario collapsed with the arrival of postmodernity and the associated post-Fordian production model. In this latter model, relationships are built up in fragmented spaces and community ritual is relegated from the physical space to a new combination of physical space and virtual reality. The traditional Sunday get-together on the town square has changed into a mix between the television get-together for prime-time programmes every evening and the various family and social get-togethers in space and time. In organising territory to meet this new situation, it should combine the traditional relationships of proximity with the new relationships of connection, but whereas before, everything used to happen in the same space, nowadays social relations look for a different meeting point for each activity. That is why the public space is questioned, or why it is at least necessary to redefine it.

An interpretation of territory using network urbanism will help us to discover this new framework of social relationships. A network structure defining the new territory has been generated by using many kinds of networks as new territorial intermediaries. These can be characterised on three levels:

- Networks of personal relations;
- Networks of production, distribution and consumption;
- Transport and telecommunication infrastructure networks.

Whereas a traditional family used to get together on the town square, go to the market and use the transport and telecommunication networks to leave the nucleus from time to time, in the case of a family in a post-Fordian territory, each of the two working members of the family would be employed in a different town in the agglomeration from where they live. Their children would go to school in a third town and the leisure and consumption spaces would also be in other towns. In this way, the post-Fordian family creates a complex network of personal relationships, conditioned by the logistic networks available, including the daily use of transport and telecommunication infrastructure networks.

Onto this scenario of territorial fragmentation and complexity, the media have imposed new discourses about the city, including that of the city as remembered, projected and imagined. Therefore, the physical perception is mixed with the image created by the telecommunication media. This has radical consequences when designing new public spaces. An example of this is the historical centre of Barcelona and the urban remodelling that has been carried out there (Magrinya & Maza, 2001).

## The public spaces in the historical quarter: Discovering the limitations of current urbanistic instruments

Ciutat Vella, the Old Quarter of Barcelona, is confronted with a process of change marked by the confluence of multiple wills that reflect the images represented by each of the social groups in the city. The historical memory, the conquest of a new space to be designed, the idyllic image of the mixing of peoples and the concentration of immigration are some of the most significant components in this confluence of projections. Its public spaces are becoming ever more complex objects in a tissue which is in the process of transformation due to significant flows of immigration. However, the urban remodelling is still based on traditional methods limited to eliminating buildings and creating spaces in the urban tissue. Few new activities have been added, and where this has happened, it has mainly been in the form of the central space being taken over by institutions that are foreign to the historical centre.

The new squares have been designed according to the idealised model of the agora, but with a certain stamp of modernity in their constructive elements. As a result, these new urban spaces do not function well as urban forms, because they have not been accepted as part of the integral fabric of the communities who live there, because they have no real utility for them.

What gives the public space its real character is how the relationships that are actually exercised there are combined, so the more complex relationships between the different social groups become, the more fragile this character will be. Other essential elements in defining these spaces are the establishments around them, and the various types of mobilities or forms of interaction exercised there. In the case of the Raval district, the design of the constructed space is just a part of the definition of that space. It is impossible to understand it without analysing its function for the neighbourhood and the agglomeration as a whole, and for its interrelations with a wider area.

Therefore, it is no surprise that although these newly designed spaces have been embellished with high-quality sculptures and construction elements, what has actually been created are ghettos and closed urban ecosystems. What has been ignored are the key parameters in redefining the public space: the weight of immigration in the interrelations between the inhabitants of an agglomeration; the co-existence of different groups within the same space; possibilities for mixing social groups; the role of the agglomeration in overseeing the central neighbourhood at night.

## The origin of the public spaces in the historical centre: Revising the urbanistic approaches that were implicit in remodelling Barcelona

If we analyse the evolution of the great interventions on the historical centre of Barcelona, we see that there are three stages, each with their own references (Ill. 21.1). These correspond with the three periods during which Barcelona was transformed:

- The Mercantile Stage associated to the Mendizabal disentailment (1835), recognisable in the markets (Boquería, Santa Caterina) and squares (Medinacelli, Real, Sant Jaume);
- The Industrial Stage coinciding with the second industrial revolution that was associated with the introduction of electricity and confirmed with the opening of the Via Laietana (1907);
- The Culturalist Stage associated with the postmodern era, with the introduction of the new post-Fordian production relations, and centred on new cultural artefacts and the relocation of existing entities (Museum of Contemporary Art of Barcelona, the Centre for Contemporary Culture in Barcelona, Foment de les Arts Decoratives, Edicions 62, Institut d'Estudis Catalans, etc.).

An analysis of the various urbanistic remodelling initiatives shows, transformations of the historical centre have been repeatedly based on creating new spaces in the form of streets or squares (Ajuntament de Barcelona, 1993), and the most recent stage (1986-2000) has been no exception (Ajuntament de Barcelona, 1999). Instead of opening up streets, squares have been taken over and new spaces have been created following a model closer to that of the Mendizabal disentailment of 1835 than to the opening of the Vía Laietana in 1907. With each remodelling initiative, a new emerging social group that has tried to take over the central sectors of the city. The last 'take-over bid' was initiated by the cultural industry, and this generated a gentrifying discourse. The central axis of this appropriation was

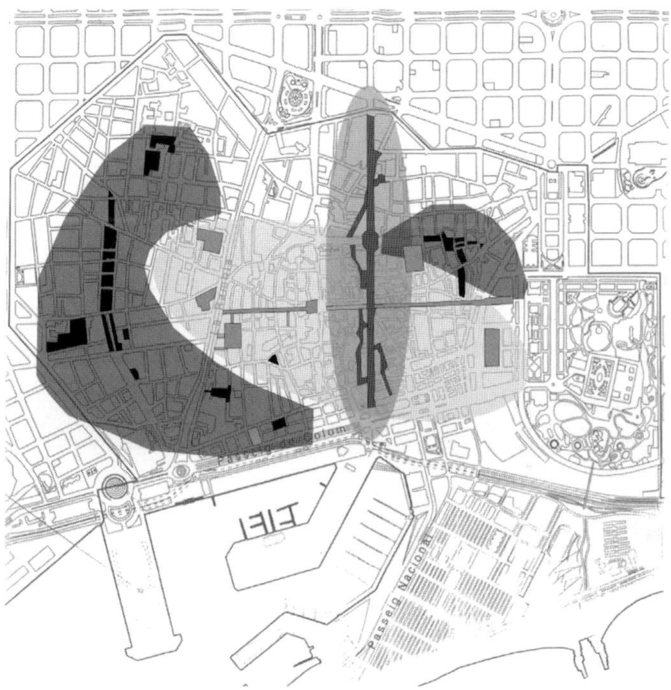

Mercantile Stage. Disentailment Mendizábal (1835)

Industrial Stage. Opening of the Via Laietana (1907)

Culturalist Stage. Public spaces and urban projects (1986)

**Illustration 21.1: Gentrification stages and associated urban actions in Barcelona historic centre**

the idealisation of urban mix and multiculturalism in public spaces. This has brought the discourse of the urban project into the design of public spaces.

However, in a postmodern scenario and with a post-Fordian production model, we can see that social relationships are no longer homogeneous. Furthermore, the dualisation of space has set in and characters and scenes that had not been foreseen in the script have made their appearance: problems such as insecurity, neighbourly strife, street patrols, and street children have arisen. From the perspective of the gentrifying discourse, none of this was foreseen for the district.

In the 80s, urban project discourse, in which the positive effect of the metastasis of specific interventions was applied (Bohigas, 1987), proved an interesting method for working-class districts with a consolidated population. At that time, and after the Plan General Metropolitano de Barcelona (PGMB) was passed in 1976, it was necessary to specify how the planned transformations would be put into effect. In the 70s, urban projects were started in working-class districts in the industrial belt of Barcelona. These districts had a stable population, but poor public facilities. Therefore, one of the aims of these projects was to improve public facilities and spaces in response to requests from the neighbourhood leaders. This was made easier by the fact that the technicians who had drawn up the PGMB in 1976 had included mechanisms to reserve these spaces, which allowed for a large margin of manoeuvre (Magrinyà, 1999). The discourse of the urban project fitted perfectly into this context. This was the instrument that fulfilled the specification of the social demands and furthered the progress from Plan to Project.

However, although the discourse of the urban project was very appropriate for the urban remodelling of the working-class districts in the industrial belt of Barcelona, it is no longer valid for the historical centre of Barcelona. During the period 1986-2000, the population of Ciutat Vella fluctuated, weakening neighbourhood organisations, so that the plans for the historical centre became separated from the wishes of its inhabitants. The urban project model has been used as a sort of stage, which architects have been asked to embellish. However, the city cannot be viewed as a stage, neither can people and spaces be treated as décor sponsored by benefactors (local government, in this case) and ethnocentric discourse, because such action ignores the reality of public spaces. It is tantamount to telling the space and the people how to act. By so doing, the authorities are laying down guidelines, always from the perspective of established power, about the relationships that people should establish with these spaces. In this regard, the historical centres have become subsidiary elements in city discourse.

The spaces have not been considered from the point of view of how the environment is occupied, under what terms (as owner/tenant), and what sort of population lives there. This means that some of these spaces are fragile. Decisions such where as to situate establishments and institutions that have office hours, how to use ground and basement floors, are much more decisive than they might appear to be at first sight. The caricaturisation of the public space as a purely architectonic intervention that adds a certain decoration to urban fittings and exercises a certain control over them is devoid of the universal urbanistic considerations, which these are essential in dealing with historical centres.

## The co-existence of different mobilities: Towards the creation of dual mobilities in central spaces

Apart from controlling the building conditions, which is essential for achieving stability in the relationships between inhabitants of urban tissue, it is also necessary to understand the mechanisms of the relationships established by users from outside the neighbourhood and how they combine with those of the everyday inhabitants of the place. If we take a look at the new public spaces in the Raval district, we will be able to analyse this mix, or combination of mobilities, in more detail. In this study, we will analyse the effects of transport and telecommunications, as significant new intermediaries, on the relationships that are established and on the new ways in which public spaces are used.

For their part, the inhabitants establish a certain degree of territorial attachment with elements of the place. Relations among inhabitants can be classified as homogeneous if they establish relationships with members of their own social group, and as heterogeneous if relationships are established among members of different groups. As for variable territorial attachment, two extreme cases can be identified. On the one hand, there are those who appreciate living in the historical centres, and those who could live perfectly well in any part of the city outskirts without having any bond with their environment. There are also two extremes in this latter group: those who only maintain relationships with those who have the same social and cultural interests as they have, and, at the other extreme, those who look for a mix of people or an unexpected meeting with members of other groups (Magrinyà, 1998). The cross between the various combinations of these two variables generates four large groups of significant mobilities: neighbourhood, dual, fragmented and centrality (Ill.21.2).

*Neighbourhood mobility* is characterised by the inhabitant's close relationship with the territory and by homogeneous relationships. A paradigmatic example of this type of mobility is that of a traditional working-class neighbourhood, where there is a single social group that is attached to its physical environment. *Dual mobility* is also characterised by a close relationship with the territory, but in this case different social groups co-exist, and heterogeneous and homogeneous relationships arise within each group.

| **Type of mobility** | **Territorial relationship** | ***Homogeneous space*** | ***Heterogeneous space*** |
|---|---|---|---|
| ***Proximity mobility*** | *Territorial attachment* | Neighbourhood mobility | Dual mobility |
| ***Connection mobility*** | *Spatial isomorphism* | Fragmented mobility | Centrality mobility |

**Illustration 21.2: Types of mobility resulting from the relationship between space and the type of relationship (Magrinyà, 2002)**

The combined large-scale introduction of transport and telecommunications generates a profound transformation of such relationships by individualising them and separating them from the physical environment. The cohesive and compact community situated on the same physical space is broken down in this way. First of all, *fragmented mobility* occurs (see Ill. 21.3). In this case, the inhabitants maintain relationships with members of the same group (family, work, children's friends, etc.), but their position in space is no longer a decisive element. Parallel to this, we have the mobility that is characteristic of large centres, where the inhabitants look for visible and easily recognisable, but usually artificial, centres in the territory, which also generate unexpected meetings with other groups. This is what we shall call *centrality mobility*. A clear example is the shopping centre. The character of a space will depend on which of these mobilities, or a combination of them, dominates (Ill. 21.4).

## The squares in the Raval district: Examples of various ways of appropriating public spaces

We will now analyse some of the new public spaces developed when the Raval district was remodelled. The first significant example is the Plaza de las Caramellas (Ill. 21.5), a new public space inaugurated in 1990, and characterised by neighbourhood mobility. This is a space created in the lower part of the Raval district after the housing there had been demolished and the residents – an autochthonous low-income population – had been displaced and rehoused in new, subsidised social housing. The design of the square meets the approval of its new neighbours because it has some interesting construction features. However, from an urbanistic perspective, it is an introvert space, closed in on itself to the extent that one of the entrances is actually closed for most of the day. Designed more as the back of a housing block, for the people who live there, who form a homogeneous group, this new square has failed to become the focus of residential mixing, neither has it become a recognised part of the imagery of the neighbourhood. It is a new space characterised by traditional neighbourhood mobility and the relationships within a low-income social group.

Els Horts de San Pau (Ill. 21.6) is another space that, given its location, should tend towards neighbourhood mobility. It is a rather large space (2 hectares) for the district in which it is situated. It was designed as a delimited space, closed in on itself, and this has been decisive in the way it is being used. This space hosts the most important historical feature of the district, the Roman church of Sant Pau, in addition to gardens, a sports centre, a school, and a car park. These four, unrelated construction elements provide services for the city, but not for the district. The car park is used by the audiences of the Paralelo theatres. The sports centre is more of a city than a district facility, and the gardens – Els jardins de l'Hort de Sant Pau – were designed to enhance the church, although this sentiment is lost on most residents and visitors. Situated next to the church, there used to be waste ground following a factory fire that functioned as a temporary football field, but it is now fenced off and disused.

In short, Horts de Sant Pau is not a space that encourages circulation, but one that is closed in on itself. It looks unsafe, divided into compartments and plots, when it could be an easy element for connecting the street Sant Pau with the Avda del Paral.lel. The autochthonous groups in the neighbourhood showed little interest in using this space, so it became a 'non place', until finally the space was taken over by the Pakistani community. Now it has also become a drug trafficking site, with the car park used as a hideout. Thus a dual mobility space has been generated, used by two well-established communities – the autochthonous group and the Pakistani community.

**Illustration 21.3: From neighbourhood mobility to fragmented mobility**

Centrality mobility (Ramblas)

Dual mobility (c/ Hospital)

Neighbourhood mobility (Rambla Raval)

Dual mobility as stage (Born)

**Illustration 21.4: Types of mobility**

As well as these rather large spaces, there are several smaller open spaces, two of which are the Jardins de Volta de Encires and the plaza Pere Corominas. These are also new places, interstitial spaces, not large enough to be considered as squares. They have been provided with some benches where the passers-by from outside the district can rest for a while, alongside those who live in the area. None of these spaces is embellished with sculpture, significant buildings, nor does any cultural initiative take place there.

By comparison, the Jean Genet and Pierre de Mandirgues squares represent two tributes to literary memory and are examples of using city discourse as a place of remembrance. The background mix is still evident in the appearance of these two spaces: they are spaces created by demolishing a building block and building new apartments, but the new squares have retained the atmosphere of the original urban setting. The recently named, Plaza Pierre de Mandirgues was once a street well-known for its pavement prostitution. Now, the trade has simply moved into the square. The Plaza de Jean Genet is a space that was converted into a 'passing' square where street patrols could deal with conflicts. The designs of these new public spaces have resulted in a simple change of scenery, but not a change in the activities that take place there.

As well as the most introvert spaces, with their combined neighbourhood and dual mobility, there are the spaces that have a stronger relationship with the city, those characterised by dual and centrality mobilities. Two examples of these are the Plaza dels Àngels and the Rambla del Raval. These two spaces are characterised by a predominantly dual mobility, but the Plaza dels Àngels has a symbolic centrality component associated with the new cultural centres of the Museum of Contemporary Art of Barcelona (MACBA) and the Centre for Contemporary Culture in Barcelona (CCCB). The Plaza dels Àngels is a square where we can see the museums and take photographs, although their position on the square appears uneasy in that sometimes they seem to dominate the square like an exhibition, and on other occasions they seem to be defending themselves from the life of the square. This is one of the places in Barcelona that has generated the most discussion. It is a space where modernity, a décor embellished by an architect, contrasts sharply with a poor district where people 'live together'. As we have seen, such idyllic sets are untrue to life and therefore prone to conflicts. The need for security guards around the Museum of Contemporary Art is visible proof of this.

In contrast, however, he Rambla del Raval, inaugurated in September 2000, is a great empty space. It was designed with the architectonic reference of spaces such as the Plaza Navona, and no thought was given to the surrounding districts. The space is located at a confluence of neighbourhoods, where there are dual and centrality mobilities. Although the autochthonous community is still dominant, the Pakistani community has recently started to establish themselves, and the social cohesion has become more fragile. Given its dimensions, the intention was that it would become a centrality space, but this has not yet occurred, and it may very well continue as a dual space. This is evidenced by the number of bars around this public space that have been appropriated by the immigrant community, and by events such as the end of the demonstration organised by the 'sin papeles' (no documents) movement.

Another paradigmatic example of a space with dual and central mobility is the Plaza Orwell (see Ill.21.7). This space was formed by eliminating a block of buildings, and generating a new environment with a statue as the central element. This space, given its proximity to the ramblas, the Plaza Real and the more run-down calle Escudillers, has generated an underground environment, which has marginalised adjacent neighbourhoods from the life of the Plaza. It is a good example of an urban mix: auto-

**Illustration 21.5: Plaza dels Àngels and Plaza de les Caramelles with the MACBA and the CCCB (Magrinyà, 1996)**

**Illustration 21.6:**
**Before and after the project realisation**

LA VANGUARDIA

VIERNES, 29 JUNIO 2001

vivir en Barcelona

URBANISMO

El Fossar de les Moreres tendrá una llama eterna

Ojo con el ojo

División de opiniones en Escudellers y la plaza George Orwell por la videovigilancia

HISTORIA Y VIDA

¿Dónde está la plaza del Tripi?

**Illustration 21.7:**
**Plaza Orwell**

chthonous groups with neighbourhood mobility, young groups of urbanites who seek to dominate the plaza, nocturnal mobility with a certain underground touch and centrality mobility from tourists wanting to experience the urban mix.

## Contemplating the container-content relationship: finding a new meaning for public space

The combination of these different mobilities allows us to describe the relationships established in Ciutat Vella from a new perspective. The first conclusion is that the idyllic image of the traditional agora has disappeared. Communities are no longer compact and cohesive and urban spaces are no longer centred solely on traditional everyday usages. Television, cinema or advertising are just as significant as everyday proximity. An inhabitant can perfectly well make use of a shopping centre made familiar by repeated advertising, but only visited once. Open spaces have become more complex because the mixture of mobilities generates different behavioural patterns, so public space as a unit where interaction can take place has become no more than an illusion.

In analysing the various spaces, we have seen that it is the ones where neighbourhood mobility is more predominant where the structures have remained minimal. Where relationships have actually been established with the public spaces in the Raval district, these have been shown to follow rules of appropriation based on the groups that inhabit them on an everyday basis. The spaces with a homogeneous composition of autochthonous groups and first generation immigrant groups, especially large spaces, such as Horts de Sant Pau or Rambla del Raval, are more fragile. There is a greater likelihood of such spaces being taken over by uncontrolled nomadic groups that generate security problems. There is also the issue of cultural ghetto settlements, such as the Pakistani community that completely took over certain sectors, such as the rambla del Raval between calle Hospital and calle Sant Pau. Such spaces then start to exist autonomously, separated from the gentrifying discourse of the area in which they are embedded.

In the spaces where there are both dual and centrality mobilities, the space is not appropriated to the same extent on a daily basis, making it easier for this type of space to be colonised by nomadic groups such as street children, drug traffickers, etc. This effect is strengthened further where centrality mobilities are predominant, because the space becomes a 'non place', making it difficult to appropriate such spaces on a daily basis, as in the Plaza Orwell. Spaces not appropriated by homogenous neighbourhood mobilities, become an attractive territory for nomadic groups such as street children or gangs of youths who take over the space temporarily, thereby weakening normal public relationships and creating security problems.

To sum up, we can say that in the postmodern era, where territory is both complex and fragmented, if public space is to be designed to meet the needs of the physical and social urban fabric, then it is essential first to analyse the predominant relationships and mobilities.

## References


Ajuntament de Barcelona, 1993, *Barcelona, space público*, Ajuntament de Barcelona, Barcelona

Ajuntament de Barcelona, 1999, *Barcelona 1979/2004. Del desenvolupament a la ciutat de qualitat*, Ajuntament de Barcelona, Barcelona

Bohigas, O., 1987, Metàstasi I estrategía; In: Ajuntament de Barcelona, *Barcelona. Espais I escultures 1982-1986*, Ajuntament de Barcelona, Barcelona

Magrinyà, F., 1998, Urbanismo de redes y planeamiento urbano, *Revista de Obras Públicas*, Urbanismo II, June

Magrinyà, F., 1999, Barcelone, du Plan Cerdà aux jeux olympiques; In: Cuillier, F., 1999, *Les débats sur la ville*, Editions Confluences, Bordeaux, pp. 111-121

Magrinyà, F., 2002, *La théorie urbanistique de Cerdà et son l'application à l'"Ensanche" de Barcelone (1859-1953): une genèse d'urbanisme de réseaux*, Ecole Nationale des Ponts et Chaussées de Paris (ENPC), Paris

Magrinyà, F. & G. Maza, 2001, Inmigración y huecos en el centro histórico de Barcelona (1986-2000), *Scripta Nova. Revista Electrónica de Geografía y Ciencias Sociales*, Universidad de Barcelona, No.94 (62), 1 de agosto (htpp://www.ub.es/geocrit/sn-94-62.htm)

# Disciplinary Language Barriers

Iwan Kriens

## An example from the practice

The south-west part of the municipality of Delft (Delft Zuid) is an agglomeration of four separate districts, each accommodating about 7000 households. The urban structure in two of them is characterised by broad avenues (the '*Dreven*') that give access to the districts from North to South. These districts lie back to back, as it were. In the years to come, the traffic structure of Delft Zuid will undergo radical changes, due to a tougher policy on cars and the construction of the A 19 national highway. As soon as this national road is open, the provincial road will lose part of its regional function, what will create new spatial possibilities. The space currently occupied by the provincial road can be completely or partly used for other uses, in order to benefit the surrounding areas. The present road can become a local access road, thereby reducing the volume of traffic (Nevalainen, 1996). The space may be used to create a quiet recreational route to accompany the new locations that will be developed. In this way, the isolated position of several of these neighbourhoods may be improved.

From a spatial planning point of view, this idea offers very attractive possibilities for helping the adjacent districts to solve some urban bottlenecks and to get space for new developments. The urban issues that may be tackled are:
- Improving the isolated position of some neighbourhoods;
- Increasing the scale of the educational institutes;
- Renovating residential property owned by housing associations;
- Meeting the demand for housing for senior citizens;
- Meeting the demand for larger playground facilities;
- Increasing the scale of health care;
- Meeting the demand for near-home neighbourhood facilities.

The possibilities for this area have been seriously studied by the Urban Planning Office of the Municipality of Delft on different occasions from 1990 onwards (Stadsontwikkeling Delft, 1991; 1998). However, the planning process has been repeatedly blocked because of the sector approach. The quantitative traffic research stated that reducing the profile of the provincial road would put additional pressure on the north–south avenues, which in turn would create an unsafe situation. The several opportunities to improve the area have been lost due to the unattainable claims of the traffic study.

This example shows the difficulties that exist between the different disciplines involved in spatial planning in the practice. One could argue that there is a language barrier between the disciplines involved. Possible explanations for this are:
- Either the disciplines involved are unable to make their contribution accessible to the other disciplines;

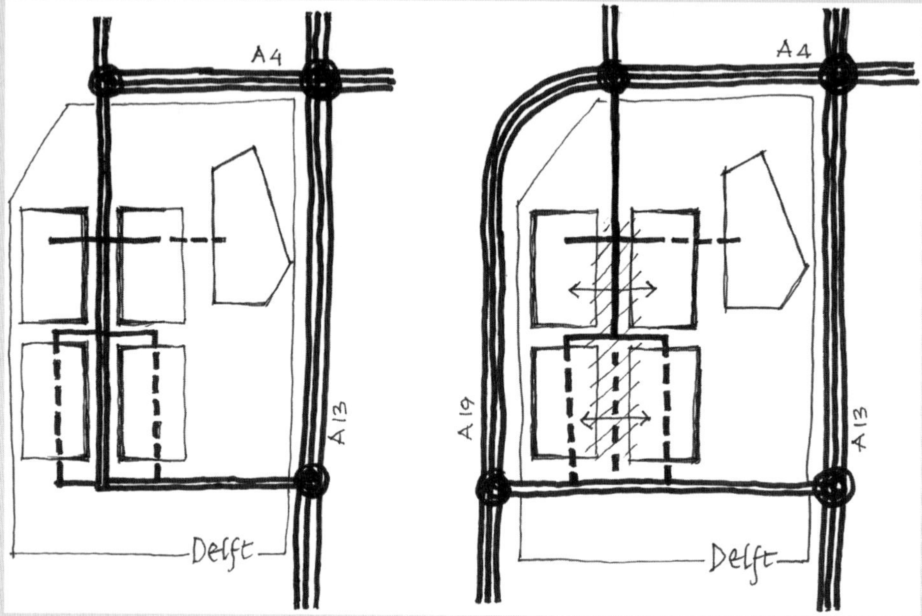


- Or the disciplines involved restrict themselves too much to serving the interests of their own profession;
- Or the disciplines are insufficiently capable of operating across the borders with other disciplines;
- Or the management of the planning process is insufficiently capable of stimulating the disciplines involved to make cross-border contributions.

These issues raise the question of how much attention should be given to good communication among the spatial disciplines in higher education.

## In higher education

For answers to the previous question at the Faculty of Architecture of the Delft University of Technology, we have to go back to 1948. In his inaugural lecture, as Professor in Urban Planning and Design Research at the Department of Architecture, Prof. Ir. Th.K. van Lohuizen stated that making urban plans could not be left to one person, but should be undertaken "by a number of practitioners with different talents, professional training and attitudes working together, the most important of which are urban architectural designers, researchers and engineers in different fields. It is of the greatest importance that by collaborating in this way, their knowledge, talents and insights will be synthesised, as though they were united in one person. To achieve this, it is necessary that all those involved have the awareness and capacity to immerse themselves in the way of thinking and needs of the others, and that they have a great respect for one another's work."

With regard to higher education, he stated: "Wouldn't it be a good idea if, during the study period already, contact could be made between future researchers into academic training, and students of urban architecture, so that they can get to know and appreciate one another's way of thinking, in the free manner that is the student's prerogative?" (Van Lohuizen, 1948). This pursuit of interdisciplinary collaboration resulted in the foundation of the first *Stedebouwkundige Studiegroep* [urban planning and design study group] in 1950, which was focused on the city of Gorinchem. The group consisted of 13 students, one from architecture, three from urban planning and design, seven from civil engineering and two from social demography (Van Hezik, Lemmers & Dijk, 1976).

The main goal 'to stimulate interdisciplinary collaboration in relation to the problems of spatial planning' was specified in 1974 as follows:

- Learning to recognise and interpret the character and cohesion of problems in the field of spatial planning, and in particular to be able to indicate their societal and scientific relevance;
- Learning to approach, or if necessary solve, these problems in a methodically and scientifically responsible way;
- Learning to collaborate with people from one's own and other disciplines in a co-operative team, in which all participants contribute optimally to the collective result, giving each other the opportunity to develop as comprehensively as possible;
- Learning to assimilate and apply knowledge, insights, methods and techniques from one's own and other disciplines in a critical way;
- Learning to establish objectives and methods independently; learning to manage a project independently, and to take responsibility for the entire project.

The first country-wide organisation of study groups started during that same year. Since then, inter-disciplinary study groups became an accepted and structured institution for project training in all Dutch universities. In 1975, the first study group focused on a foreign issue began to function and many others followed (Dijk, Geraedts, Van Woerden & Van Zonneveld, 1999). However, the situation changed completely in 1981, when a new law obliged the Dutch universities to reduce the length of study in scientific education from five to four years. Since the study groups functioned in de fifth year, they had to be suspended, despite the favourable opinions of students. A quick survey among graduates (Bertels, 1982) showed that 80% of them emphasised that participation in a study group was a valuable aspect of their studies. Despite this and the recommendations of a conference organised to discuss this issue (Bertels, Rijpma & Van Zwieten, 1986), the inter-university study groups were definitely abolished in 1984.

## Statement

The organisation of inter-disciplinary, inter-university study groups was always a complex task. The participation in the groups was not easy for the students. But they acquired useful experiences in dealing with the complexity of the various spatial and social disciplines and with the differences in disciplinary languages and concepts.

Evidently, there is no proof that the elimination of the *Inter-universtaire Studiegroepen Planologie* (as they were renamed in 1979) has been the direct cause of the stagnation in mutual understanding between the disciplines in practice. But the 1980s research mentioned earlier stressing the importance of inter-disciplinary communication in education gives an indication of what is missing in the present context.

Research on the influence of (the lack of) inter-disciplinary skills in the planning process should be stimulated to overcome the present problems between the spatial disciplines in the practice. The outcomes will surely underline the significance of leveling out language barriers between the disciplines, and the need for re-introduction of academic opportunities to put the students in contact with other spatial disciplines during their study.

## References


Bertels, F.G.M., 1982, *Een blik vanuit de beroepspraktijk op de interuniversitaire studiegroepen planologie-Delft*, Technische Hogeschool Delft, Delft

Bertels, F.G.M, J. Rijpma & K. van Zwieten, 1986, *Interdisciplinair Projektwerk, grenzen aan ISP en ISP over grenzen*, Technische Universiteit Delft, Delft

Dijk, W.J., H.G.M. Geraedts, W.M. van Woerden & J.B.F. van Zonneveld, 1999, *Innovatief Interdisciplinair Projectonderwijs*, Faculteit der Civiele Techniek, Technische Universiteit Delft, Delft

Nevalainen, M., 1996, *Transformatie Provinciale weg 15*, DN Urbland, Delft

Stadsontwikkeling Delft, 1991, *Structuurschets Delft*, Afdeling Stedebouw, Gemeente Delft

Stadsontwikkeling Delft, 1998, *Ontwikkelingsvisie Delft*, Sector Ruimte en Groen, Gemeente Delft

Van Hezik, M.L.A.M., L. Lemmers & W.J. Dijk,1976, *Vijftig Vijfenzeventig, 25 jaar stedebouwkundige studiegroepen*, Technische Hogeschool Delft, Delft

Van Lohuizen, Th.K., 1948, *De eenheid van het stedebouwkundige werk*, Inaugural Lecture, Technische Hogeschool Delft

# Part IV

## Strategies

# Introduction to Part IV

In the first three parts of this book, the contributions are primarily about substantial aspects of spatial planning. In Part IV, *Strategies*, the procedural side of spatial planning is brought to the fore. Planning implies achieving a stated aim in a structured and effective way, while strategic thinking is about ascertaining what should be achieved, and remaining focused on how it can be realised. Strategic thinking and working is a prerequisite for developing spatial plans. A deep awareness of the feasibility of the decisions that have to be taken provides the best chance of succeeding. Both the academic and professional sides nurture strategic thinking and working: the scientists offer theories and research instruments to improve spatial planning, and the practitioners analyse problems as they encounter them, on the basis of which, they formulate outlooks and solutions. Both substantial aspects of planning, and strategic factors are influenced by societal and technological developments.

Part IV starts with a chapter by Jan Vogelij *Towards Strategic Spatial Planning in Europe*. Vogelij explores the future role of spatial planners and the importance of educating them both academically and professionally. This must include a regional strategic orientation, and spatial development and planning in Europe. As president of the European Council of Town Planners (ECTP), he refers to the guidelines laid down in *The New Charter of Athens*, published by the ECTP in 2003. This is followed by *The Spatial Planner's Changing Roles*, an analysis by Hilde Blank and Iwan Kriens of Blank's experiences working as a practitioner in the Netherlands and in the United States of America. Described are the spatial planner's different contributions and the changes that have taken place in planning practice, largely due to policy changes.

Jan Bredenoord's chapter *Municipal Planning and Implementation Strategies*, concentrates on the specific possibilities offered by a strategic structural plan on the local level. He views this plan as an instrument for forming and maintaining strategy, and for this, proper communication on the local level proves to be a vital element. He bases his proposals on experiences in both Dutch and Latin American municipalities. Robin Houterman and Edward Hulsbergen discuss the local level further, and especially the importance of local initiatives, in *Neighbourhood Initiatives: Time for Bottom-Up*. Bottom-up initiatives from residents and users are often a direct response to local needs, but they are seldom linked to, or integrated with, municipal plans. Participation, empowerment and strategy are key concepts here. Crucial for the success of bottom-up initiatives is the (political) space and support given by local planning authorities in general, and spatial planners in particular.

In *Urban Mobility; Towards the Conditions for the Planning and Design of the Mobile City*, Remon Rooij explores the important role played by the new urban-transport technologies in the development and transformation of the socio-spatial environment. User movements on an urban scale reveal the functioning of urban areas and can be translated into strategies for seamless multimodal mobility in vital cities. In *Carrying Structures: Urban Development Guided by Water and Traffic Networks*, Sybrand Tjallingii describes the background of his 'two network strategy', a conceptual tool for urban planners, which utilises as a starting point the two spatial networks of water and traffic. The tool combines the ecologically sound management of these networks with contrasting conditions for built-up and green areas.

On the regional scale, Marc Jacobs's *Building a Multimodal Metropolis: A Short Guide* offers insight into planned and unplanned urbanisation. The development of a multimodal urban system, originating from two or more urban systems, is analysed and the spatial actions required to achieve such a system are identified. Jacobs stresses the importance of using an approach that is underpinned scientifically.

Finally, in *Olympic Thoughts in Urbanism: The New Charter of Athens*, Rypke Sierksma scrutinises the phenomenon 'charter'. The New Charter of Athens, approved in 2003, has figured in several chapters of this book, but there are other charters as well. Sierksma concludes that the function of both modern and post-modern charters is mainly ideological; and that they fail to identify the true object of town planning. He asks whether there isn't a gap between what systems theory actually performs, and the practical demands and expectations of spatial planners who use its terms. He closes with the thought that 'the town planner is left in good faith to muddle through, hoping that not too much will go wrong'.

# 22 Towards Strategic Spatial Planning in Europe

Jan Vogelij

## Introduction

The future role of spatial planners is an important determinant of the way planners will have to be educated. Most practicing spatial planners, who work for municipalities, regional bodies and consultancies, conduct local orientations. In future, however, a wider, more regional strategic orientation will be needed.

In this contribution, first the changing context of cities in Europe will be discussed, and how the increase in scale and the need to position the city within the context of networks is influencing the scope of the planner's work. Secondly, attention will be paid to a variety of activities related to spatial development and planning in Europe. These consist of community initiatives, member-state activities and transnational local and regional projects. Many individual spatial planners are already involved in these activities and many more will be involved in the future. Finally, an attempt is made to describe some educational and practical requirements that planners will need in the future. These include a wider, more strategic, orientation towards spatial developments.

## The changing context of European cities

The unification of Europe after the Second World War resulted in an increase in scale. In the highly fragmented European territory, this was not only a spatial increase, but also a mental increase in scale. Apart from some internationally oriented European cities and regions, most of the others experienced a widening perspective, from a provincial orientation to one of trans European network co-operation. Nowadays, cities and regions have to define their strategic position within that wider context. This forces them to critically review their identity and their qualitative attractiveness for economic activities.

The move to unify Europe has lowered the barriers between nations. This has created more direct international competition between cities and regions and a general widening of the context in which spatial developments have to be considered. With the exception of some large cities of global or European importance, such as London or Paris, it used to be enough to consider the role and position of a city within its national context. However, the surrounding region no longer offers the true spatial context. Nowadays, cities operating in the same functional or thematic networks, but located further away, also need to be considered.

Since the enlargement of the EU in 2004, the spatial context has grown even wider. The implication of this wider context is that, in future, cities will become more confronted with competition and so will be directly challenged with their own genuine qualities and competencies. National legal, fiscal and cultural differences will play a lesser role. On top of that, economic developments have also drastically changed the situation. Not only have traditional production industries moved to Asia or Eastern Europe, but essential parts of ICT service industries have also found cheaper locations.

In order to keep a certain level of prosperity in Europe, especially at this time of gentrification and decreasing population, Europe needs to reconsider its economic strategies. The questions that must be addressed are: What are the real comparative advantages of European cities and regions? For what economic activities can European cities and regions provide the natural location? Which of the specific newly developing economic sectors do fit in especially well with specific qualities of European cities and regions?

In the *Rise of the Creative Class*, Richard Florida (2002) describes the importance of creativity and the creative class for economic innovation. He has also analysed Europe against his creativity index. One of Europe's strengths is its large cultural diversity and its creative culture. New economic activities in Europe, apart from general production and services, should be based on these strengths. In line with this, cities need to analyse the qualities of their offer in relation to new economic activities and to the requirements of specific groups of the creative class. More than before, city managers will have to search for strategic opportunities. The winners will be those who capitalise best on their qualities.

The city's own identity is the result of historical developments in its specific topography, and it should be considered as a part of its capital. The physical expression of that identity, the spatial quality (spatial planners' concern) is important to its external image as well as its capability to attract new activities. The new, wider context in which European cities are required to operate, require them to critically consider their position in Europe-wide city networks. These may be physical and thematic networks, and an individual city needs to identify its position in both of them. For instance, the European high-speed railway system forms a coherent network. Every city in this system with a station lays a specific role in that network, as well as within its surrounding region. Next to its other endogenous qualities, a station in a network such as this, contributes to the city's capital for future development.

The same applies to inland waterway systems. A system that only functions with sufficient interdependent port facilities in various cities. But at the same time, the ports themselves add a quality to the regions in which they are situated, a quality which distinguishes them from cities without a port. Waterways cross different regions, so a port city in a wine-producing region will also be part of a network of places in the wine production chain. These cities are part of those networks of inland ports that stock wine and of those that act as market places for wine, but also of the networks of stocks and markets of other products shipped in other harbours. Spatial developments should be geared to enhancing their distinct positions by improving certain qualities and covering-up certain weaknesses. Therefore developments should be selective, aimed at distinctive specialisation, and strengthening the identity of individual cities.

In addition to the above mentioned examples of physically connected transportation networks and industrial networks, in which complementary industrial activities are clustered for instance for car building or food processing, those with a touristic theme are also relevant. Some networks on touristic themes build strongly on cultural historical connections, some of which are physically connected by routes (e.g. the pilgrim route to Santiago, wine routes, routes connecting fortified cities in former boundary regions, or for instance specific touristic routes following traces of Celtic Culture).

The Council of Europe initiated the European Heritage Network (HEREIN) with support of the European Union. In 1998 the Council of Europe entrusted to the European Institute of Cultural Routes the monitoring of the cultural routes programme (http://www.culture-routes.lu). Touristic themes offer possibilities for even small and medium-sized towns to become part of an international system which will help them to become more widely known. The challenge will be for cities to identify their qualities more precisely within the context of a larger group of more or less similar cities. It will be important to be able to identify and specify what exactly constitutes a distinct quality compared with those of others in the various networks, and to enhance those qualities in future developments.

The increased importance of the regional scale should be emphasised here as well. The more agricultural production decreases, the greater the urgency to reconsider the future of the regions where this is occurring. To combat international competition between regions, caused by vanishing national barriers, cities and smaller regional centres should identify their endogenous qualities as part of their region. Co-operation between cities and regions in networks also includes acknowledging each other's strengths and specialisations. Being complementary to each other in a coherent system may also raise city's awareness of the interdependency of their developments.

## The European Union and spatial planning

Although the member states have attributed important competencies to the European Union, the responsibility for future spatial development still rests with the individual nations. The European responsibilities for regional economic development and environmental protection have resulted in the combined policy aiming at sustainable development. Informal meetings of the national ministers responsible for spatial planning have resulted in an agreement concerning the long-term objectives for spatial development in Europe. This is known as the European Spatial Development Perspective. Although Interreg projects, as well as the European Spatial Planning Observatorium Network (ESPON) initiative, explicitly focus on spatial planning in the sense of distributing activities in the European space, and although the draft Constitution includes territorial cohesion, spatial planning is still not one of the EU's competencies.

The history of the unification of Europe is strongly influenced by the common experience of the Second World War. First there was the general wish to avoid the re-occurrence of war. Secondly Europe wanted to ensure the availability of strategic materials, such as steel and coal, and thirdly there was the wish for the European population to become self-sufficient in supplying their own food. Meeting these wishes required the strong co-operation of the European nations. Although this co-operation began in the 1950s, spatial development became never an explicit part of the co-operation.

The European Community firstly directed its attention towards securing the availability of coal and steel, despite the fact that these economic sectors were already shrinking at that moment. The second important aspect of European co-operation was the Community Agricultural Policy (better known as the CAP). For about half a century, the European Commission stimulated efficient food production by subsidising unit product prices. During that same period, the economies of mining regions gradually collapsed as the mines were closed down, bringing in its wake an urgent need for urban and landscape regeneration and suburbanisation in these areas. Coupled with high unemployment and social degeneration, these old industrial towns became unattractive places to live in. The agricultural subsidies mainly benefited farming on the north-western plains of Europe, bringing high prosperity there, but at the same time increasing the scale of farming operations and reducing variety in the rural landscapes. Small rural centres and middle-sized towns flourished and increased suburbanisation.

The EU was being confronted with the environmental impacts of its prospering economy. As an aid for balancing conditions for economic production in a level playing field, the European Commission gained responsibility for environmental policy as well. This resulted in a wide range of European environmental directives, which addressed all components of the environment. The confrontation between economic development and environmental protection also brought about a sectoral confrontation between two different Directorates General of the European Commission: DG XVI (Regional Development) and DG XI (Environment, Civil Protection and Nuclear Safety). The competing interests of the economy and the environment came together in the common policy on sustainable development. This policy sets requirements for spatial development, although neither of the two Directorates General is responsible for spatial development as such. In 1996, however, the variety of initiatives culminated in the policy paper *European Sustainable Cities*, which offered an outlook on sustainable spatial planning (European Commission, 1996).

Although still not an official responsibility of the European Commission, even more attention was given to spatial planning by programmes for co-financing local and regional initiatives through structural funding. The so-called Interreg programmes entitle local and regional authorities to financial support when carrying out transnational or cross border co-operation in spatial development (Ministerie van Volkshuisvesting, Ruimtelijke Ordening en Milieubeheer, 2000). The Interreg programmes are organised from six regional offices.

At the same time, the Ministers of the Member States responsible for spatial planning were working on what amounted to a plan for the whole of the European territory: the European Spatial Development Perspective (ESDP). After a decade of preparation, the document was adopted at the informal meeting of these Ministers in Potsdam in 1999 (European Commission, 1999). One of the requirements of the Interreg programmes is that the spatial planning objectives of the ESDP have to be incorporated in Interreg projects. In these projects, the initiatives of local and regional authorities are subsidised if they co-operate transnationally in spatial planning and in exchanging experiences. Hundreds of initiatives, throughout Europe, involving planners on the local and regional levels have been carried out, and all were obliged to address aspects of sustainable development and ESDP objectives. The number of planners involved exceeds that of the projects that were awarded subsidies. This is because project applications require extensive and detailed preparation, but not all of them were awarded subsidies. Interreg projects are organised according to super-regions, like the Baltic Sea Region, the North Sea, North West Europe, and the Atlantic Area. Spatial Visions have been prepared for most of those regions, aiming at a more regionally focused elaboration of ESDP objectives. These spatial visions were prepared to evaluate the individual Interreg projects by planning consultants for spatial vision groups composed of government representatives from the countries in each of the specific super-region.

An amazing result was that even governments that had never before prepared a coherent vision of their own national spatial development, now helped in preparing these large-scale spatial development visions. For instance, Belgium, France, Germany, Ireland, Luxembourg, the Netherlands and the United Kingdom prepared the mapped spatial development vision for the North-Western Metropolitan Areas (NWMA), later named North West Europe (NWE). The Spatial Vision for North West Europe contained some rather basic maps, illustrating envisaged developments. Of the involved countries, only the Netherlands had a long tradition in national planning. In Belgium, separate plans for Flanders, Wallonia and Brussels were prepared in the late 1990s. In Germany, spatial planning is a competence of the Länder, while the national government formulates the main strategic objectives. In France, La Délégation d'Aménagement du Territoire et d'Action Régional (DATAR,) the high competency Interministerial organisation, focuses on development planning through the 'contrats état-région'. In the United Kingdom, National Planning Principle Guidances are issued. Luxembourg and Ireland recently prepared strategic spatial development plans on the national level in 1999 respectively 2001. Ireland seems to have been really inspired by the role their representatives have played in preparing the Spatial Vision for North West Europe. It is quite noticeable that mapped national plans seem to be more feasible for the smaller countries than for the larger ones. Feasible or not, in most of the countries involved, there is a noticeable revival of interest in spatial planning. So it could be argued that EU initiatives have been instrumental in stimulating spatial planning activities in Member States.

Although prospective maps were produced when the European Spatial Development Perspective was being prepared in its first official draft in Noordwijk in 1997 (European Commission, 1997), at its adoption in Potsdam in 1999 a version was tabled that was back to the level of policy objectives, without maps depicting locations of developments. The policy objectives cover a wide range of issues and levels of scale. For instance, they address suburbanisation, trans-European infrastructure networks, new urban-rural relations, creative management of natural and cultural heritage, and polycentric urban development (European Commission, 1999).

At the pan-European level, the objective of polycentric urban development is especially interesting: it argues for spreading economic activities over the enlarged territory of Europe. This objective emerged from the notion that too many economic activities (40%) are concentrated in North West Europe, within too small an area (20%). This concentration occurs within the area known as the Pentagon, because of the five major cities, London, Paris, Milan, Munich and Hamburg, at its corners. This area partly overlaps the area that has been identified in an early DATAR study as the European economic core area (Brunet, 1989). This area consisted of a zone from London over the Benelux, the Ruhr area, along the Rhine valley towards Milan. The shape of this zone inspired to its name: it was known as the 'Blue Banana'. The DATAR study aimed at bringing under French government's attention the importance of the connection of Paris with the developments which took place around it, in the 'Blue Banana'. It was important for Paris to be one of the key cities in Europe's economic core area. Not surprisingly, Germany found also the 'Blue Banana' to be too limited, and wanted to include their cities of Hamburg and Munich. As a result the Pentagon including Paris, Hamburg, and Munich has been defined in the European Spatial Development Perspective as the economic heart of Europe.

The European Spatial Development Perspective objective to enhance polycentric urban development addresses similar political strategic considerations: for the social and economic coherence of Europe other, more peripheral, parts of the European territory should play a more important role. In particular, cities with high development potentials, such as Barcelona, Madrid, Berlin, Copenhagen, Stockholm, Prague, Warsaw and Budapest, should be stimulated as growth nodes outside the Pentagon. However, it is still an open question whether this political objective for the spatial distribution of activities is feasible in a period of economic decline. A set of decisions was made concerning the implementation of European Spatial Development Perspective. One of these was to establish the European Spatial Planning Observatorium Network (ESPON). This EU initiative explicitly addresses spatial development and spatial planning on the European level, covering 29 countries. Part of the ESPON programme is to set up a database, establish statistical evidence for spatial developments, and formulate suggestions and recommendations for spatial planning at the European level. In relation to the polycentric urban development discussed above, one of ESPON's lines of investigation is to analyse how capitalising on comparative regional advantages can best be coupled with the polycentric growth-node policy. Finally in the draft European Constitution the new policy objective territorial cohesion was introduced. Together with the other two long existing policy objectives of social and economic cohesion, all three aspects of sustainable development were now to be addressed by the main European official policy (European Commission, 2004). The developments described here may look like steps towards official European involvement in spatial (or territorial) planning or co-ordination.

It may be noticed, that although Europe has no competencies with regard to spatial planning, a long history of related activities address spatial planning aspects.
Through regional development, environmental policies, sustainable development became an accepted objective. The ESDP policy, the Interreg programmes and the ESPON initiatives ere further stimulants to spatial development attention on the EU level. Since territorial cohesion is an official part of the general EU cohesion policy, Europe's involvement with spatial planning is rather close. This poses new challenges to spatial development and planning.

## New challenges for spatial planners

The strength of the spatial-planning practitioner is at the same time a weakness. Their distinct design orientation, combining research and processing capabilities, sometimes leads to too great a focus on small-scale architecture. The changing context of cities and the spatial development within the EU require spatial planners to make an important strategic contribution. Recent planning activities in the EU, in connection with the Interreg and ESPON projects, have heightened expectations about the capabilities of planners. To meet these expectations, higher education should be preparing them better for the different roles that they will be required to play in future.

Traditionally, when they are involved in expanding or regenerating a city, spatial planners are oriented towards the spatial organisation and quality improvement of parts of a town or city and its public spaces. All proposals for buildings and for housing, economic activities and services are confronted with the possibilities and limitations of spatial and environmental legislation. In the Netherlands, in particular, spatial planning activities on the local level have been dominated for decades by the expansion of cities for housing purposes.

The economic activities of towns and cities used to be based more or less historically on their significance for the surrounding hinterlands. For instance, they acted as a market place for agricultural products, as a fortress to defend an area, as a regional capital, as a logistic centre for the transportation of goods, or as a place where raw materials were mined or manufactured. Together with providing services and provisions for the population, cities were more or less complete entities. Their economic basis did not come into discussion, especially during times of prosperity.

Recent developments, also on the European level, have changed this traditional pattern. The times have become more dynamic, traditional industries have disappeared, production has moved to cheaper parts of the world, European agriculture has to compete on the world market, the increase in scale of economic activities has detached these activities from stable localities. Prosperity is no longer self-evident in this part of the world.

To remain prosperous and to use prosperity to attract new economic activities, the decision-makers of European cities need to re-define the position of their cities within this new context. They need to specify their city's role within relevant regional and transnational networks, because their specific positions within those networks are important determinants for their relative strength and attractiveness, in the competition with other cities. In this task, the professionalism of spatial planners can be put to good use. Although accessibility and other physical aspects, and social aspects, such as the composition and educational level of the population, will remain generally important, in future, the competition between cities will be different from that encountered at the present time. It will be more specific and qualitative. The city's identity, based on its specific topography and the cultural historical development within it, will also be more important, because the physical expressions of that identity are part of a city's spatial quality.

Although modern economic activities are generally described as foot-loose, the capability of a city to attract economic activities will depend on new broader site qualities. When considering relocating, managers are in a position to choose between different sites of more or less equal quality, so it is other aspects of those sites that gain weight in making a decision. These are qualities that previously have been qualified as soft factors: such as the beauty of the town or city, excellent service levels, cultural richness, good schools, a civic centre with buildings of distinctive monumental identity, or a lively intellectual climate.

What is needed is a form of qualitative sustainable spatial development that connects the past with the future. Here the New Charter of Athens 2003 offers a way forward. In this document, the idea of the 'Connected City' is worked out in more detail (European Council of Town Planners, 2003). This concept of social economic and environmental connectivity strongly emphasises the importance of the internal as well as the external cohesion of the range of systems of which the urban complexity is comprised. The concept of the Connected City aims at strengthening the different identities of European cities so that they not only can be matched as effectively as possible with new economic activities and sites, but also offer new innovative milieu's, attractive for creative groups. By doing so, spatial planning and design contributes to the European innovation policy aiming at the position of the most competetitive knowledge economy: the Lisbon/Gothenburg agenda (European Commission, 2003).

Spatial planners' capabilities are needed for combining specific requirements and demands with site qualities. Research and design capabilities relating requirements for economic activities to specific spatial qualities should be enhanced, as should the specification of environmental requirements for creative groups. Richard Florida's finding that economic activities tend to follow the creative class, is relevant in differentiating creative groups according to specific economic segments associated with a city or a region. The different spatial preferences of those groups may be indicators of the spatial qualities that could offer opportunities to certain cities. Spatial planners, be they researchers, urban designers, mediators or urban managers, should be strongly aware of the city-specific spatial qualities that offer opportunities for future developments. Urban design should be oriented away from fashionable architecture , more towards strategically enhancing some times hidden identities that are related to specific economic activities.

In summary, the changing context of European cities as described in the second section, together with the consequences of EU policies described in the third section, result in enhanced requirements to spatial planner's capabilities. They regard a better knowledge of the strategic innovative economic aspects of territory specific spatial qualities considered within the wider context of the regional scale. This has important consequences for the training of spatial planners. The ECTP's New Charter of Athens 2003 states that spatial planners have to fulfil different roles. They can be researchers, designers, mediators and urban managers within a new, wider urban context that takes the spatial developments in Europe into account. To do this, spatial planners will need to place more emphasis on the long-term and wider spatial aspects of regional planning and design, gain a better understanding of economic developments in relation to spatial aspects, gain more knowledge of the relation between (historic) endogenous qualities and new economic activities, and define relevant groups of the creative class and their preferences more effectively.

## References

Brunet, R., 1989, *Les Villes "Européennes"*, Rapport pour DATAR, La Documentation Française, RECLUS/DATAR, Paris

European Commission, 1996, *European Sustainable Cities*, Report by Expert Group on the Urban Environment, Brussels

European Commission, 1997, *European Spatial Development Perspective, First Official Draft*, Office for Official Publications of the European Communities, Luxembourg

European Commission, 1999, *European Spatial Development Perspective Towards Balanced and Sustainable Development of the Territory of the European Union*, Office for Official Publications of the European Communities, Luxembourg

European Commission, 2003, *Competitiveness, sustainable development and cohesion in Europe. From Lisbon to Gothenburg*, European Commission, Directorate-General for Regional Policy, Brussels

European Commission, 2004, *A New Partnership for Cohesion, convergence competitiveness co-operation*, Third Report on Economic and Social Cohesion, Office for Official Publications of the European Communities, Luxembourg

European Council of Town Planners, 2003, *The New Charter of Athens 2003, The European Council of Town Planners' Vision for Cities in the 21st Century*, Lisbon, http://www.ceu-ectp.org/e/athens

Ministerie van Volkshuisvesting, Ruimtelijke Ordening en Milieubeheer, 2000, *A Spatial Vision for North West Europe: Building Co-operation*, National Planning Agency, Ministerie van Volkshuisvesting, Ruimtelijke Ordening en Milieubeheer, Den Haag

Florida, R., 2002, *The Rise of the Creative Class: And How It's Transforming Work, Leisure, Community and Everyday Life*, Basic Books, New York

**website**

http://www.culture-routes.lu

# 23 The Spatial Planner's Changing Roles

Hilde Blank and Iwan Kriens

## Introduction

Since the Housing Act (1901) was passed in the Netherlands, spatial planning has been linked strongly with socio-societal questions and with the corresponding political-administrative field of influence. In the previous century, there were various distinct periods during which the character of the discipline changed; this was caused by either an economic recession or by prosperity, or by a change in the political climate. Thanks to the dynamics that is a hallmark of the discipline, spatial planners have proved time and again to be able to develop a new characteristic profile and to contribute to the debate about the city and urban form. However, in the transition to the 21st century, the spatial planner's domain does not seem so self-evident anymore. This is enhanced by the changing role of the government, the influence of private parties and the increasingly complex societal questions. The field of the discipline is widening and socio-economic policy sectors are determining urban development to an increasing degree. Urban design is no longer the exclusive reserve of the discipline of urbanism (urban planning and design), architects and landscape designers are also becoming more and more involved in urban planning and design assignments. This chapter will give a sketch of the changing role of the spatial planner from the early 90s onwards, and concludes with key concepts for redefining the discipline.

## Process and content: Widening the discipline

In the 80s and 90s, in the Netherlands, there was a rigorous separation between urban planning and design studios and project management offices. The studios drew up development plans [*bestemmingsplan*nen] and gave advice on policy, and were left undisturbed to focus all their attention and energy on finding the ultimate design solutions, while the project management offices organised the decision-making process with its accompanying programme of participation and implementation activities. Process and content were viewed as two separate courses of activity. Because of the tension that often arose between the designer and the project manager concerning the societal and economic feasibility of the plans and the monitoring of time and budgets, many creative design solutions came to nothing (Nooteboom, 1999).

To link process and content together, a new (spatial) planning form was needed, a plan that no longer tried to achieve a pre-planned final image, but one that came into being by the interaction of process and content. The spatial plan came to mean something different, and so was used for another purpose. It was used to chart problem areas and to find solutions on a higher level of scale. The existing planning-juridical instrument, the Structural Plan, was considered to be unsuitable for this. These new planning forms were introduced with the help of terms such as structural vision, development outlook, urban vision or quality atlas. They enabled municipalities to establish integral outlooks as a policy instrument. The plans served as frames of reference for new initiatives and assignments but also as development frameworks for directing the spatial development of towns and cities. As far as content and level of abstraction were concerned, this new planning form was strongly related to the structural plan, but the advantage was that time was gained, as no spatial-planning decision-making procedures were needed any more. What is striking is that many municipal services did not develop these integral outlooks themselves, but relied on the assistance of external consultancies. Someone had to take responsibility for these integral outlooks and be accepted as the integrator who could prepare policy choices, independently. Where municipal urban designers failed, the external consultancy succeeded. This was because, on the one hand, an external advisor acting as integrator has a greater air of authority, and on the other, because there was a demand for another approach to the discipline of urbanism. A sense of political-administrative relationships, creatively combining wishes and developments in the fields of economy, infrastructure and living environment, and developing an implementation strategy at an early stage were difficult to realise for spatial planners in the civil service, because they themselves were part of the municipal machinery. The space that was offered to allow or to enable someone to act as an – external – integrator was largely dependent upon the decisiveness of the civil-service directors and council, or the alderman whose responsibility it was. Cities such as Maastricht, Groningen and Rotterdam, the administrations of which were very responsible politically in those days, with councils that were strongly oriented towards content, made a visible leap qualitatively in the 90s, in the field of urbanism and architecture.

In order to make structural outlooks, it was necessary to operate not only from one's own discipline, but to allow integral thinking to gain precedence. The challenge in doing this was not to slip back into compromises between the various sectors, but to grasp the opportunity to look across each other's borders. Working through the scales together resulted in revised concepts, ideas and solutions. In this way, the possibility arose to make work with work and to bind (external) parties together, by which the products that feature in the outlook are not only appealing, but also implementable. The accompanying process approach demanded of the municipal authorities a new culture and work attitude. Working on integral outlooks has been strongly educative for the local authorities and has often been behind civil service reorganisations.

By the end of the 90s, the spatial planner's role had widened into a combination of integrator, visionary designer and manager.

## The guardian of spatial quality at a higher level of scale

After the turn of the century, the administrative political climate in the Netherlands changed. Because of the political shift from left to right, the emergence of many local-interest parties, the dual system (Albeda, 2004), the economic recession, and the fact that there has been no new Government Memorandum for spatial planning as yet, the need for grand, compelling structural outlooks has diminished. Long-term outlooks are often theoretical and abstract, and appeal less to the imagination, while the 'New Politics' wants to see quick results and demands plans oriented towards implementation, without an interactive planning process. It is a reaction against the previous period that was characterised by the 'polder model', which was an endless process that continued until consensus was reached.

Nowadays, spatial planning has a lower ranking on the political agenda. Issues such as safety, welfare and the economy are becoming, or are already, more important. More attention is being given to large-city policy, and outlooks are becoming less and less part of the spatial planning domain. The influence of social and economic developments and, consequently, also the influence of these sectors of policy, is getting stronger, and the (municipal) spatial planner's task is changing from directing to following. Under the influence of these changes, it might seem as though the planning process would be more difficult to manage; the town councillor and the top civil servants are becoming executors rather than preparators of policy. There is a growing awareness that the government cannot manage on its own, so municipal services are seeing their role, as visionaries and strategists, disappear. Instead, they are being allotted the roles of planning assessors, budget guardians and regulators of spatial planning procedures. A new balance in roles and responsibilities is being sought within both the government (state, province and municipalities) and market parties (developers, builders, speculative and long-term investors), but also between the public and private sector.

Traditionally, the private party's role has been that of implementor, and they only came into the picture once the outlook had been developed, and when there was an approved development plan on which implementation could be based. Because of the shifts mentioned above, the private party now becomes involved at an earlier stage in the planning process, more frequently putting themselves forward as initiator. The private sector sometimes adopts the role of the public sector, and with that, risk-bearing moves to the beginning of the planning process. The private sector commissions consultancies in urbanism and designers or employs its own concept developers. The private sector takes on, as it were, part of the public sector's task, and, in so-doing, is given a larger decision-making role in the spatial layout of the Netherlands. As a result, the public sector has to find a new role, that will more frequently manifest itself as the guardian of qualitative cultural heritage and landscape. To help it perform its new role well, the public sector is developing new planning instruments.

A striking example of this is the image quality plan at urban and regional structural level, such as the one that was made mandatory for the area 'Noord-Holland North' by the province of Noord-Holland (Provincie Noord-Holland, 2005a). A development image will be created that will offer both local government and private enterprise a lot of room for manoeuvring. The image quality plan is meant to stimulate planners, when taking on new developments, to be inspired by the individual character of the area, the qualities of the landscape and valuable cultural elements. The role of the provincial government is shifting from that of empowered authority and assessor, to that of knowledge broker. In this way, the provinces stimulate thinking about spatial quality on the regional level. This manner of working is an illustration of the search to find the correct balance between allowing the private sector and local authorities more planning freedom, and determining the rules for safeguarding the basic quality of the layout of Noord-Holland. To increase its grip on spatial quality, the province of Noord-Holland is even going to appoint a provincial architect (Provincie Noord-Holland, 2005b). More and more urbanists are also appearing on the local level. On the national level, the *Rijksbouwmeester* (Gouvernement Architect) has appointed special Rijksadviseurs (Gouvernement Advisers) for landscape, cultural history and infrastructure (Ministeries van Onderwijs, Cultuur en Wetenschap; Volkshuisvesting, Ruimtelijke Ordening en Milieubeheer; Landbouw, Natuur en Voedselkwaliteit; Verkeer en Waterstaat; Buitenlandse Zaken; Economische Zaken en Defensie, 2005).

A natural tension can be observed between the various governmental authorities (national, provincial, municipal). Together, they represent the public interest, but they appear to have different opinions about how to go about this. At first sight, spatial planning in the Netherlands appears to be regulated very well, and to have a range of instruments to fit the various levels of scale, such as the regional plan, the structural plan and the development plan. In actual practice, however, this set of instruments often causes irritation, because a lot of time is lost in lobbying, and in finding creative interpretations to increase ones own influence. It sometimes seems as though policy documents are designed as point from which to deviate, rather than that they are used as a framework for assessing and developing. Deregulation in the Netherlands offers the chance of being able to review the procedures for spatial planning thoroughly. However, the question is whether this will stimulate decisiveness, solve the stagnation in building production, and whether it will contribute to the spatial quality of the layout of the Netherlands. Maybe it would be wiser to re-evaluate the existing planning instruments to get rid of all sorts of extra regulations that have been added in the course of time. This is pre-eminently a research assignment that pertains to the field of study of the spatial planner; the university environment offers the possibility to develop and test experiments in this field.

Because of the increasing complexity of spatial planning and the increase of actors who play a role in it, there is currently a strong demand for taking spatial quality up again as a spearhead, especially spatial quality on a higher level of scale. The spatial planner is becoming the guardian of spatial quality and the developer of new planning instruments that will anchor this quality in spatial planning policy.

## Area development as the key

The growing interrelation between public and private parties can certainly help to accelerate the realisation phase of projects, but the question is whether intermixing divergent interests should also take place at an earlier stage in the planning process. After all, there is a considerable difference in interests between public parties serving the public interest, and private parties with their own business interests. Solutions for desired, but financially impossible, developments will frequently be found by placing the assignment within the context of area development, where private parties are allowed to take the initiative. However, although this will cover specific situations, there will always be those which will remain partly unprofitable (Adviescommissie Gebiedsontwikkeling, 2005).

The differences and similarities with between the US (California) and the Netherlands are remarkable. California has a form of spatial planning that is strongly managed by the private sector, while the Netherlands have been managed, from way back, by the government. In both situations, it is noticeable that the concept of area development is used. In the Californian situation, this means that the areas that are developed are those that are not interesting for the private parties, while in the Dutch situation, it means that because the Dutch authorities cannot complete the projects on their own, they are looking for development partners in the private sector (Blank, 2004).

In California, there is an institution called the 'Redevelopment Agency' (California Redevelopment Association, 2002). A redevelopment agency helps the local authorities in revitalising the city by stimulating new developments, creating employment and organising collaboration between local authorities and private parties. Stated simply: the tax revenues arising from increases in real-estate values all go to the area itself. Banks loans and other financial arrangements are contracted in this basis. A redevelopment agency can be set up if at least 50% of an area is struggling to combat unfavourable physical or economic circumstances. This must first be approved by the municipality or the elected representatives of a community. The redevelopment agency is a separate organ in the municipal organisation and is accountable to the regular authorities. The agency is set up with a clear aim in mind, and as soon as that aim is reached, the agency will be discontinued again. A redevelopment agency makes a plan for the area for which it is appointed, that contains a spatial outlook and economic and social targets. Its role is clearly directive, both in the initiating phase of projects and in the implementation phase. The redevelopment agency acts as an intermediary between private and public parties and has an important means of directing, namely, money. Its role as an intermediary is effective because all parties remain recognisable; each having its own role to play and taking its own responsibility. Examples of areas that have been developed by setting up a redevelopment agency are the Gaslamp Quarter in San Diego (San Diego Redevelopment Agency, 2000), the centre of San Jose (San Jose Redevelopment Agency, 2000) and Mission Bay in San Francisco (San Francisco Redevelopment Agency, 1998).

In the Netherlands, area development is viewed as a new possibility for realising seemingly unprofitable spatial developments. Area development stands for integrality, in which it is important that developments are found that reinforce each other ('win-win' situations). Collective targets must be formulated that are directed towards reaching spatial-functional, economic and socio-societal quality demands. Area development can only succeed when there is co-operation between complementary parties; when each party does what they are good at, and when parties complement one another in expertise. By organising finances and implementation on the area level in such a way that an equalisation between the various developments can take place, unprofitable parts can be included in the realisation. A prerequisite is that a collective plan is developed, based on a collectively felt urgency. Examples of area development are the Wieringerrandmeer (www.wieringerrandmeer.nl) and the Bloemendalerpolder (www.noord-holland.nl/projecten/Bloemendalerpolder).

To accelerate the planning processes and to keep the desired ambition going to the end of the realisation phase, the spatial planner will acquire the role of development planner. This will mean a further widening of the area of knowledge to include real-estate development and strategy development within the area of financing.

## Less control: More demand for strategy

The presentation of the *Nota Ruimte* [National Spatial Strategy] in March 2004 (Ministerie van Volkshuisvesting, Ruimtelijke Ordening en Milieubeheer, 2004), marked a turn. It gave clarity to the spatial layout of the Netherlands and it made clear what the national government would deal with and what would be left to other parties. Municipalities and provinces will be offered more space to manoeuvre, and private parties and the societal middle field will be given a more prominent role in the future spatial organisation of the Netherlands. The State and the provinces will of course not relinquish control over spatial organisation in the Netherlands. This has created a lot of uncertainty. As a result, there has been a revival of the integral outlooks in which frameworks are created for spatial developments, ambitions are established, priorities are set and investment flows are directed. The outlooks are seen as an instrument for lobbying other authorities and as the basis for making arrangements and agreements of intention with private parties. Both the State and the provinces are considering how to deal with the spatial development of the country and the region and what sort of method is required. As a complement to the more traditional authorisation planning, development planning (Dammers, Verwest, Staffhorst & Verschoor, 2004) is entering the scene; a phenomenon, that has been around for much longer, known as 'integral area development', now brought to the fore again by the National

Spatial Strategy. In the provinces, 14 pilot projects have been designated (www.2.vrom.nl/notaruimte/p.html) in which, by using new forms of co-operation between the parties and sectors involved, the development possibilities of areas will be enhanced and their inherent development strengths will be used to accelerate the planning process. This requires another approach to spatial planning, regulations and planning procedures. As far as content is concerned, the layer approach will make an appearance (Ministeries van Volkshuisvesting, Ruimtelijke Ordening en Milieubeheer; Landbouw, Natuur en Voedselkwaliteit; Verkeer en Waterstaat; Economische Zaken, 2004), in which areas to be developed will be analysed from the moment they were opened up, to their present use and appearance. As far as process is concerned, public and private parties, but also other interested parties, will enter into co-operative agreements at an earlier date, in order to combine various initiatives and energetically and simultaneously pursue their realisation. Within the total area, possibilities for equalisation will be sought, so that valuable elements or necessary, though unprofitable, facilities can be retained and materialised. Large-scale projects, such as *Ruimte voor de rivier* ['Room for the River'] (Ministeries van Onderwijs, Cultuur en Wetenschap; Volkshuisvesting, Ruimtelijke Ordening en Milieubeheer; Landbouw, Natuur en Voedselkwaliteit, Verkeer en Waterstaat, Buitenlandse Zaken, Economische Zaken en Defensie, 2005), coastal protection, dike reinforcement and National Landscapes will be linked to new building assignments. The underlying idea in this is that the qualitative approach and the synergy between different interests will eventually lead to innovative, renewing and realisable planning and design solutions.

The developments on national level and the large on-going assignments in the Netherlands require a strategic approach of the discipline. Combining the various spatial, often necessary, assignments in an intelligent way offers opportunities for realising innovative design solutions. For this, the spatial planner's field of study will be required to develop in the direction of strategic urban planning and design.

## Conclusion

The urban societal complexity and the political-administrative context that formulate the frameworks within which this complexity has to be tackled and unravelled is threatening to submerge the spatial planner's identity. At the same time, the field of study of the urbanist is no longer restricted to that discipline, but is often annexed by architects or landscape architects. However, because of the above-mentioned complexity, a need has arisen for the spatial planner to have a clear profile. The broadening scope required in practice, offers the institutes of higher education the possibility of developing the discipline further and of starting co-operative links with comparable fields of study both at home and abroad. In his article about what sort of urbanism will there be in the future (Drewe, 2005), Paul Drewe has called on the Faculty of Architecture of Delft University of Technology not to let urban planning and design become part of architecture, but to broaden the discipline to 'Urban Studies and Planning'. In this way, the TU Delft can express its societal and scientific mission to strengthen the significance of spatial planning. In his article, Drewe also refers to the New Charter of Athens (European Council of Town Planners, 2003), in which the European Council argues four roles for the spatial planner:
1. Scientist;
2. Designer and visionary;
3. Political advisor and mediator;
4. Urban manager.

In practice, it has become clear that the economic and social policy sectors cannot take over the spatial planner's role. From those sectors, in particular, there is a need for the spatial visualisation of future assignments. These assignments take place in very different domains, such as those of the ageing, the transformation of manufacturing industry, shrinking cities and changes in mobility, and can be actualised at specific locations. As the analysis in this chapter has shown, this is why an extra two roles can be added to the four listed above:

5. Strategic urbanist;
6. Development planner.

The strategic urbanist plays an important role in the initial phase of a project, in which the aim is to get a good picture of the assignment by researching while designing, through all scale levels and all policy disciplines. The emphasis lies on his or her analytical capacity, rather than on the capacity to find solutions. The strategic urbanist also looks for the interventions that will act as a driving wheel for achieving the implementations. The development planner is an expert in the field of area development and serves as an intermediary between public and private parties (including societal organisations) with the aim of speeding up planning processes and advancing their realisation, without losing sight of the planning ambitions.

Strategic urbanisme and development planning will be key concepts in redefining the spatial planner's field of study.

## References


Adviescommissie Gebiedsontwikkeling, 2005, *De Praktijk van gebiedsontwikkeling*, Lysias Consulting, Amersfoort

Albeda, H.D., 2004, *Doelmatig en doeltreffend, controle door de gemeenteraad in het duale stelsel*, Stichting Rekenschap, Amsterdam

Blank, H., 2004, Silicon Valley, op zoek naar publiek-private samenwerking BVR-NL; In: Eric Pasveer (ed.), *BVR, Ruimte en Regie*, Uitgeverij Thoth, Bussum (Neth), pp. 114-130

California Redevelopment Association (CRA), 2002, *The Community Guide to Redevelopment*, California Redevelopment Association, Sacramento Cal.

Dammers, E., F. Verwest, B. Staffhorst & W. Verschoor, 2004, *Ontwikkelingsplanologie, Lessen uit de praktijk*, NAi Uitgevers, Rotterdam

Drewe, P., 2005, Welke stedebouwkunde van de toekomst?, *Atlantis*, 16.2, pp. 39-43

European Council of Town Planners, 2003, *The New Charter of Athens 2003, The European Council of Town Planners' Vision for Cities in the 21st Century*, Lisbon, http://www.ceu-ectp.org/e/athens

Ministeries van Onderwijs, Cultuur en Wetenschap; Volkshuisvesting, Ruimtelijke Ordening en Milieubeheer; Landbouw, Natuur en Voedselkwaliteit, Verkeer en Waterstaat, Buitenlandse Zaken, Economische Zaken en Defensie, 2005, *Aktieprogramma Ruimte en Cultuur*, Den Haag

Ministerie van Volkshuisvesting, Ruimtelijke Ordening en Milieubeheer, 2004, *Nota Ruimte: Ruimte voor ontwikkeling*, Ministerie van Volkshuisvesting, Ruimtelijke Ordening en Milieubeheer, Den Haag

Ministerie van Verkeer en Waterstaat, 2005, *Ruimte voor de rivier*, Den Haag (www.ruimtevoorderivier.nl/)

Nooteboom, C., 1999, *Nooit gebouwd Nederland*, V+K Publishing, Blaricum

Provincie Noord-Holland, 2005a, Beeldkwaliteitsplan, Inspiratie voor kwaliteit, *Ontwikkelingsbeeld Noord-Holland Noord*, Provincie Noord Holland, Haarlem

Provincie Noord-Holland, 2005b, Ontwikkelen met kwaliteit, Ruimtelijke samenhang op uitvoering gericht (Streekplan), *Ontwikkelingsbeeld Noord Holland Noord*, Provincie Noord Holland, Haarlem

San Francisco Redevelopment Agency (SFRA), www.sfgov.org/site/sfra, project Mision Bay (1998)
San Diego Redevelopment Agency (SDRA), www.sandiego.gov/redevelopment agency, project Gaslamp Quarter (2000)
San Jose Redevelopment Agency (SJRA), www.sjredevelopment.org, project Down Town (2000)

# 24 Municipal Planning and the Strategies for Carrying it out

Jan Bredenoord

## Introduction

This chapter is about strategic spatial planning on local government level. In the Netherlands, municipal development plans are called *structuurplannen* (structural plans) because they are concerned with the underlying structure of the city or region as a whole, but in this chapter the term 'master plan' is used. Before these master plans can be realised, a municipal planning system and an implementation realisation strategy are needed. The most important areas of activity in this context are municipal prioritising, municipal organisation and monitoring and directing external actors. We can say that there is a municipal development plan, when it is actually being carried out. The plan is put into practice by carrying out strategic (territorial) projects and making strategic investments using private funding, and the longer-term investments of entrepreneurs and, for social housing developments, housing corporations. Almost every city or municipality is working on developing ideas for the future. In most cases, this is a periodic activity, in which policy for the physical environment (connections, functions, land use, etc.) is augmented with the main points of the socio-economic policy. The best way to do this would be to upgrade the master plan into a municipal development plan.

In this chapter, attention will be given to master planning and how it is practiced within some municipalities in the Netherlands. This will be compared with similar developments in municipalities in Latin America. Of particular interest is the North–South co-operation between the Dutch municipality of Utrecht and the Nicaraguan municipality of León, which has resulted in the development of a planning system for León. Bringing this priority urban development project into a form in which it can be carried out, is a point of attention in both countries at the present time.

## About master planning

Master planning means the development of spatial ideas for the future development of municipalities, cities and (urban) regions. The best way to do this is to work interdisciplinarily. The participation of the local population and their organisations, other social organisations and private investors is indispensable. Every planning level has its own actors, who will participate in different ways, and the municipality is very dependent upon the influence of those external actors (hereafter called the participants) for enabling the realisation of the ideas for future development. So, in order to make the right decisions concerning municipal ambitions, and to establish the right priorities, the municipality must obtain appropriate knowledge about the participant's potentials, and within the municipality's organisation, there needs to be a well-structured planning practice, in addition to the master plan. However, the municipality's organisation must show the pros and cons of relevant spatial, ecological, socio-economic and financial interests. To be able to carry out ideas for future developments, additional instruments are needed, such as an investment strategy and a plan for realising it. The relation between the long-term master plan and short-term implemative activities should always be clearly differentiated.

## The importance of master planning

Holding onto the main principles of municipal policy for a long period of time, which is a necessity where master plans are concerned, is viewed as being politically problematic. In daily practice, new issues arise which attract the attention of (some) politicians, who may or may not, depending on their own agenda, take into account the consequences for current priorities of dealing with these issues.

The effectiveness of the planning depends to a large extent on the quality of the municipal management (the officials as well as the politicians). The effectiveness of the municipal organisation, the extent of governmental financial resources available and the municipality's role in directing the activities of participants in 'the market' will be discussed below. Ideas for future development are worked out alongside strategic projects and by means of long- range sectoral investment programmes. If the local government decides to undertake a too large number of strategic projects at the same time - which is the rule rather than the exception – both attention and investments become too widely spread. Setting and maintaining adequate priorities is one of the most difficult tasks in master planning. It is essential to strengthen local government in this matter.

## Some uses in the Netherlands

Most larger municipalities in the Netherlands have a master plan or ideas for future developments. Some of them, such as Amsterdam, Groningen, Breda, Apeldoorn and Tilburg, have a long tradition of structural plans. In 2004, Zonneveld and Tjallingii carried out an analysis of 25 years of master planning in Breda (Zonneveld & Tjallingii, 2004). In this analysis, they underlined the necessity of master planning and of designing ideas for the city of the future, but noted that the domain within which municipalities can act has become limited, due to modern (scale) dynamics. Nevertheless, at the same time, new functional connections and relations are coming into existence, which should help to counteract that trend. It can be assumed that Zonneveld and Tjallingii's conclusions for Breda also apply to the planning dynamics in other municipalities in the Netherlands, and that these dynamics make ongoing monitoring and planning inevitable. The municipality of Breda maintains its own ideas for future development (taking into account societal ambitions and those of the city council) by basing the annual *Kadernota* (an overview report) on them, in addition to the municipal investment plan, and urban renewal and greater city projects. An urban monitor, an urban management model and urban development programmes are used as complementary instruments.

The municipality of Nijmegen recently stopped using a formal structural plan and opted for a refreshing new approach, a flexible steering instrument for spatial development known as the *Book of urban opportunities*. This book of opportunities gives insight into all the municipal developments, their coherence, new development outlines, and their possible effects on programming, policy and realisation. The status, being free from obligations, is however under discussion.

**Illustration 24.1: Structural vision for the municipality of Utrecht 2015-2030 (Dienst Stadsontwikkeling Gemeente Utrecht, 2004)**

In 2004, the city council of Utrecht announced a new master plan, one of the main features of which was to determine the relation between the existing city and the large expansion area Leidsche Rijn. This *Structuurvisie Utrecht 2015–2030* (Dienst Stadsontwikkeling Gemeente Utrecht, 2004) is, above all, a spatial plan based on sectoral plans and on existing plans for neighbourhoods and urban districts (Ill. 24.1). Spatial policy, however, is often implemented along sectoral lines, which makes integral programming difficult and sometimes impossible. The Utrecht master plan is a good urban development tool, with site-focused projects. However, its status is somewhat open (free from obligations) and there is no clear linkage between monitoring and actualisation. Because Utrecht is the central city of an urban region, for the spatial development of that region, it co-operates with nine other municipalities. All the participating municipalities have their own master plans or are busy establishing them. These plans make up the substance of the regional master plan. The *Regionaal Structuurplan Utrecht* (Bestuur Regio Utrecht, 2003) is by no means free from obligations. This is because the agreements with the central government and between the municipalities are like implementation contracts. The regional master plan reached its final stage at the beginning of 2005 and an *additional strategy* for implementing the main targets is currently being formulated. In this case, the participants are the ten municipalities in the urban region and the provincial and central governments. Whenever strategic projects are being determined, they try to find and involve private participants to actually carry them out.

Another example of a master plan for an urban region is that for the *Network City of Twente* (Regio Twente, 2002). This plan gives insight into the future collective development of four municipalities (Almelo, Borne, Hengelo, and Enschede) around the two most important development concepts. The first of these concerns the public transport system, and the second, the design of the surrounding structure of the landscape and its connections with the green urban areas. The urban and regional coherence between all areas with regard to housing and economic functions will be strengthened by linking these physical structures and basic qualities, such as landscape and urban centres, and, at the same time, the functional specialisation will be improved. However, despite the consensus on paper, the harmonisation of all these activities, and the co-ordination of the different municipality's policies, the regional and provincial government is insufficient as yet to carry out completely the collective ideas for urban regional development. An ongoing discussion point contributing to this seems to be deciding on the functional specialisation of the participating municipalities. The municipality of Enschede, as participant in the regional plan, has its own master plan called *Ruimtelijke ontwikkelingsvisie 2015-2030* [Spatial development vision] (Ill. 24.2). This was announced in 2001, for implementation in 2005. This plan, used as input for the network-city plan, establishes the preconditions for urban developments at the lower planning levels. Its status is comparable with that of the Utrecht master plan.

It can be concluded that municipalities and urban regions work with master plans or structural plans, but that, in a number of cases, their status, in the Netherlands at least, is rather weak, and an adequate planning system, with monitoring, prioritising and actualisation is not always available. In an attempt to avoid the 'free from obligation status' and, in some cases, an isolated sectoral approach, some municipalities try to find ways of developing an implementation plan as part of the master plan.

## Some uses in Latin America

Because of the necessity of giving strong impulses to urban development through housing and economic development plans, a large number of municipalities in Latin America are actively setting up municipal development plans (or strategic plans as they are sometimes called). In many Latin American countries, local economic development is strongly needed to combat urban poverty. The main target is often to promote well-balanced development within an ecological, social and economic setting, with the aim of it eventually leading to sustainable development. The municipal development plans of the district of Villa el Salvador (in the southern part of the metropolis of Lima), Peru, and the municipal district of Ventanilla (a satellite city in the northern part of Lima), are examples of modern municipal master plans. The Pachacútec Pilot Project, a huge urban expansion plan, lies in the municipal territory of Ventanilla, but it is being developed in a rather isolated way and facilitated by the central government. The only option for this project is for it to be integrated eventually within the municipality of Ventanilla. The Villa el Salvador development model has attracted a lot of international attention because it incorporates a number of interesting advantageous aspects such as urban participation, participative urban budgeting and concrete economic facilitation from the municipality, all of which are rather unique in these countries.

The strategic municipal development plan of León, the second city in Nicaragua, is the next interesting example. This municipal planning system has been developed during the last few decades in close co-operation with the municipality of Utrecht in the Netherlands. In Nicaragua, the national government established the National Development Plan (*Plan Nacional de Desarrollo, PND*; Government of Nicaragua, 2003) and the municipalities and regions were invited to contribute to it by defining their own potentials. The government is trying to use this national plan to promote sustainable economic developments through co-operation within local and regional public-private partnerships. Practically every municipality in Nicaragua now has a strategic development plan with economic elements. On

**Illustration 24.2: Master plan municipality of Enschede 2015-2030 (Regio Twente, 2002)**

the one hand, the economic potentials of León can be used to enable the realisation of some of the main targets of the PND (e.g., the cities of both León and Chinandega in North-Western Nicaragua), and, on the other, the PND can help to further local/regional plans (such as those for León). This is now all in progress.

## Strategic planning: some conclusions

The municipal master plan and the development ideas for urban regions embody a variety of approaches: the managerial approach, the physical/spatial approach and the public finance approach. The importance of these types of plans is increasing, because they form the municipal integration framework for the traditional sectors. Cities are compositions of networks of infrastructure, green structures, functional and socio-economic structures and components. Besides that, they form part of larger regional, national and international networks. Relations between the different levels should always be clear. Municipal master planning should have a strongly defined outlook and be focused on the main aspects of policy. Urban policy is not just a matter of dealing with threats; good opportunities also present themselves. These offer development chances, and the challenge is to incorporate them. When the regional level is under discussion, co-operation is more and more with governmental participants. One can have direct democratic influence on the municipal level, but it is essential to harmonise the different levels of planning, which is a municipal task in the first place.

The most concrete planning levels are those for he neighbourhoods and projects, but at municipal level all interests converge. This is why in each municipality in Nicaragua a Municipal Development Committee is being established. This is a legal requirement. Besides the municipality (acting as coordinator and facilitator), all important actors in the community, private actors and neighbourhood representatives are participating. In Dutch municipalities, an equivalent could be realised; a kind of (non political) 'forum for municipal development', in which all participants are equal.

## The city-wide approach

When strategic planning at local level is under discussion, spatial aspects become important. The city-wide approach combines the physical qualities of the city and the technical networks of infrastructure with the social, economic and cultural aspects, and underlines the coherence of urban society, despite all the differences. This broadened approach brings together all activities, investments and potentials, so that it is the inhabitants, entrepreneurs and other participants who become the most important actors in self-activation and participation. The function of the municipal planning system should then be to support them and make space available so as to create opportunities for all (acceptable) activities. For example, it could create special zones for new economic initiatives and new forms of housing, and houses with workplaces in or near residential areas. On the local level, formalised co-operation between the local planning board and housing corporations (which are very important in the Netherlands) could be explored and long-term contracts for managing housing stock and public space could be drawn up. Within neighbourhoods, it is more important to directly involve residents and entrepreneurs. If local management supports good and realisable local initiatives, this can lead to new housing and economic projects, so that the municipality can concentrate on redesigning public space, thereby improving the local economy and the quality of life.

A city-wide approach can be supplementary to the traditional municipal steering models. In Tilburg, for example, the 'city programme' is an integrative instrument for sector programmes, and from that starting point, they also determine neighbourhood budgets. Budgeting for neighbourhoods (for future developments rather than normal maintenance) is based on quality data and on the possibility of supporting favourable initiatives. Municipal management is acting more and more as a facilitator of urban processes by establishing preconditions, applying for subsidies, co-ordinating and stimulating.

Local government can indicate which municipal territories have been designated as consolidation areas. Some residential areas are vital urban territories that must be correctly preserved. Other urban territories will be transformed to another type of neighbourhood, for example with more economic activities or a higher density, or for other target groups. The monitoring of the quality of urban areas, in relation to urban planning is a very important activity. Although some municipalities have developed a city monitor (e.g., Breda and Utrecht), more monitoring systems would be desirable. In the city-wide approach, the municipal master plan can be expanded into a real Municipal Development Plan, including an implementation strategy.

In these paragraphs, a number of aspects concerning strategic planning have been described. However, if one talks of an ideal municipal planning system, a further description is needed. A selection of the good practices of a number of municipalities is given below. It is preferable to base local strategic planning on the city-wide approach, to gain maximum participation at all levels. The most important items needed to develop municipal planning systems are:

1. Instruments, e.g. a structural plan, a master plan for future development, a book of urban opportunities, a municipal development plan or implementation plan;
2. Urban monitoring and research as an ongoing activity if necessary with some research assignments carried out by external parties, with representatives of these activities occupying a strategic position on the municipal planning board;
3. The main aspects of municipal policy should be integrally and strategically developed and presented in the final municipal development plan;
4. Strategic urban management must be integrative as far as the main aspects are concerned so that realisable short-term goals can be achieved. Municipal managements continuously up-date their ideas for future developments, implementing them periodically;
5. The possibilities for local government to invest in the city are limited; many investments, such as subsidies or projects, come from other government levels;
6. The urban participants, their interests and investment potentials must be investigated. For example, in the Netherlands, the housing corporations own a large proportion of the housing stock. Investments of other private parties are also very important;
7. It would be advantageous to set up a municipal forum for urban development;
8. Attention should be given to participation, informing the public and the active participation of reliable parties.

Besides using an integral approach to solve urban problems, the housing, economic and other sectors will also be important areas of attention. In the housing sector one could think of activities such as investigating housing demand and other housing market aspects, sector programming, public-private consultations with participants in the housing sector and setting up sector agreements. In a comparable way, one could think of public-private consultations in the economic sector, for example with professional groups and the Chamber of Commerce. Such consultations should always be based on the municipal development plan.

## The urban planning system in León

From 1994 onwards, the city to city co-operation between León and Utrecht gave way to the setting up of a municipal planning system in León. This is one of the first two strategic development plans in Nicaragua; the other one is for the municipality of Estelí.

### The start of structural planning in León

Structural planning has been carried out in León since 1985/1986. The first overall planning instrument at that time was the EUDOF (*Esquema Urbano de Desarrollo Fisico*). It was set up by MINVAH, the Ministry of Housing and Human Settlements, and the municipality accepted the plan. This plan had a planning term of 15 years and three 5-year plans for its detailed implementation. Because local public influence was limited and there was a lack of financial and organisational support, this plan unfortunately had limited impact, especially as far as developing an infrastructure was concerned.

## The León Master Plan 1996

In 1994, therefore, the municipality of León chose a more fundamental approach. With help from the Netherlands, they formulated a Master Plan for León (*Plan Maestro Estructural de León*; Municipality of León, 1996) (Ill. 24.3). This was a physical and economic development plan for the entire urban area, including 862 $km^2$ of rural land. The increasing demand for space for the fast growing population was one of the most important items. The León Master Plan consists of two parts:

1. Ideas for the identification and long-term development of space sufficient for new housing developments and economic activities near the city centre, and the identification of a number of small development centres in the rural area around the city;
2. The description of 13 strategic projects and recommendations for carrying them out.

The León Master Plan identified the district of South-East León as the only urban expansion area for the next 25 years. This is because there are good possibilities for integrating this area with the existing city and for connecting the district with the main road to Managua. The municipality would try to improve the local economy, in co-operation with labour unions and local entrepreneurs, and would prepare new employment areas to give entrepreneurs opportunities to create jobs. However, since then, the development of new industrial parks and parks for offices has been limited, with some exceptions in a Free Trading Zone, which has attracted a number of foreign companies. The Master Plan also recommended developing high-quality public spaces and a green image to encourage tourism. The other main targets were to conserve and reconstruct the historic structure of the city, to preserve the colonial architecture and set up a new system of cycling tracks. After establishing the León Master Plan, they organised conferences with the government and potential donors and investors, such as non-governmental organisations (NGOs), and established links with foreign cities. A limited number of strategic projects, such as the South-East León Development Plan, were chosen for implementation. The municipality of León gained a lot of experience in local strategic planning and they have a well-equipped planning department.

## The 1999 Strategic Plan

A new Strategic Plan was made for 1999 onwards (Proyecto Estratégico de León, 1999). This was not formulated within the municipal organisation, although the municipality took part in, and facilitated, the process. All social and local organisations, the education sector and the commercial sector also participated, and the result was a collective strategic plan. This plan outlines the need for public provisions, infrastructure, and quality of life. However, there was no implementation plan, and a set of priorities, to accompany it. Nevertheless, the participative element is still there. In 2004, a *Committee for Municipal Development* (representing all the actors) was established to harmonise all important local activities. The 1999 Strategic Plan office is still in operation, acting as a platform for consultancies with, and between, local actors.

## The 2004 Municipal Development Plan (PEDM)

Since 2001, the municipality has been actively formulating a Municipal Development Plan (Municipality of León, 2004), by integrating the plans mentioned earlier: the 1996 León Master Plan and the 1999 León Strategic Plan. The new municipal development plan was announced by the city council in 2004. The approach is thematic, enabled by the integrative approach of the physical aspects and the

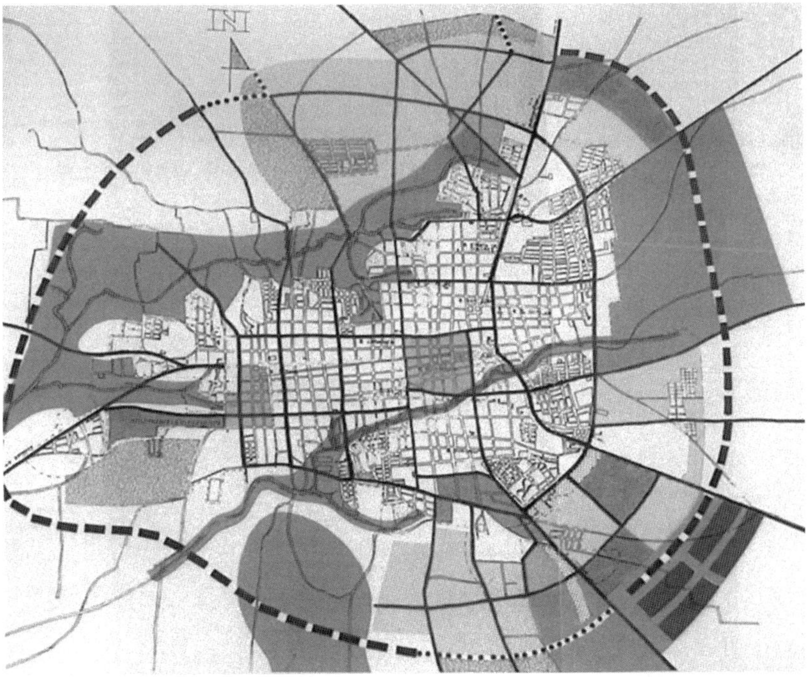


**Illustration 24.3: Master plan municipality of León (Municipality of Léon, 1996)**

participate way of working. This PEDM embodies a large number of aspirations, activities and projects. The most difficult part, however, is setting priorities when municipal resources are very limited. An estimated 50% of all municipal investments come from the links with foreign cities. The main strategic projects for the time being are South-East León (in co-operation with Utrecht), the North-Eastern rural development project (in co-operation with Zaragoza), and improving the markets (in co-operation with Hamburg). There are also active links with other cities and NGOs in this municipality.

As a result of the municipal development plan, it will be possible to improve the quality of life and work in existing neighbourhoods and industrial areas. The PEDM will also help to determine urban transformation processes and arrears, so the importance of this planning instrument will grow in the future. The Committee for Municipal Development can play a steering role in identifying urban problems, and coordinating the activities of actors in (new) strategic projects. Participative budgeting is a legal requirement for Nicaraguan municipalities. These activities have already begun in León. The planning experiences of this municipality between 1990 and 2005 are described in the book *Urban development strategies in León Nicaragua* (Bredenoord, 2005).

## Participative urban planning in VES

The municipality of Villa el Salvador (also called VES), with its 400.000 inhabitants, is part of metropolitan Lima. VES has an adequate Municipal Development Plan. The development of Villa el Salvador as a huge new residential area began in about 1971. Initially set up by the government, from an urbanisation model, this new residential area grew very fast. The layout of Villa el Salvador is based on one spatial design for a huge neighbourhood. This spatial model was then repeated many times, as one can see on the city map. The central government facilitated the urbanisation process by constructing the main infrastructure, such as access roads. Within the framework of this urbanisation model, families could obtain small parcels of land (plots of 100 $m^2$) on which to build their own houses. In fact VES was an enormous laboratory for self-help housing and for developing small-scale industries in close collaboration with inhabitants and entrepreneurs. Many families succeeded in constructing a decent house within 5 or 10 years. Some families completed their house bit by bit over a period of about 20 years. Other families did not succeed so well and still only have a small, ground-floor house. One can see these differences in almost every street. So Villa el Salvador is still under construction, and this process will probably continue for many years as families continue to enlarge and improve their homes and workplaces. In many cases, the house is also the workplace.

Besides the space to build houses, the municipality has also provided space for an industrial area, and is creating a very active economic policy. They are including in this plan a large, existing industrial area, with special zones for the different types of production, such as carpentry, furniture-making, metal-working, etc. This municipality has been progressive and has focused on providing services to inhabitants and (micro) entrepreneurs for a long time already.

## The strategic policy

The policy is for inhabitants and entrepreneurs to participate throughout the period of the urban development and to continue participating in the management of public areas. Social institutes co-ordinate these activities and collaborate with the municipality in looking for co-operation with private parties and higher levels of government. There is close co-operation between entrepreneurs' unions and the municipality, and there is a special organisation for promoting economic development.

The municipality of VES has its own spatial development plan: *Plan Integral de Desarrollo Villa el Salvador 2010*. The participation ideology of this municipality is embedded in this plan and communication with the population is always based on it. Communication takes place on two levels, the first with institutions, entrepreneurs and representative organisations and the second with the population (the street consultations). Each block of housing selects a leader to represent that block at neighbourhood meetings. In this way, residents and entrepreneurs maintain communication with local municipal institutions, utility companies (water and electricity), and non-governmental organisations (NGOs). The municipality of VES has had experience of working with participative budgeting for some years already.

## Some conclusions

The municipality of VES has a good practice of strategic planning and strong local participation. This is not the case, though, in many other municipalities in Peru. In general, the financial resources of local government in Peru are too limited to be able to invest in social housing and economic development, but steering family activities and other participants such as NGOs is very well possible, and necessary. This role involves stimulating, coordinating and bringing participants together. To facilitate these activities, urban management needs to be strengthened because this is the only municipal level that can bring about area-focused coordination. In all cases a *Municipal Development Plan* is necessary for determining the main aspects of spatial, economic and cultural development. Many municipalities are busy establishing such a plan, as all this is very necessary in Peru. Hence, the role of Peruvian NGOs is also very important, some of which will also be active in formulating urban development programmes. International co-operation has already opened up possibilities of strengthening local administration.

## The role of participants in strategic planning

In describing the role of participants in planning it is necessary to indicate the level of planning, because every planning level has its own type of participant.

### The municipal level

This is about participating in developing ideas for future improvements, and coordinating spatial investments, governmental as well as non governmental. The possible results for the municipality as a whole are that urban districts with priority status, consolidation areas, conservation areas and transformation areas will be identified. The participants at this level can be housing NGOs, other NGOs, the Chamber of Commerce and organisations representing various groups of professionals, other umbrella organisations and representatives of neighbourhoods and industrial areas. These bodies can be coordinated through a municipal coordination team, and steered by means of a forum for the urban development. With its formulated vision on future developments, the municipality provides the process and ensures that this is compatible with the higher levels of planning (e.g. the regional development plan and the plans of the government).

### The neighbourhood level

On the neighbourhood or urban district level, planning is focused on developing certain territorial sites such as housing projects, community and commercial centres, etc. Participants on this level make lists of urban arrears and opportunities and work out perspectives (plans) for entire areas, such as neighbourhoods. To apply for central municipal aid, project proposals have to be made. Clearly, at city level, all wishes and proposals cannot be honoured at the same time. Projects focused on territorial sites and programmes always have to be developed integrally, so this demands an adequate project organisation and very often project teams are needed for this. The actors on this planning level are, among others, residents, entrepreneurs and their organisations, umbrella organisations, social organisations, project developers and housing corporations.

### The role of housing corporations and NGOs

The role of housing corporations is very important in the Netherlands. These organisations construct houses, manage the housing stock and let units out to people who need accommodation, especially to those in social target-groups. In some neighbourhoods, they let out large numbers of housing units. Like the municipality, they also have to work on at least two levels. In their case, the first is on the municipal level (strategic housing stock management), followed by the project and neighbourhood levels. Close co-operation with the municipality is then possible on both levels. The different housing corporations usually have stock in the same area, but it is necessary that the different corporations co-operate together so that the municipality can operate as an intermediary for the whole group.

In Latin America, there is nothing comparable to these housing corporations, but there are NGOs, some of which specialise in housing. A very good example of one of these is the non governmental organisation Centro de Estudios y Promoción del Desarrollo (DESCO) in Lima, which devotes a lot of its time to helping social target-groups in urban neighbourhoods. In most cases, they help to support families by giving them loans and technical assistance to help them construct their own house in a proper way. In fact, this is structured self-help housing support. When NGOs do this on a large scale within projects focused on territorial sites, they have to participate in the integral development of the entire neighbourhood. On the higher municipal level some specialised NGOs can also participate in municipal development planning and help the municipality to prioritise similar activities elsewhere in the municipality. In some cases, specialised NGOs can act as consultants for communities or municipalities.

## Strategic planning and education

Strategic planning on municipal level demands well-educated and well-equipped personnel. In the present system, there are two types of municipal manager: the project manager and the urban manager. The project manager is occupied with projects focused on territorial sites and works integrally in every possible way with the local participants chosen for those projects. This project manager is not always a municipal official.

The specialised municipal official, occupied with urban planning and urban management is the urban manager. In big cities there is always a local planning board to carry out these tasks, but in smaller municipalities this knowledge is not always available. This is especially the case in the municipalities of Latin American countries, where there is very little adequate knowledge about urban planning and management. This is not surprising when you look at university-level education in those countries and the inadequacies, both in number and quality, of the urban management programmes. This could be improved by devising new programmes and establishing a set of *best practices*. Urban management needs the support of an educational and research programme to train a new generation of urban mangers, most of whom have started their careers as architects, urban planners, or lawyers, to become real urban managers. These new educational programmes need to include municipal legislation, municipal directing systems, urban planning (technical infrastructure, urban landscape, urban space, land use), socio-economic planning, urban monitoring, making long and short-term visions, municipal investment potentials and the search for reliable private participants. Good examples of planning systems are already available (e.g. in Breda and in Villa el Salvador).

In some Latin American countries progress is being made in formulating municipal planning systems. The research project *Self-help housing in Lima* (Bredenoord, 2002) describes an educational aspiration for urban managers, originally architects, civil engineers or lawyers, to be given special training in capacity building. An example of such a training is the Municipal Training Programme for Nicaragua, that was carried out in 2002 by the Dutch Association of Municipalities. An important item was public-private partnerships, because they did not have much experience in this field of co-operation. Internationally operating organisations are also active in this field of training and education. In the Netherlands, we have IHS Rotterdam, ITC Enschede, and VNG-International and the Netherlands Habitat Platform in The Hague. Nicaraguan counterpart for urban management programmes are the national network of municipalities and the universities. In Peru, besides several universities, a number of the NGOs undertake this work, for example, DESCO and Foro Ciudades para la Vida.

## Final remarks

This chapter promotes the idea of upgrading the structural plan to a municipal development plan, which is itself part of a strategic municipal planning system consisting of the following main elements: city monitoring, formulating outlooks on future development, updating, prioritising, involving participants, setting up municipal planning teams and steering committees, and formulating implementation plans. The municipal development plan must have a cyclical character and it must be kept up-to-date. In the Netherlands, there are spatial master plans for all big municipalities, and, in some cases, they have started to set up municipal management systems. The main point of attention is to establish priorities.

In Peru and Nicaragua, municipal development plans are very much in the making. The cases of León and VES, described in the chapter, are interesting examples of these activities. The interchange of experiences and knowledge is the most effective when international links have been established with a number of cities outside the region. Where the Netherlands is concerned, Dutch municipal experience with public-private co-operation and contracts between local government and housing NGOs would be of help to Latin American municipalities. In turn, Dutch municipalities could learn from their Latin American counterparts how they could increase the involvement of the local population and social and economic actors in urban planning and management (e.g. through municipal participative budgeting).

On the project level, there are interesting concepts in connection with self-help housing and the self-maintenance of complexes. Strategic planning can be developed whenever there is an interchange of knowledge and experiences.

### References


Bestuur Regio Utrecht, 2003, *Regionaal Structuurplan 2005–2015*, Utrecht

Bredenoord, J., 2002, *Investigación Autoconstrucción de viviendas en Lima*, Amersfoort, the Netherlands

Bredenoord, J., 2005, *Urban development strategies in León Nicaragua*, Dutch University Press, Amsterdam

Dienst Stadsontwikkeling Gemeente Utrecht, 2004, *Structuurvisie Utrecht, 2015-2030*, Utrecht

Government of Nicaragua, 2003, *Plan Nacional de Desarrollo*, Managua, Nicaragua

Municipality of León, 1996, *Plan Maestro Estructural de León*, León, Nicaragua

Municipality of León, 2004, *Plan Estratégico de Desarrollo Municipal de León*, León, Nicaragua

Proyecto Estratégico de León, 1999, *Plan Estratégico de León*, León, Nicaragua

Regio Twente, 2002, *Netwerkstad op koers, Ruimtelijke ontwikkelingsvisie 2030*, Regio Twente

Zonneveld, W. & S. Tjallingii, 2004, Perspectieven voor Planningprincipes. Structuurplanning in Breda 1975-2005, *S&RO*, 85/4, pp. 56-59

# 25 Neighbourhood Initiatives: Time for Bottom-Up

Robin Houterman and Edward Hulsbergen

## Introduction

The continuation of urbanised areas as living environment for a socially, economically and culturally diverse population is a thought-provoking field of study. On the one hand, the urban spatial-functional complexity is a playing field for sectors and actors, and, on the other, a major challenge to find effective problem definitions and policies. On the regional level, growth, investments and prosperity go side by side, spatially, with decay, neglect and poverty. Some districts and neighbourhoods may be known for good living, others may be designated as problem areas.

Post 1945 Europe can be seen as one large laboratory of urban (re)generation, and, as such, an almost unlimited possibility for learning. The question, however, is: what are the lessons? The multitude of cases and casuistry; the local, national and European efforts to generalise experiences and findings are tokens of the variety of problem definitions, approaches and results on offer (e.g. European Commission, 1999). Overviews of the development of approaches and policies in Great Britain can be found in *The Handbook of Urban Regeneration*. From the 1950s to the end of the 20th Century, the strategies and orientations, in terms of sectors and actors, changed from reconstruction and extension to comprehensive and integrated approaches (for a summarising table, see Roberts & Sykes, 2000:14).

In the last decade, attention became more focused on encouraging the participation of residents and users in regeneration processes, and the idea that residents' and users' participation is vital is gaining a stronger and stronger foothold. Participation is viewed broadly, not only as being active in formulating the plan and in taking part in the decision-making process, but it is coming to be seen as an essential asset in the regeneration process itself. Within this broad view, bottom-up initiatives have become a significant form of participation.

In line with this view, a recent report of the Scientific Council of the Dutch Government [*Wetenschappelijke Raad voor het Regeringsbeleid, WRR*] argues that residents with their low-scale social ties can be key actors in contributing effectively towards the liveability of their neighbourhood, but that this also requires an appropriate attitude on the part of the municipality (Wetenschappelijke Raad voor het Regeringsbeleid, 2005). This latter point is the crucial aspect of participation pin-pointed in this argumentation: the way in which planning authorities deal with and accommodate bottom-up initiatives. Participation can also be, and has often been, exercised as a predominantly top-down activity, where policy-makers expect local initiatives to fit in with their own schemes. However, this neglects the potentials of cyclical interaction between top-down and bottom-up initiatives on the neighbourhood level.

In this chapter, we explore the importance of bottom-up initiatives as a current and future instrument of urban policy and planning. We set the scene by giving a selected overview of trends and outlooks and continue with an overview of three important aspects related to bottom-up initiatives: strategy, participation and empowerment. We conclude with our view that bottom-up initiatives are an indispensable form of citizen participation on the local level of urban regeneration, which, as such, need space and support from planning authorities in general and spatial planners in particular.

## Trends and outlooks

What explains the present strong policy interest in bottom-up initiatives? Three lines of reasoning can be distinguished: past experiences, counterbalancing a globalising society, and democratic policy-making.

Past experiences with urban regeneration projects have shown that a prerequisite for success is community involvement, and this has gradually become a recognised fact. The market-led, top-down strategies - focusing on physical development in the form of property-led regeneration, and so common during the 80s - have proved inadequate to combat the multi-faceted urban problems that neighbourhoods have to face. The British Office of the Deputy Prime Minister sees lack of joint working at the local level as one of the key reasons why so little progress has been made in delivering sustainable economic, social and physical regeneration, or in improving public services to meet the needs of local people. This is one of the reasons for the current British policy focusing on the establishment of Local Strategic Partnerships (LSP). LSPs are "single non-statutory, multi-agency bodies, which match local authority boundaries, and aim to bring together at a local level the different parts of the public, private, community and voluntary sectors to tackle the multi-faceted, deep-seated problems in neighbourhoods" (ODPM website). The key assumption is that these LSPs are better able to define initiatives that respond to the demands of local society. The (renewed) focus on the demands of local society is mainly based on practical experience, which has shown that "initiatives in urban regeneration are implemented most successfully when programmes sensitively respond to local people, including those with special needs and problems" (Jacobs & Dutton, 2000:127). Internationally, this attention to the needs of local people is not a new starting point for initiatives. For example, in the Netherlands, attention to the needs of local people was already a common issue in the approach to urban renewal of the 1970s: 'building for the neighbourhood'. The difference was that the renewal strategy at that time focused mainly on physical aspects, while recent regeneration approaches are largely focused on social and economic aspects.

The second line points at the central position of the neighbourhood in urban research and policy. This might appear curious in a society where globalisation and individualisation are major forces. The importance for policy and research of the neighbourhood can possibly be explained by three arguments.

The first sees it as a response to the "new information technologies, a new virtuality in social networks and a greater fluidity and superficiality in social contact, which are eroding the bonds of spatial proximity and kinship" (Forrest & Kearns, 2001:2126). Forrest and Kearns argue that this assumption may reflect a very particularised experience of social change; that of the 'wired intellectual'. According to Pahl (cited by Forrest and Kearns) most people (still) live in narrow *gemeinschaftlich* worlds of neighbourhood and kin. Thus, for many of us, as we spend most of our time in our home environment, the neighbourhood still forms our principal spatial context.. This is especially true for social groups such as the elderly, children, ethnic minorities and the unemployed. For these groups, the neighbourhood even seems to have gained importance, in recent years (Meegan & Mitchell, 2001:2174).

The second argument is the view that social cohesion and solidarity are diminishing because of current institutional changes. Forrest and Kearns point at different factors assumed to be contributory to the social-cohesion 'crisis': the breakdown of Keynesian capitalism, the end to the progressive recruitment of households to the lifestyles and living standards of the traditional middle classes, growing inequality and social fragmentation, and a perceived decline in shared moral values. However, within this supposed context of a less cohesive society, it is argued that the neighbourhood can act as a "potentially important site for rebuilding cohesion from the bottom-up, with active, empowered citizens practicing mutuality and reciprocity" (Forrest, 2004:6). And, indeed, there is ample evidence to show that bottom-up initiatives are often successful in bringing residents and other local actors together.

The third argument for placing the neighbourhood at the centre of the urban debate is its (continuing) importance for residents as a place offering personal opportunities. Kearns and Parkinson (2001:2106) argue that "in an increasingly competitive and uncertain world in which people seek to establish themselves either alongside or over and above others, the neighbourhood can play an important role in people's personal and social identity and social position, but with highly variable outcomes". Kearns and Parkinson list a number of reasons why the neighbourhood is significant. In their view the neighbourhood is "a reservoir of resources into which we can 'dip' in pursuing our lives, an influence upon our lifestyle and life-outcomes and a shaper of who we are, both as defined by ourselves and by others" (*ibid.*:2109).

Much research on the effects of the neighbourhood's physical, social and economic environment on individual residents focuses on deprived neighbourhoods, by asking questions such as: "do poor neighbourhoods make their residents poorer?" (Friedrichs, 1998). Recent empirical research does not provide a clear and general answer. Neighbourhood effects do have many variables (individual and household characteristics, endogenous and exogenous neighbourhood variables; metropolitan area characteristics, welfare-state regimes), each with implicitly different spatial scales (Friedrichs, Galster & Musterd, 2003).

Finally, a third line of reasoning rests on the generally increasing attention being given to democratic policy-making in the context of urban governance. Though democratic policy-making has been a topic for decades, what is new is that local democratisation is supposed to be a prerequisite for sustainable urban development. For example, the European Commission states that one of the challenges faced by cities in formulating and implementing strategies for more sustainable urban development today, is extending democracy and local empowerment and involving all stakeholders - including citizens (European Commission, 1998). The European Council of Town Planners (ECTP, 2003) and UN Habitat are also placing emphasis on more democratic urban development. Tools such as the *Instrumentenwijzer* (www.instrumentenwijzer.nl) and the *Toolkit citizen participation* (www.toolkitparticipation.nl) promoted by Habitat Platform in The Hague are practical elaborations of this concern for democratisation in urban governance.

## Bottom-up initiatives and related aspects

In this and the following paragraphs, we will focus on the connection between bottom-up initiatives, strategy, participation and empowerment. Bottom-up initiatives are those that originate from inside the community to improve the physical, economic and social conditions of that community's own living environment. They usually originate from the needs of local society and, because of this, can contribute to strategies which try to address those local needs.

The cases, briefly described in Boxes 25.1-3, illustrate the possible diversity of bottom-up initiatives and their potential contribution to simultaneous physical, social and economic development on different levels of scale. Illustration 25.1 represents where the three cases stand in relation to participation, empowerment and bottom-up initiatives, and whether they are part of the municipal strategic framework.

### Strategy

Strategy might play an important role in connecting participation, bottom-up initiatives and empowerment. In the current debate on urban and regional development and regeneration, an important issue is strategic outlook, which creates the framework in which the longer-term goals, aims and objectives of individuals, organisations and areas can be realised (Carter, 2000:42).

Bottom-up initiatives might be the most effective when they are part of a strategy aiming at increasing the liveability of an area. What can be learnt from the three cases is the importance of municipal support for bottom-up initiatives, although they should not be dependent on that support. In this context, it is worthwhile examining the potential contribution of information and communication technologies (Kennisnetwerk SQM, 2004:69).

A major difficulty is how to connect top-down and bottom-up initiatives and responsibilities, so that they complement each other. Top-down decision-making and implementation is risky without the support of those who have to live with the consequences. Bottom-up initiatives alone are not able to respond to the complex urban problems with which neighbourhoods are confronted. A strategic framework could provide a bridge between the top-down and bottom-up approaches (Carter, 2000:42).

### Participation

As outlined in the introduction, we see bottom-up initiatives as a form of broad participation. Participation is important in terms of efficiency; residents and users know what the local dynamics are, and may have a sound grasp of which services will work and which will fail due to lack of interest. This is important for three reasons (Hull, 2001:306). Firstly, regeneration needs to build on the local networks, harnessing the resources and enthusiasm of residents. Secondly, local people understand the sensitivities and pull of local personalities, who could present barriers, if not included in the deliberations. Thirdly, with resident involvement, services (times, locations, and costs) can be provided to that suit the residents.

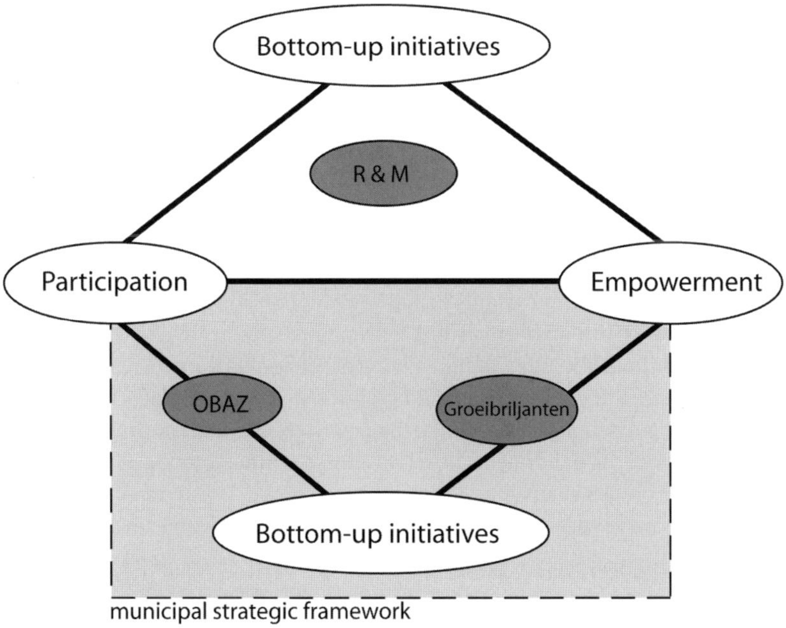


**Illustration 25.1: Representation of the cases in relation to bottom-up initiatives: strategy, participation, and empowerment**

The *groeibriljanten* case in Rotterdam was organised by the municipality of Rotterdam in the Netherlands. Its aim was to stimulate private initiatives in deprived areas of the city. The goal was to stimulate cohesion within communities by involving them more in their living environment. Initiatives could be put forward by private initiators (or groups), who could compete for financial support from the municipality.

The initiatives were judged from the following aspects:
- To what extent is it a private initiative?
- What is the spatial-economic potential?
- Does the project contribute to social cohesion?
- What significance does the initiative have for the wider city?
- To what extent is it an innovative approach?

The initiatives put forward showed a great diversity in nature, purposes and initiators. With 73 initiatives, the contest seemed to be rather successful.

Their initiators now have to continue the projects by, first working out what support is needed in terms of money and professional assistance. After this phase, the realisation of the projects will be started.

**Box 25.1: *Groeibriljanten* [growthdiamonds] (source: www.groeibriljanten.nl)**

The success of participation, and thus of bottom-up initiatives, seems to largely depend on how professionals and volunteers collaborate. In this sense, the attitude adopted by both groups is important, because they must want to collaborate and be willing to understand each other's perspective. This attitude, however, might be partly based on prejudices. For example, a large degree of participation might delay the regeneration processes. The assumption is that consensus building through partnership-based processes is time-consuming compared with processes involving only a few actors. The opposite might also be true. Time investments at the beginning of a process might save time during the implementation phase, as public support for initiatives would then have been achieved.

Participation, as such, does not necessarily secure a democratic regeneration process. This remark is closely related to one of the main critics of participation; it can provoke certain 'representatives' to exercise their power unjustly. This might even (re)produce social inequality in a neighbourhood (Jones, 2003:582). Indeed, the equal representation of both residents and users is an aspect of special attention. The cases where people just defended their own interest, and not the needs of other groups, cannot be neglected. If participation aims at serving the common interest, it is necessary to ensure that the different social groups in an area are equally represented. This is not always easy, however. At least two social groups need special attention to get them involved: ethnic minorities and temporary foreign urban residents.

The involvement of minority groups is usually difficult (Blair & Hulsbergen, 1996). Language barriers and illiteracy, among other factors, may be obstacles against involvement (Mzalassi & Kleinhans, 2004). A report of the Department of Environment, Transport & the Regions (1997) provides a list of key factors for involving ethnic minorities in regeneration. The demands of temporary urban residents also need specific attention. These foreign residents are usually either highly skilled professionals or low-skilled foreign workers, who reside in urban centres for shorter or longer periods. Their presence might become more important in the future, especially in the larger cities (European Council of Town Planners, 2003).

## Empowerment

Community empowerment is one of the central issues in the debate on the regeneration of deprived neighbourhoods. In a broader sense, empowerment can be obtained by providing residents with better education, job training, and support for economic initiatives, and by increasing the accessibility of local services and facilities. In the context of participation, empowerment can, in our view, be understood as both a means and a goal. As a means, empowerment stands for the extension of the community's ownership of programmes and projects, which provides local people with involvement and decision-making responsibilities. Empowerment can become a goal when participation is seen as a strategic asset to increase residents' skills. By letting people participate on management boards, and in steering groups or executive bodies, they will gain skills that could help them to find regular jobs. Perrons and Skyers (2003), however, argue that these kinds of strategy to address deprivation always need to be combined with transformative redistribution strategies that battle against the processes leading to social inequalities.

Bottom-up initiatives can be seen as a means of self-empowerment, because residents and users create their own responsibilities. In this context the observation of Lyons, Smuts and Stephens (2001:1237) about the relation between empowerment and sustainability seems useful. In relation to empower-

*'Onze Buurt Aan Zet'* was a national programme initiated by the Dutch national government. In certain neighbourhoods, municipalities were given the opportunity of experimenting with community participation and decision-making.. The goal was to give residents space for their own initiatives and to bring residents in closer contact with their own living environment. The essence of the programme was that residents could decide on spending part of the municipal budget for improving their living area. The decision on which initiatives the budget would be spent was made in different ways. In Schiedam, for example, decision-making was organised with representatives of the community following a brain-storming session, while, in Zwolle, the residents proposed ideas, which were then put to the vote among the residents of that neighbourhood. Part of the budget was then reserved for the voted-for ideas, and their implementation was supported by professionals. The initiatives that were taken by residents were mostly focused on improving the liveability of the neighbourhood, an important item being measures to improve security in the neighbourhood. Social activities were also initiated and funded.
This was a 4-year programme, but some of the municipalities have decided to continue it without the national government's support.

**Box 25.2: *Onze Buurt Aan Zet* (OBAZ) [Our neighbourhood's turn] (sources: Van der Graaf, Oudenampsen & Wentink, 2004; www.onzebuurtaanzet.nl)**

R&M is a resident organisation aimed at improving the liveability of Gouda Oost (5,500 residents), in addition to the formal organisations. Gouda Oost faced increasing environmental decay and marginalisation because of growing unemployment and poverty, while the cultural and ethnic diversity added to the reduction of social ties. R&M was initiated in 1997 by two residents, who cared about the lack of perspectives for young Moroccan residents, in particular. The main problem was to find a location for activities. After years of discussions with the municipal and professional social organisations the R&M Foundation was established in 2000 as a corporate body positioned somewhere between the municipality and an activity centre. From 2000 to 2003 many activities were organised for different age groups (leisure as well as those geared towards education and employment; housing wardens and environmental reporters). To contribute to the coming regeneration of the district, a design studio for the central area was organised. However, at the end of 2003, the municipal's social policy changed, and with that the financial support to R&M. In 2004, the board decided to dissolve the foundation, but to keep the activity centre open and activities going as far as was financially possible. Now, in 2005, relations between the centre and municipality have improved, financial support continued, and residents are working on establishing a neighbourhood association as a corporate body.

So far, R&M has been managed by volunteers among the residents, but this might change. Although the majority of initiatives and projects have been successful, R&M is not explicitly part of the development plans for the neighbourhood. Though, for some activities, use has been made of the centre' s space and organisation capacity, R&M has so far not been explicitly seen as a strategic asset for the neighbourhood.

**Box 25.3: R&M Gouda-Oost (sources: Hulsbergen & Vellinga, 2001; Galesloot, 2000; www.activiteitencentrumgouda.com)**

ment, they define sustainability as "the ability acquired and held by communities over time, to initiate and control development, thus enabling communities to participate more effectively in their own destiny". It is here where empowerment can be connected to (national) subsidiarity, with respect to the European Union. The question is, on which level of decision-making are residents and users competent to participate. Bottom-up initiatives might be an answer to this question. Finally, from research in South Africa, Lyons et al. conclude that: "if, in the process of participatory development, people receive training in usefully transferable skills, empowerment will develop on three levels: personal, project and community. Fully empowered people, projects and/or communities are then able to contribute towards the sustainability of development projects which, in turn, contribute towards the broader notion of sustainable development" (*ibid.*:1249).

## Conclusions

In this chapter, we have explored, theoretically and practically, the importance of bottom-up initiatives with respect to current and future urban policy and planning. We have looked at their possible contribution in solving complex problems on the neighbourhood level, and have come to the conclusion there are four ways in which bottom-up initiatives can be a valuable instrument for urban policy and planning.

Firstly, bottom-up initiatives lead to new interpretations of the problems in neighbourhoods. Residents are frequently considered to be 'part of the problem'. But, as Angela Hull (2001:304) argues: "rather than portraying the residents as the perpetrators of a 'problem', that we, professionals have to deal with, we should recognise their agency in the solution to the problem of deprived neighbourhoods". Providing more space for bottom-up initiatives is a recognition that residents are part of the solution, instead of part of the problem. This awareness might induce new ways of defining problems in urban neighbourhoods, and lead to new, innovative regeneration strategies.

Secondly, bottom-up initiatives can be true investments in urban areas. The cases described in the boxes suggest that residents are willing to spend a vast amount of time and energy in making their direct living environment more liveable for themselves and others. Bottom-up initiatives can be a tool for people to modify the physical, economic and social aspects of their own living environment. In this way, they could, and probably do, stimulate social cohesion, diminish social exclusion, and, through empowerment, offer local residents more opportunities.

Thirdly, at the same time, bottom-up initiatives can address physical, social and economic problems. One of the main topics in the regeneration debate in the Netherlands at the present time is the integrated approach on the local level. Much can be expected when top-down strategies are linked to bottom-up initiatives, in a complementary way.

Fourthly, bottom-up initiatives can help to enhance the democratisation of (local) urban governance. If new forms of local democracy develop, these could also be a means of securing the interest of the many residents and users in a city, in an era in which public authorities, plagued by financial constraints, leave parts of their responsibilities for the collective interest to the free market (European Council of Town Planners, 2003). It is assumed that democratising local urban governance can help to make our cities, and their development, more sustainable.

Another conclusion might be that bottom-up initiatives should be judged on three issues: firstly, on how they contribute to the common interest in a neighbourhood; secondly, on how they contribute to the long-term objectives of a regeneration strategy and whether they are complementary to top-down initiatives; and thirdly, on how they contribute to the sustainable empowerment of residents and users and lead to wider-scaled subsidiarity in the future.

Finally, one aspect is crucial for the success of bottom up initiatives; how much (political) space and support do they receive from local planning authorities in general, and spatial planners in particular. If bottom-up initiatives can rely on both groups of professionals, then we view them as a necessary extension of participation in planning and implementation.

## References


Blair, T.L. & E.D. Hulsbergen, 1996, Conclusions and practical guidelines; In: Council of Europe, Directorate of Social and Economic Affairs; *Area-based projects in districts of high immigrant concentration*, Community relations, Council of Europe Publishing, pp.13-26

Carter, A., 2000, Strategy and Partnership; In: Roberts, P. & H. Sykes, *Urban Regeneration: A Handbook*, Sage Publications, London

Department of Environment, Transport & the Regions, 1997, *Involving Communities in Urban and Rural Regeneration: A Guide for Practitioner's Second Edition*, DETR, London (summary available on www.odpm.gov.uk)

European Commission, 1998, *Sustainable urban development in the European Union: A framework for action*, European Commission, Brussels

European Commission, 1999, *Guidelines for the new URBAN community initiative 200 ??? -2006*, European Commission, Brussels

European Council of Town Planners, 2003, *The New Charter of Athens 2003, The European Council of Town Planners' Vision for Cities in the 21st Century*, Lisbon, http://www.ceu-ectp.org/e/athens

Forrest, R. & A. Kearns, 2001, Social cohesion, social capital and the neighbourhood, *Urban Studies*, 38/12, pp. 2125-2143

Forrest, R., 2004, *Who cares about neighbourhood?*, CNR paper 26, ESCR Centre for Neighbourhood Research, Bristol

Friedrichs, J., 1998, Do poor neighbourhoods make their residents poorer? Context effects of poverty neighbourhoods on residents; In: H.-J. Andress (ed.) *Empirical Poverty Research in Comparative Perspective*, Ashgate, Aldershot (UK), pp.77-99

Friedrichs, J., G. Galster & S. Musterd, 2003, Neighbourhood Effects on Social Opportunities: The European and American Research and Policy Context, *Housing Studies*, 18/6, pp. 797-806

Galesloot, H., 2000, *Vraag het ze persoonlijk: inspiratiebronnen voor veelkleurige bewonersparticipatie*, Instituut voor publiek en politiek (IPP), Amsterdam

Hull, A., 2001, Neighbourhood renewal: A toolkit for regeneration, *Geojournal*, 51, pp. 301-310

Jacobs, B. & C. Dutton, 2000, Social and community issues; In: Roberts, P. & H. Sykes, *Urban regeneration: A handbook*, Sage Publications, London, pp. 109-128

Hulsbergen, E.D. & B.M.K. Vellinga, 2001, *Buurtvernieuwing door actief burgerschap*, Stichting Habitat Platform, Den Haag

Jones, P.S., 2003, Urban regeneration's poisoned chalice: Is there an impasse in (community) participation-based policy?, *Urban Studies*, 40/3, pp. 581-601

Kearns, A. & M. Parkinson, 2001, The significance of neighbourhood, *Urban Studies*, 38/12, pp. 2103-2110

Kennisnetwerk SQM, 2004, *ICT en sociale cohesie in buurten en wijken*, Kenniscentrum Grote Steden, Den Haag

Lyons, M., C. Smuts & A. Stephens, 2001, Participation, empowerment and sustainability: (how) do the links work?, *Urban Studies*, 38/8, pp.1233-1251

Meegan, R. & A. Mitchell, 2001, It's not community here, it's neighbourhood: Neighbourhood change and cohesion in urban regeneration policies, *Urban Studies*, 38/12, pp. 2167-2194

Mzalassi, F. & R. Kleinhans, 2004, Inspelen op sociale netwerken: Allochtone bewoners over de Van der Spekbuurt, *VHV-bulletin*, 31/5, pp.22-24

Perrons, D. & S. Skyers, 2003, Empowerment through participation? Conceptual explorations and a case study, *International Journal of Urban and Regional Research*, 27/2, pp. 265-285

ODPM, www.neighbourhood.gov.uk, March 2005

Roberts, P. & H. Sykes, 2000, Evolution, Definition and Purpose; In: Roberts, P. & H. Sykes, *Urban Regeneration: A Handbook*, Sage Publications, London, pp. 9-36

Van der Graaf, P., D. Oudenampsen & M. Wentink, 2004, *Onze buurt aan zet onder het vergrootglas*, Ministerie van Binnenlandse Zaken en Koninkrijksrelaties, Den Haag

Wetenschappelijke Raad voor het Regeringsbeleid, 2005, *Vertrouwen in de buurt*, WRR-rapport 72, Amsterdam University Press, Amsterdam

# 26 Urban Mobility
## Towards the Conditions for the Planning and Design of the Mobile City

« Nusquam est, qui ubique est. »
« He who is everywhere, is nowhere. »
Seneca, *Epistualae Morales*, no. 2

Remon Rooij

## Introduction

The role of mobility in today's city and that of tomorrow is one of the major strategic design and planning tasks for the field of urbanism. At the interface between urbanism and transport, it is widely agreed among experts in both fields that the new urban-transport technologies are important in developing and transforming today's socio-spatial environment, although it is still very unclear how these new technologies should be interpreted in terms of planning and designing the network city.

This chapter presents guidelines and recommendations for planning and designing seamless multi-modal mobility in the network city. It focuses firstly on the actual and potential role of this form of mobility for the vitality of cities and the functioning of urban areas, and secondly on the changes that have taken place in spatial planning due to the rise of the mobile city.

### The city and mobility

When thinking about the future of cities and their desirable and possible level of mobility, the field of spatial planning is confronted with the notion of a city that has changed significantly during recent decades and centuries. Traditionally, a city could be defined as the concentration of buildings around a centre – a uni-nodal conglomerate of functions, but urban developments of recent decades have not followed this pattern. Brand (2003) traces the recent shifts in urban development from the traditional city, to the urban district, to the polynuclear city, and the urban field (see Ill. 26.1).

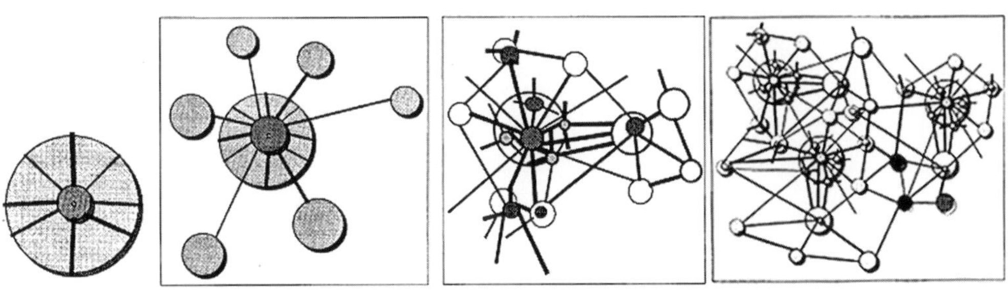

**Illustration 26.1: Recent stages in urban development from the traditional city, the urban district, the polynuclear city, to the urban field (Brand, 2003)**

Bertolini and Dijst (2003) reason that "...the lives of people and the workings of organizations are increasingly independent of physical and administrative urban boundaries..." (*ibid.*:27) and that "... not only the spatial reach of people has increased, but also and the diversity in activity and travel patterns." (*ibid.*:28). These processes have led and still lead to "... an increasing disentangling between human activity patterns and the physical city. Increasingly, each individual can create his own virtual city; one that has no set physical and administrative borders, but a rather specific, changeable combination of activity places, connected by transport, information, and communication networks, within definite socio-economic and behavioural constraints." (*ibid.*:29)

So, it is networks that have become prominent in daily life, and these have increasingly proved to have their own dynamics. Each network contributes to the overall economic, social, cultural, political, and spatial developments (Boelens, 2000). The dynamics that accompany this world-wide network society put not only the city and rural areas under pressure, but also socio-economic and socio-cultural interactions, and the present structure of government and governance.

With the increase of motorised mobility, people's action space has increased enormously during the last 100 years, and especially during the last few decades. The multi-nodal and hierarchical infrastructure network has to accommodate the movements of personal transport. In accommodating for this movement, the spatial planner should always try to retain and improve the vitality of urbanised areas. In fact, this is the core business of spatial urban planners and designers. Based on a wide variety of urban and transport research, the author holds the view that this vitality is endangered most by:

1. The amount of space consumed by infrastructure and vehicles;
2. The increasing inaccessibility of urban areas;
3. The lack of coherence between the hierarchical levels of the urban web.

As mentioned above, Seamless Multimodal Mobility (SMM) is widely considered to be a possible solution for some of the problem fields in spatial planning (see also Bovy, 2002; Van Wee, 2002). First of all, it is discussed as a congestion and space consumption mitigation strategy, and as a strategy for increasing liveability in cities and city centres, and for increasing the multimodal accessibility of activity places.

Furthermore, a proper hierarchy in the multimodal network, with seamless transfer points in both time and space might produce optimal conditions for a vital city (Rooij, 1999). It is suggested in the TRAIL research programme on seamless multimodal mobility that "...seamless multimodal trip chaining on the regional scale could be the answer to growing restrictions in using the car for personal travel. Sophisticated information and communication technologies, as well as real-time transport process control, could offer the means for effective and efficient multimodal transport services exhibiting an unprecedented high level of quality, able to compete with the car..." (TRAIL, 1999).[1] Supported by high-tech information and communication technologies, the traveller can compose and direct his own trips by combining several separate transport modes within one trip.

1 TRAIL research school for TRAnsport, Infrastructure and Logistics; see http://www.rstrail.nl/smm

## Problem definition and chapter objective

The role of mobility in today's city and that of tomorrow is one of the major strategic design and planning tasks for urbanism as a whole (see e.g. Calabrese, 2004; Van der Hoeven, 2001; Van der Spek, 2003; Rooij, 2005). It is widely agreed among urban and transport experts that the new urban technologies, such as transport technologies, are playing an important role in the development and transformation of the present socio-spatial environment, but how these new technologies should be interpreted in planning and designing the network city is still very unclear. The main objective of this chapter therefore is to present guidelines and recommendations, based on the findings for *The Mobile City* (Rooij, 2005), for incorporating seamless multimodal mobility into the planning and designing of the network city.

In the next section, planning objectives for the mobile city of the future are formulated. The third section describes the shifts in the discipline of spatial planning resulting from the rise of the mobile city, and those that can be expected in the future. Finally, conclusions are drawn regarding the role of mobility for the functioning of the city.

# Formulating planning objectives for the network city

« planning: big plans require strong foundations »

In formulating planning objectives for the network city, this section follows the three spatial-planning problem fields identified in the first section of this chapter.

## Space consumption

If infrastructure and vehicles consume an excessive amount of space, liveability in urban areas is endangered with respect to:

- Traffic safety;
- Emissions, noise, stench;
- Parking;
- Barriers in the city;
- Green-structure intersections;
- Congestion.

Collective forms of transport use space more economically, and are therefore suitable for use in dense urban areas. Moreover, the more that collective transport is used, the more it will improve over time, because higher demands will lead to a more frequent services. However, one of the limitations of collective transport is that its availability is restricted to timetables and transfer points (e.g. bus stops).

It can be reasoned that most means of transport have certain characteristics that are better suited to particular spatial circumstances, and so, conversely, they will probably function relatively badly under other spatial circumstances. From this, several 'ideal' methods of travel for different types of trips can be identified (Ill. 26.2). For certain types of trips, it might be best to combine travel modes, thus introducing the option of multimodal travel.

Bovy (2002) reasons that if a traveller comes from a rural area, destined for a large city located some distance away, then we can assume that the person would prefer to start the trip by car and reach their destination by tram or bus. For longer distances, an Intercity train could be best for the journey in between. So, the traveller has a choice between making the whole trip by car, or making a multimodal trip (car-train-tram). In many cases, the car is a moderately good to good alternative. However, this example shows that if the traveller's demands with respect to the transport service differ in the various sections of that trip (speed of bridging the distance, facilities required while travelling, departure and arrival times), then multimodal travelling becomes a viable option to the ever-present option of using the car.

Multimodal travel can be attractive for spatio-geographical reasons (Bovy, 2002) for example for trips:
- Between areas with a (very) high urban density and those with a low urban density;
- Over long distances;
- Between areas with average urban densities.

And for other reasons (very often in combination with the previous ones):
- At times of the day when the alternative offers insufficient quality (e.g., during peak hours, congestion or hindrance);
- Combining other activities, such as the need to work while travelling;
- Which have other specific traveller demands.

From this perspective of spatio-geographical reasoning, it can be concluded (Bovy, 2002; Van Nes, 2002; Adviesraad Verkeer en Vervoer, 2002) that unimodal travelling is suitablefor most trips and that multimodal travel is only preferable for limited niche markets. Its share in total mobility is therefore limited; at the present time only about 3 % of all trips are multimodal, and the potential percentage in future is expected to be only about 6%. A multimodal trip is defined here as a trip (movement between two activity places for one specific motive) made by using two or more means of transport. 'Walking' is not included here, so trips following the sequence walking-car-walking, walking-bus-walking, walking-tram-walking, walking-train-walking, walking-bicycle-walking etc. are all considered to be unimodal trips.[2] Multimodal travel can be seen to be crucial for certain market segments, and has the potential to increase even more there in the future, especially for long-distance trips between high-density urban conurbations. If typical types of trip are considered, then the 3% share increases. For instance, of all interurban trips to the main cities in the Netherlands, the multimodal share is 20%, and of all train trips, then the multimodal share is 80% (Goeverden and Van Nes, 2000). At the local level, therefore, multimodal travel can be quite significant (Van Nes, 2002) and its contribution to liveability (and accessibility) in urban areas can be essential.

The research results presented here, refer to nationally oriented figures based on the Dutch National Travel Survey. Of course, multimodal travelling can be, and certainly is, very competitive on the international and intercontinental levels of scale, but unfortunately, there is little in-depth research available about activity-travel behaviour on an international level of scale. It can be expected, however, that travelling over those larger distances will take place even more frequently in future than it does today, and that, because of this, the share of multimodal travelling (train-aeroplane-taxi, or car-aeroplane-rented car, etc.) will grow along with the globalisation of activity patterns.

2 This is due to the fact that in (inter)national research and statistics on people's travel behaviour, 'walking' is very often not considered to be a separate means of transport.

| Under these circumstances... | the traveller has wishes for his or her transport service... | ...and thus would like to travel with... |
|---|---|---|
| Urban environment, Medium long distance | I want to move freely and I don't want to have to park | Metro, train, bus |
| Urban environment Short distance | I want to move freely and I don't want to have to park | Bicycle |
| Long distance | I want to walk, eat, pee and I don't want to drive | Intercity, HST, airplane |
| Non-urban area | I want to move freely and I don't want to wait for ever | My car |
| In a hurry | I want to move fast and with certainty | Sometimes the car, sometimes the train |
| Shopping | I want to be able to take stuff with me and I don't want to drag my luggage | Car |
| I would like to work | I want a quiet environment and I don't want to drive | Train (1st class) |

**Illustration 26.2: Examples of 'ideal' transport modes for different sets of circumstances (Bovy, 2002)**

Van der Hoeven (2001:165) relates infrastructure, and the problems accompanying it to the necessity "to use space more intensively and with more quality." "... Large-scale infrastructural links use more and more space, and also a lot of indirect space, in order to cope with more and more traffic. Large zones near those links can only be used to a limited extent, because of the poor quality of the living environment around." In searching for solutions for this problem, Van der Hoeven suggests that "... constructing infrastructure underground has two important advantages:

- Due to the construction of a second ground level above the rail or road infrastructure, the space taken up by that infrastructure can be put to a double use;
- At the same time, that double bottom absorbs the hindrance that one usually gets from using that specific infrastructure.

In this way, the indirect use of space by specific infrastructure can be optimalised. This is important because motorways' or railways' indirect claims on space are usually much larger than their direct claims. The direct claims on space of a major connection is seldom wider than 100 metres. On the other hand, the zones that are exposed to all kinds of hindrance can be over a kilometre in width: a multiple of the actual amount of space used." (*ibid.*)

## Accessibility

Urban areas, especially city centres, bundle all kinds of functions into a relatively small area, so that all kinds of efficiency advantages can be achieved, such as the proximity of functions and of people. At the same time, densely urban areas are usually increasingly difficult to access or leave by car. The metaphor of the fish-trap is well known. The fact that this situation has worsened during recent decades and is continuing to worsen, is mainly due to disintegration, i.e. the dispersal of activities; a process

3 Bovy (1993) has shown that a completely congestion free car network is undesirable for a society; the costs are simply too high to prevent congestion from happening completely. He speaks about an optimal level of congestion, where the marginal (societal) costs and marginal (societal) benefits are in balance.

that takes place because of the accessibility problem. The spatial constellation and organisation of the Netherlands makes it impossible to guarantee perfect accessibility by car,[3] i.e., without congestion, nor is it possible to exploit an attractive and competitive public transport system everywhere. Moreover, the spread of activity places, even drastically worsens accessibility by bicycle.

This line of reasoning clearly shows why multimodal transport to and from cities is very important, not only for the Netherlands, but also for many other countries with similar accessibility problems. Multimodal accessibility makes it possible to maintain and even to improve the urban structure(s), and so offers possibilities of bettering public transport facilities. In line with Bovy's (2002) conclusions, we can therefore conclude that multimodal passenger mobility is important, above all, for the good functioning of high-density urban areas: maintaining and raising the level of essential urban-centre functions, such as retail, culture (museums, theatres, cinemas, etc.), governmental machinery (municipal, judiciary, etc.), the hotel and catering industry (restaurants, cafes, etc.), public spaces (library, churches, squares, etc.), leisure (discotheques, sports arenas, etc.), tourism, and, in the case of the largest city centres, the head offices of multinationals. For this, the city has to be accessible at least by collective means of transport, but the overall accessibility of urban areas improves if the traveller is provided with multiple options: unimodal, multimodal, individual, and collective. Improving the multimodal transport system, i.e. by improving the subsystems (unimodal, multimodal, individual, collective) and the coherence and connectivity between them, should not be interpreted as a move to bring about a so-called modal shift, but as a way to bring about better, more efficient and effective accessibility to and from cities.

In this book's chapter on the 'Urbanism of Networks', the multidimensional Christaller scheme was introduced to conceptualise the shift from the traditional (medieval) city centre to today's reality of urban complexes with several urban centres, each with their own characteristics. A key concept here is selective accessibility. Today's network city, with its diversity of urban centres, requires a diversified approach to its accessibility. As discussed above, high-density city centres may profit from a multiple option accessibility. Urban centres organised under specific time and/or space conditions (e.g., opening hours, events on specific days, etc.) may profit from varying their accessibility, in line with those specific temporal-spatial conditions.

## Coherence

Interaction between the different elements of the urban fabric – streets, shops, houses, offices, green areas, pedestrian zones, etc. – increases the vitality and success of cities. A city's vitality largely depends on its geometry; its connections and networks. The processes that generate the urban web can be summarised in terms of three principles, expressed as follows (Salingaros, 2005:19):

1. *Human-activity nodes.* The urban web is anchored on nodes of human activities (activity places), the connections to which make up the urban web. There are distinct types of activity nodes: home, work, church, museum, restaurant, etc.
2. *Connections.* Infrastructure and/or public space can link nodes of activities. Empirical observations verify that the stronger the connections, and the more substructure there is in the web, the more lively the city. However, if there are too many coinciding connections, that overload the channel's capacity, then this may result in non-vitalising deconcentration.

3. *Hierarchy.* A hierarchy can rarely be established at once. When allowed to do so, the urban web organises itself by creating an ordered hierarchy of connections on several different levels of scale. It becomes multi-connected but not chaotic. It organises itself in a strict order starting from the smallest scales and progressing upwards to the higher ones. A web becomes pathological if a connective level is missing.

A complex process of organisation connects the different nodes of the urban web. The connections enable easy movement to any point, preferably by many different paths (Salingaros, 2005:21). An ordered urban environment that is strongly connected might look highly irregular from the air. On the other hand, geometrical regularity in the plan might not be directly experienced on the ground. A mathematical theorem says that two points can only be connected by a straight line in one way, but they can be connected by curved lines in an infinite number of ways. Following this reasoning, if we want the maximum possible number of connections between urban nodes (*ibid.*), then we cannot just insist on direct connections.

Based on De Jong and Paasman's (1998) table (see Ill. 26.3), Van Nes (2002) referred to these composition principles for complex interacting systems as a natural phenomenon – "a hierarchy in (transport) networks by scale factor 3". The urban network as a whole becomes unstable if the transport network takes leaps within this hierarchy. An example is Immers' suggestions for improving the road network along the south wing of the Randstad area in The Netherlands. To improve the functioning of the road network (flow of traffic, the accessibility of urban areas, congestion), Immers suggests that the secondary network should be extended. The primary, motorway, network is currently being used for several activity-travel goals, some of which (e.g., intra-regional travel) explicitly belong to the inadequate secondary network.

When discussing hierarchy in levels of scale within the city, it is worthwhile taking a look at some new insights on networks from fields of expertise other than urbanism and transport engineering. Rules of hierarchy in complex networks have been studied by physicst Barabasi (2002). When we take into account the 'rules' for the form and structure of complex networks, as presented by Barabasi, we can come up with a number of starting points for designing and planning a a network of transfer points. Barabasi shows that the complex networks that function well are, in fact, the so-called scale-free networks; the ones that are ruled by the law of power. The largest (best connected) node is closely fol-

| Level of scale | R=...km | Density [km/km$^2$] | Connection every...km | Spacing in km |
|---|---|---|---|---|
| National | 100 | 0.02 | 30 | 100 |
| Regional | 30 | 0.07 | 10 | 30 |
| Agglomeration | 10 | 0.20 | 3 | 10 |
| City | 3 | 0.70 | 1 | 3 |
| City district | 1 | 2.0 | 0.3 | 1 |
| District quarter | 0.3 | 7.0 | 0.1 | 0.3 |
| Neighbourhood | 0.1 | 20.0 | 0.03 | 0.1 |

**Illustration 26.3: Level of scale and infrastructure network (De Jong & Paasman, 1998)**

lowed by two or three somewhat smaller hubs, followed by dozens of even smaller ones, and so on, arriving eventually at numerous tiny nodes. Barabasi argues that the scale-free topology is a natural consequence of the ever expanding nature of real networks. It is up to the field of urbanism to take a closer look at the concept of the city as a scale-free network, and its related rules. Many complex networks (e.g. the human body) grow 'naturally' or 'organically', without very much human intervention. As far as the city is concerned, post-war urban design and planning has not succeeded in creating vital new urban areas. It can be hypothesised that the rules of scale-free networks do not apply here, but the truth of this will have to be proved or disproved by in-depth research.

## Shifts in spatial planning

« Nothing endures, except for change. »

"Confronted by this whirlwind of social and spatial transformation, the intellectual categories that constituted the foundation of planning in general, and of city planning in particular, have been made obsolete. Yet, the issues treated by city and regional planners are more important than ever, and the stock of skills accumulated in the field, both in the profession and in the academic institutions, are absolutely precious. What is at issue is the ability of city planners to renew their thinking, their framework, and their method, while departing from the world that is left behind: a world centred on the welfare state, on rigid zoning, on the belief in models of metropolitan growth, on the predictability of social patterns, on the legitimacy of national governments, on the long-term benefits of economic growth without social and environmental constraints and on the view of the world from patriarchalism as a way of life.

The danger for the profession is to face this transformation defensively. As in all major processes of social change there are extraordinary opportunities to be seized, but also serious costs for those institutions and individuals unable or unwilling to adapt. Thus, there is an obvious danger of digging the trenches of cultural resistance and resisting change by refining old concepts, or by embarking on a process of self-reflection in which planning itself becomes the goal, rather than the means. While in the professional world, the harsh reality of bureaucracies, politics and markets will leave little room for intellectual escapism, in the academic planning field the building of fantasy worlds made of abstract categories, or the attempt to justify planning by inventing a new academic discipline around an ad hoc theoretical foundation, could substitute for the harsh task of reinventing what to do out there, in an increasingly complex world. It must not be." (Castells, 1997).

Castells further describes in his course 'the education of the city planner in the information age' that, in future, city and regional planning will have to deal with a wide range of issues, but of these, the following will be the foremost:

1. The overarching issue of sustainability;
2. The planning of urban and metropolitan infrastructures;
3. The reconstruction of cultural meaning in spatial forms and processes;
4. The shift towards local and regional governments as decisive instances of governance.

The policy of sustainability has been widely adopted. Also in spatial planning, economic, ecological, and social aspects should be considered and weighed at the same time. The reconstruction of cultural meaning in spatial forms matches the reconciliation of the global with the local, and thus the humanisation of public spaces.

Boelens (2000) elaborates on Castells' work and states more generally that spatial planning will have to deal more and more not only with the changed notions of time and space, but also with a new steering philosophy. Space is not just determined by place anymore, but by its position and function in boundless networks. The same applies to the notion of time. The revolution of telematics has resulted in a kind of timeless time, that can hardly be measured. After all, we can communicate with the other side of the world in a split second (see also chapter 4 by Luuk Boelens).

But there are dangers. Increased (physical and virtual) mobility and network developments open up new possibilities for segregation: the Fourth World. In contrast to the First World (booming Western economies), the Second World (the former communist countries), and the Third World (developing countries), the Fourth World cannot be localised, but is present in all the other three worlds. The Fourth World consists of people who live (voluntarily or not) outside the new dynamics associated with the information '(r)evolution' – a revolution that not only stimulates segregations spatially, but also socially. For spatial planning, this puts the public domain on the agenda, not only as a spatial, but also as a social dilemma, and perhaps even as a cultural dilemma as well.

By discussing the role of mobility environments in the network city, Bertolini and Dijst (2003) hope for a better integration of mobility and accessibility considerations into urban planning and design. They present four steps for arriving at an urban development strategy based on the concept of mobility environments:

1. Different sorts of mobility environments are identified within an urban-regional context.
2. The public actor sets its investment priorities. To promote the emergence of a wide range of nodes connected on multi-spatial scales, one of the main criterions should be to encourage different sorts of mobility environments to specialise and complement each other's activities.
3. The public actor becomes increasingly dependent upon private actors. Forms of governance appropriate to the scope and scale of the strategy need to be devised.
4. Urban planning and design professionals work out how they can contribute to achieving these broadobjectives at the project implementation level.

In Bertolini and Dijst's opinion (2000), urban planning and design strategies, in which mobility environments are a leading approach, can be effective in influencing spatial developments in an increasingly mobile society.

Spatial planning has so far derived its steering power largely from the nation state. This state, and the Netherlands is a good example, consists of a hierarchical structure of departments: many local and a few regional governments, 12 provinces and one national government. In a network society, the government is not the only relevant actor and factor for spatial planning; it is not even the most important anymore, the primus inter pares. In fact, nowadays, the government stands between the other actors, instead of above them, and in the future, its role can be expected to be predominantly that of defining, and controlling the conditions, under which developments can take place.

# Conclusions

## What are the potential advantages of SMM for the city?

Firstly, the main (potential) advantage of a seamless multimodal transport system is that it is expected to improve the effectiveness and efficiency of transporting people into and out of relatively dense urban areas, thus increasing the city's multimodal accessibility. Seamless multimodal mobility reduces transfer and total travel times on individual and/or aggregate levels, thus making travelling easier. Furthermore, seamless multimodal travelling increases the efficiency of the transportation process, partly because of sophisticated (demand-responsive) information and communication technologies, but also due to the seamless physical connection of different transport subsystems (i.e. transport nodes and transfer points). In this way, societal costs per travelled kilometre can be minimised, and the number of people transported into and out of cities maximised, so that infrastructures and vehicles, and also direct and indirect space, are used more efficiently.

Secondly, multimodal transport is very well suited to long-distance travel (> 30 km). Connecting urban conglomerations multimodally, on a large scale will improve the accessibility of city areas. Traditional multimodal means of transport, such as the aeroplane and high-speed rail are indeed the most effective and efficient means of transport for (inter)continental and cross-national travelling at the present time. But there also seems to be a role for the car in (inter)national long distance, multimodal trip-making, both as a method of accessing and leaving a place, and also as a main means of transport, especially for travel motives, such as recreation and tourism, where a degree of flexibility is required.

Thirdly, a good multimodal transport system combines transport-system differentiation (speed, degree of collectivity, degree of time-space flexibility) and organisational coherence between the subsystems. Individual activity-travel patterns are more diffuse than ever, personal action spaces have grown, and most probably will grow even larger with economic growth, whether sustainable or not. The large variety (in both time and space) of trips requires a range of transport alternatives and services. The more car access is restricted at specific times or places, as we see happening today, the more good alternatives are needed, also for the traveller, in order to maintain and improve the accessibility of cities. Every transport system has characteristics that are better suited to certain spatial, social or other circumstances. In the case of spatial circumstances, it seems wise to connect subsystems at the point where those circumstances change. Then it is clear to the traveller that it would be an advantage to shift modes, because the systems to which they transfer are much better adapted to the spatial conditions that will be encountered during the next part of their journey. This physical connection of different spatial circumstances, mostly connecting different levels of scale, also has the potential to improve the spatial coherence of the urban web.

Fourthly, intermodal transfer points, as urban focus points in the city, open up possibilities for new mobility environments, places interesting for both their transport connectivity and their urban programme. Specific connectivity between transport (sub)systems (individual versus collective; slow versus fast methods) offer different potentials for different urban programmes.

Regarding the transfer points and especially the process of choosing a location, we should be aware that transport engineers and urbanists are not the only actors involved. Customers, governments (politicians), financers, transport providers, developers and contractors are all parties who might have a say in these kinds of complex processes. So, multimodal transport is not only a matter of technology, but also one of inter-organisational optimisation; a decision-making process requiring management and negotiation.

## What can SMM not do for the city?

Seamless multimodal mobility is a potential way of enriching people's mobility choices, and of improving the efficiency and effectiveness of the transport system as a whole with a minimum of societal costs per travelled kilometre. Transport history reveals that the introduction and adoption of new, faster, more efficient or effective transport technologies results in more travelling, i.e. in more kilometres travelled. For personal travel, the law of constant travel time applies. Only economic recession seems to have a negative impact people's mobility. We have to disappoint those who expect or hope that seamless multimodal mobility will reduce personal mobility. We expect the opposite to happen, because better, new alternatives usually lead to more travel; supply creates demand. However, SMM will bring about an improvement in the efficiency and effectiveness of the transport system as a whole, and so an improvement in the accessibility of cities and in people's reach.

Nor is seamless multimodal mobility a one-way ticket to solving the congestion problem of most industrialised countries. Although this may seem contradictory, congestion is a sign of a well-functioning network. In fact, it can be reasoned that there is an optimal level of congestion for all kinds of networks – a point at which marginal societal costs (that have to be made to prevent congestion from happening) are balanced by the marginal societal benefits (the savings in time and money if there is no congestion). However, seamless multimodal mobility is advantageous for the environment. Transport systems with characteristics that fit best in certain spatial circumstances are better for the environment because they put less pressure on it. Where there is a lot of pressure on space, e.g. in dense city centres, transport systems (infrastructure and vehicles) that use less space should be promoted. Where pressure on space is low, individual means of transportation should be the predominant form.

## Mobility in 21st century society

In today's socio-spatial environment, well-known geographical notions, such as the centre and the periphery, are no longer as significant and meaningful as they used to be. The linear movement from A to B has been largely replaced by the more or less permanent participation of people in extensive networks of movements and activities. Castells (1991) views this as a transition from a sense of place to a sense of flow. The notion of travelling has changed significantly. The traveller has evolved from homo viator, someone who travels on his own, into a homo transportandus, someone who is transported (mechanically), and a homo transportans, someone who transports (goods, information, data, energy).

The present and expected developments in personal transport do not just imply blessings. Increased mobility in general and the large increase in individual motorised transport has also caused large social problems of a different nature. As both individually and collectively motorised travellers, we have made a growing claim on the physical and social environment, causing ecological, social, and economic damage (Steg, 1996).

A reduction in people's mobility seems both socially improbable and impossible in the near future. It may even be socially undesirable. It seems to be culturally determined in most societies for economic growth, whether sustainable or not, to be at the top of the list of priorities, and we have learnt from recent history that the positive relation between economic growth and increasing mobility seems irreversibile. As long as the collective concepts on economic growth are imbedded in our society, what the fields of transport planning, spatial planning, and urban design are (confronted with is the problem of facilitating and accommodating the increasing demand for mobility as sustainably as possible. So, the challenge is to manage and control mobility rather than reduce it.

## References


Adviesraad Verkeer en Vervoer, 2002, *De markt voor multimodaal personenvervoer. Onderzoek naar de markt- en beleidspotentie van multimodaal personenvervoer*, Report no. 1, Literature study, Ministerie van Verkeer en Waterstaat, Den Haag

Barabási, L., 2002, *Linked. A New Science of Networks*, Perseus Publishing, Cambridge Mass.

Bertolini, L. & M. Dijst, 2000, Mobiliteitsmilieus. Ankers voor het vluchtende stedelijke leven; In: Boelens, L. (ed.), *Nederland netwerkenland. Een inventarisatie van de nieuwe condities van planologie en stedebouw*, NAi Uitgevers, Rotterdam, pp. 35-45

Bertolini, L. & M. Dijst, 2003, Mobility environments and Network Cities, *Journal of Urban Design*, Vol. 8, No. 1, pp. 27-43

Boelens, L., 2000, Nieuwe uitdagingen. Een virtueel interview met Manuel Castells; In: Boelens, L. (ed.), *Nederland netwerkenland. Een inventarisatie van de nieuwe condities van planologie en stedebouw*, NAi Publishers, Rotterdam

Bovy, P.H.L., 1993, *Weg met congestie?*, Inaugural Lecture, 10 December 1993, Delft

Bovy, P.H.L., 2002, Multimodaal personenvervoer: soms marginaal, maar essentieel; In: Adviesraad Verkeer en Vervoer, *De markt voor multimodaal personenvervoer. Onderzoek naar de markt- en beleidspotentie van multimodaal personenvervoer*, Report no. 5, Ministerie van Verkeer en Waterstaat, Den Haag

Brand, A., 2003, *Het stedelijk veld in opkomst. De transformatie van de stad in Nederland gedurende de tweede helft van de twintigste eeuw*, PhD Thesis, Faculty of Social and Behavioural Sciences, Universiteit van Amsterdam

Calabrese, L.M., 2004, *Reweaving UMA. Urbanism, Mobility, Architecture*, PhD Thesis, Delft University of Technology

Castells, M., 1991, *The Informational City: a Framework for Social Change*, University of Toronto

Castells, M., 1997, The Education of City Planners in the Information Age, *Berkeley Planning Journal*, No. 12

De Jong, T.M. & M. Paasman, 1998, *Het metropolitane debat : een vocabulaire voor besluitvorming over de kaart van Nederland*, Stichting Milieu en Stedelijke Ontwikkeling MESO, Zoetermeer

Goeverden, C.D. & R. van Nes, 2000, Karakteristieken en ontwikkelingen van multimodale verplaatsingen, *Colloqium Vervoersplanologisch Speurwerk*, Delft

Rooij, R.M., 1999, Coherent Multimodal Couplings in the Urban Web for Successful Cities; In: Bovy, P.H.L. (ed.), *Proceedings Part 1, Five years "Crossroads of Theory and Practice" TRAIL Anniversary Congress*, 5th TRAIL Annual Congress 1999, The Netherlands TRAIL Research School, Delft

Rooij, R.M., 2005, *The Mobile City. The Planning and Design of the Network City from a Mobility Point of View*, PhD Thesis, T2005/01, TRAIL Thesis Series, The Netherlands

Salingaros, N.A., 2005, *Principles of Urban Structure*, Design/Science/Planning Series, Techne Press, Amsterdam

Steg, E.M., 1996, *Behaviour Change with Respect to the Decrease of Car Use. Theoretical Analysis and Empirical Research about Problem Consciousness, Willingness of Reducing Car Use, and the Evaluation of Policy Measures*, PhD Thesis, University of Groningen, Groningen (Neth.)

TRAIL, 1999, *Seamless Multimodal Mobility. The research program*, Delft University of Technology and Rotterdam Erasmus University, TRAIL Research School for Transport, Infrastructure and Logistics, Delft/Rotterdam

Van der Hoeven, F., 2001, *RingRing*, PhD Thesis, Faculty of Architecture. Delft University of Technology

Van der Spek, S., 2003, *Connectors. The Way beyond Tranferring*, PhD Thesis, T2003/01, The Netherlands TRAIL research school, Delft University Press, Delft

Van Nes, R., 2002, *Design of Multimodal Transport Networks. A Hierarchical Approach*, PhD Thesis, The Netherlands TRAIL Research School, Delft University Press, Delft

Van Wee, B., 2002, Personenvervoer: dood paard of volop kansen? Een essay over multimodaal vervoer vanuit milieuoptiek; In: Adviesraad Verkeer en Vervoer, *De markt voor multimodaal personenvervoer. Onderzoek naar de markt- en beleidspotentie van multimodaal personenvervoer*, Report no. 5, Ministerie van Verkeer en Waterstaat, Den Haag

# 27 Carrying Structures: Urban Development Guided by Water and Traffic Networks

« The good city is one in which the continuity of complex ecology is maintained while progressive change is permitted. »
Kevin Lynch, *Good City Form*

Sybrand Tjallingii

## Introduction

### The setting: Devising a planning tool for sustainable urban development

The Dutch cartoonist Stephan Verwey once pictured a group of deer running from a forest to a wall of high-rise buildings, hesitating, and then running back at high speed, back to nature. The message is clear, the city is no place for wildlife; nature and the city are enemies. In this chapter I will argue that this idea is a widespread and deeply rooted myth in our culture, but it is a myth that is at odds with our modern understanding of urban systems and ecology. Nature does not stop where the city begins. The city itself is an ecosystem that generates new and modified cycles of water, minerals and energy. Moreover, the city provides a variety of urban habitats for humans, plants and animals. Rather than presenting a solution, the 'enemy' myth causes problems for both resource and habitat management. They arise if human activities, in both town and country, fail to take ecological processes into account, and they persist if ecologists confine their activities to nature reserves or to wildlife species. In the long-term, conditions will only improve if ecologists and urban planners work together on the issues of making urban development more ecological. If that is the challenge, then the *Two Network Strategy* is one of the 'tools' that can be used (Tjallingii, 1995, 1996). In this context, 'tools' are planning principles for guiding the making of 'good city form' (Lynch, 1981). The Two Network Strategy is a planning tool that uses the spatial networks of water and traffic systems as carrying structures for making good plans for sustainable urban development.

This chapter starts with some fundamental reflections underlying the Two Network Strategy. Then, I will discuss separate guiding models for water and traffic networks before presenting the Two Network Strategy in a case study about the north wing of the Randstad. The final section discusses some relevant questions.

### The myth about nature and the city

Before sharpening central questions in this article, it is worth analysing the motives underlying the myth surrounding nature and cities in the professional cultures of ecologists and urban planners. It is tempting to delve into the rich history of thought about these issues, a history that goes back to the beginning of the 20th century, and to look at the leading roles played by planners and theorists such as Patrick Geddes and Lewis Mumford (Hall, 1996:136-174; Van Schendelen, 1997). This would go far beyond the scope of this chapter, however. Here, the relevant issue is the influence that the myth about nature and the city being enemies has had on present-day discussions concerning urban and regional planning projects.

Viewing cities and nature as enemies is deeply rooted in our culture. Not surprisingly therefore, the 'enemy' image has had a major impact on the way landscape ecologists have looked at cities. When landscape ecology emerged as a branch of biology and physical geography in the second half of the 20th century, researchers in this new field had another good reason to by-pass the urban environment: the new science was conceived as a natural science, so what was more natural than to look first at pristine nature, and how that works, before looking at how man has interfered with the natural processes? From that, the idea matured that cities only disturb nature. This idea still dominates the approaches taken by many modern landscape ecologists. Take, for example, Cook's work on the ecological networks of the urban landscape of Phoenix, Arizona (Cook, 2000). Cook turns to ecological theory to explain how the population dynamics of species is influenced by ecological networks that connect patches of habitat. Understanding the importance of these patches and green links then led to a plan to establish green networks in urban areas. The emphasis of the study is on the role of green networks in defending nature against urban disturbance. In this context, the hydrological cycle was only used as an assessment parameter for wildlife habitats. Yet, in the desert climate of Phoenix, water is also a basic condition for the human habitat. Natural areas may play an important role in water conservation for the benefit of the city. The focus on defending nature against the city implies that less attention was given to the potential synergism between conditions for wildlife and for vital human aspects of urban life. The Two Network Strategy addresses this very synergism by looking at the city itself as an ecosystem.

Following a different line of reasoning, many urban planners and designers came to a similar mythical view. For a long time, up to the beginning of the 20th century, urban form was shaped by the technical need to adapt cities to potentials offered by the landscape, such as existing slopes, flood plains and local climate. As technology developed, however, planners, designers and other decision-makers distanced themselves from the bonds of nature. Leading architects came to regard tuning urban forms and processes to nature, or to the local landscape, as a backward idea. They viewed the city as the place where human beings are in control; they viewed it as being the centre of culture, and therefore the opposite of nature. While urban development threatens to destroy the countryside and its nature, natural forces, such as earthquakes and floods, pose a threat to the city. With this understanding at the back of their minds, urban planners tried to contain urban development by introducing green-belt and compact city policies. In the Fifth Memorandum on Spatial Planning *Ruimte maken, ruimte delen* [Making Space, Sharing Space] (Ministerie van Volkshuisvesting, Ruimtelijke Ordening en Milieubeheer, 2001) so-called contour lines are drawn to futher contain urban expansion; in order 'to save nature'.

## 'Working with nature', a process-oriented approach

A different approach in urban planning is to 'work with nature'. So instead of protecting natural ecosystems against the city, the aim then is to make the city itself more ecological. This approach draws on the experiences of using the 'design with nature' line of thought in urban planning, following the work of McHarg (1969), Hough (1984, 1995), Spirn (1984, 1998) and others. The Two Network Strategy is a tool that can be used in exploring this approach further.

The myth that the city and nature are enemies expresses basic concerns of ecologists, urban planners and many others about the quality of life for both people and wildlife. The 'working with nature' approach addresses these concerns by taking the ecological processes of the city itself as a point of departure. At this point it is appropriate to distinguish between 'object-oriented' and 'process-oriented' approaches to ecology and planning (Tjallingii, 2000). The objectives in the first approach were protected areas or desired species, and the planning task was then to organise the means to achieve these goals. The process-oriented approach takes existing ecological processes, such as the hydrological cycle, as the starting point. The planning task then becomes creating sustainable conditions for resource management and multifunctional urban land use. The Two Network Strategy was conceived as a tool for use in this process-oriented approach; a tool for 'working with nature' (Tjallingii, 1995, 1996). Both water and traffic are basic 'flows', essential processes for urban development. Moreover, they both form spatial networks that may act as carriers for a contrast between dynamic and quiet types of urban land use. As a planning strategy, the combination of these two networks may create an environmental and spatial quality frame that leaves freedom for flexible infill.

### Questions

The general discussion in the preceding sections leads to the following questions:
1. What are the goals and means of the Two Network Strategy?
2. How can this strategy be used as a tool in actual planning processes?
3. To what extent is the 'process-oriented' approach an alternative for the defensive 'object-oriented' approach?
4. Why the Two Networks of water and traffic? Why not a green network?
5. How does the Two Network Strategy cope with the antagonism between city and nature?

## Guiding models for water networks

This section of the chapter will elaborate the 'working with nature' idea. Making ecologically sound plans for urban areas starts by translating basic ecological principles into guiding models that can serve as useful tools for planners when making a specific plan to fit a local situation (Tjallingii, 2000). The Two Network Strategy itself is a guiding model, but it combines a number of more detailed guiding models for water and traffic flows. These will be discussed first.

The *general guiding model for water* (Ill. 27.1) is based on fundamental ecological principles that may be summarised as: 'keep it clean and keep it longer'. From an ecological point of view, this sounds like commonsense, but in practice, in managing rainwater and surface-water, it used to be the practice to 'make it clean and get rid of it quickly'. However, both the practice of incomplete purification, and the quick removal of rainwater, used to create pollution problems and flooding further downstream. The guiding model, therefore, indicates the need to make management plans for each water system. River valleys, polders and other manageable catchment basins have to be treated as unities in water-quantity planning. The basic unity for quality planning is the house, the firm or farm where human activity is the starting point for preventing pollution. Sediment management focuses on working with nature by using vegetation to prevent erosion and by using the river flows and tidal movement to keep channels open and coasts safe. The guiding model addresses the general situation of river valleys and a delta with polders.

The *connection model* (Ill. 27.2) addresses itself to urban regions which have different land-use functions. Here the underlying ecological principle is to create a stable gradient by allowing water to flow from clean to polluted, or, for that matter, from nutrient-poor to nutrient-rich conditions. The guiding model incorporates two strategies: a series connection and a parallel connection. In both cases, the aim is to attain the best mutual adjustment between the water systems and the land use.

The *circulation model* (Ill. 27.3) is a guiding model for the R block (residential areas) in the connection model. The general 'keep it longer' principle leads to a cascade-like approach. In a relatively flat area, the steps of the cascade may provide the space and opportunity for surface water to circulate before it goes one step further down. The underlying ecological principle is to use the area's own water resources optimally by retaining the water and purifying it within the system. Retention includes seasonal storage, i.e. retaining rainwater in the winter season, when surpluses can be expected, so that it can be used during summer shortages. In the flat Dutch polder area, the normal practice is to pump off the clean rainwater in winter. In summer, water is let into the system to maintain water levels in the polder ditches. However, the water that is let in is from more polluted rivers. So solving a quantity problem creates a quality problem. The circulation model, with its seasonal storage, provides an ecological answer to this problem and suits the conditions of many urban areas in the Dutch polders., An increasing number of plans for new developments and for existing cities in the Netherlands use this guiding model (Van Eijk, 2002).

The details and backgrounds of the connection and circulation models were discussed in Tjallingii, 1995, and 1996. One of the water systems based on the circulation model is at Morra Park in Drachten, in the Dutch province of Friesland. This was monitored for over 10 years, and the results are positive. Phosphate content in the urban waters is 0.1 mg/l, compared with 0.4 mg/l in the polder outlet waters of the surrounding countryside. Other positive developments noted during this assessment were a rich wildlife and residents' appreciation of the water system (Van Eijk, 1997).

## Guiding models for traffic networks

The *general transport guiding model* (Ill. 27.4) embodies the principle: create contrast and avoid fragmentation by concentrating transport into traffic corridors. This principle is important because it improves the ecological conditions for both wildlife and people. These corridors create more contrast between quiet areas, both rural and urban, and the dynamic axes of traffic and transport-related activities. The concentration of infrastructure creates conditions for joint investment in noise barriers, run-off water treatment facilities, tunnels and bridges. These measures would be very expensive, if applied to each road separately. Concentration makes it easier to protect the environment and to alleviate the barriers for wildlife, pedestrians and cyclists. In the Dutch situation, where there is a high density of infrastructure, the concentrated corridors are planned in distances of 10-100 km.

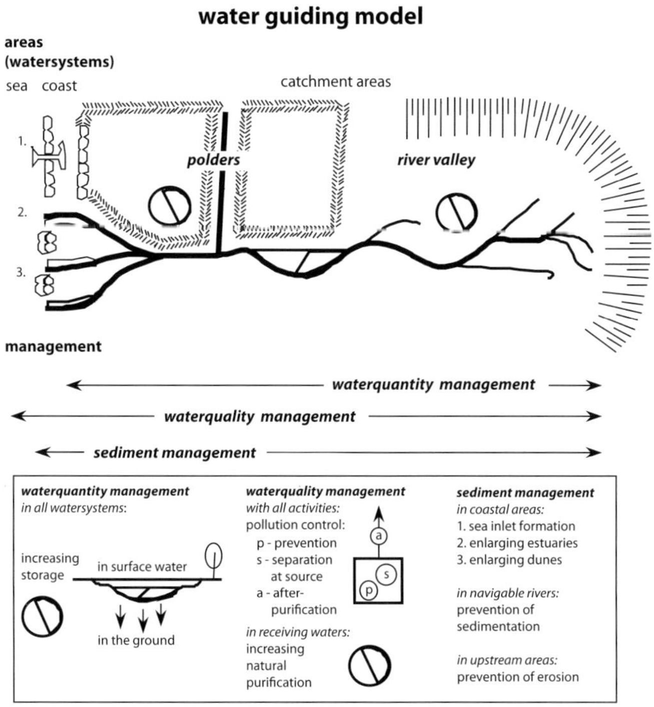


**Illustration 27.1:**
**The general guiding-model for water systems**

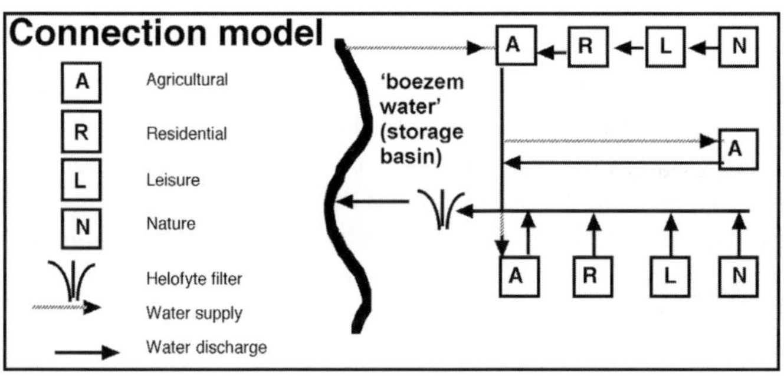


**Illustration 27.2:**
**The guiding model for regional land-use and water planning in lowlands**

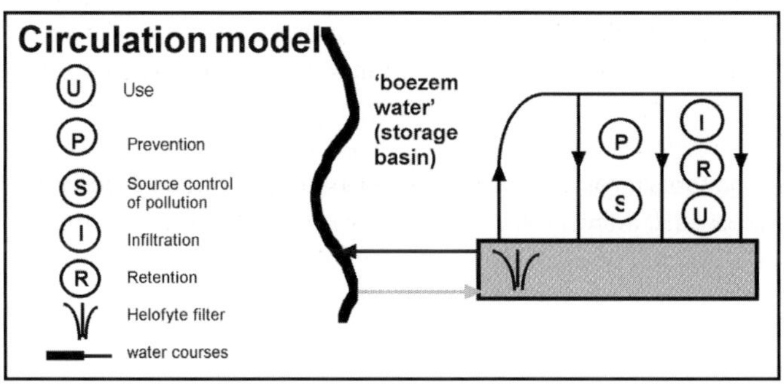


**Illustration 27.3:**
**The guiding model for planning with water in polder-cities.**

Within the corridor area of 100-200 m, for safety reasons, there can be a separation of transport lines. Between the nodes, the multi-modal structure offers different options – roads, railways and waterways. Thus, collective transport, or new intermediate-distance, individual and collective, means of transport are possible options. Around the nodes, however, individual transport will prevail.

The *lobe city and string of beads model* (Ill. 27.5) is a guiding model for the interaction between transport planning and urban development. This model can guide planners when looking for alternatives for scattered sub-urbanisation and criss-cross road building brought about by people's preference for living in a green setting and going into town by car. The idea behind this model is a new version of the 'lobe city' (Tjallingii, 1996:225), in which urban expansion takes place along lobes and there is a central axis for public transport and trunk roads, with attractive cycle tracks through the green wedges. Further urbanisation is accommodated in smaller satellites along public transport axes, like strings of beads. This idea follows suggestions made by Breheny and Rookwood (1993).

By increasing the length of the border between built-up and green areas, the guiding model provides a spatial structure that allows a maximal number of people to live along the edges of green areas. The model rejects the popular ring roads, because ring roads destroy the quality of this border. Where there already is a ring road, instead of letting the road contain further urban growth, development should jump the ring in the form of extra lobes, again with attractive edges that can be made even more attractive by constructing water-retention lakes there.

## Working with the Two Network Strategy

### The tool

The *Two Network Strategy* guiding model (Ill. 27.6) combines elements of the guiding models for water and traffic flows with an ecological zoning principle. So, for instance, the gradient between quiet and dynamic traffic would coincide with the transition between upstream clean water and more polluted water downstream. The zoning is a translation of the *connection model*. The *circulation model* is applied along the edges of residential areas. The *general transport guiding model* is applied to the traffic corridor that carries the industrial zones and intensive agricultural areas where good logistics and reliable transport are primary necessities. The *'border length and string of beads' model* is applied to the underlying idea of regional development.

The Two Network Strategy is a tool that guides the design of carrying structures rather than form. To use the tool, planners are required to use their skills to explore local hydrology and other ecological potentialities of the landscape as well as any elements of the existing cultural landscape that may contribute to the new plan.

A case study can best illustrate how the tool works. The Two Network Strategy has been used on the national level to make scenario plans for 2030 (Ministerie van Volkshuisvesting, Ruimtelijke Ordening en Milieubeheer, 1997) and to make a number of plans for new urban developments (Zonneveld & Dubbeling 1996, Aalbers & Jonkhof, 2003). Recently, the strategy has guided the making of a plan for redesigning and renewing the Schalkwijk district of Haarlem, a residential district with 33,000 inhabit-

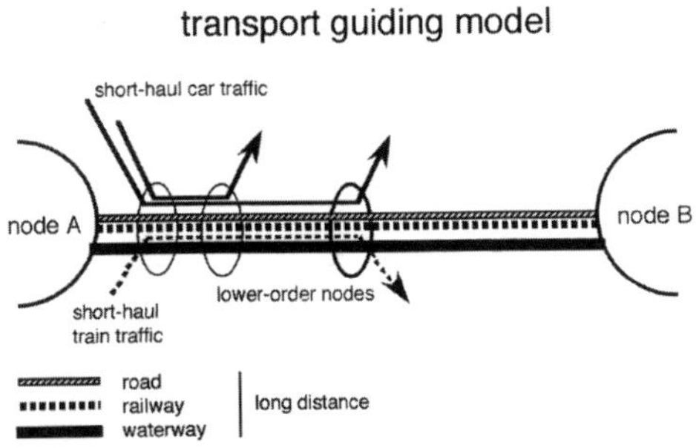


**Illustration 27.4:**
**The general guiding model for transport corridors**

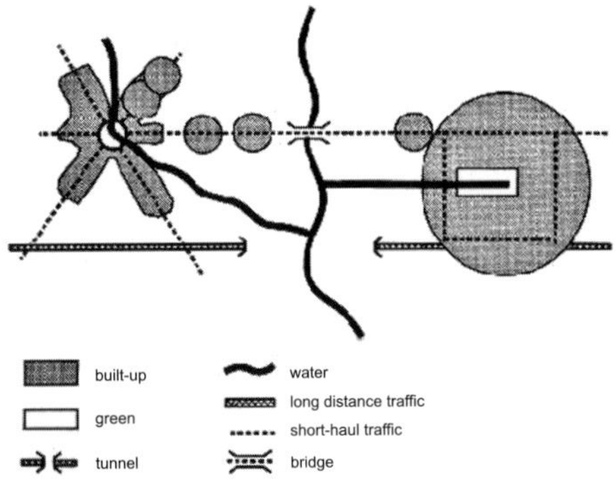


**Illustration 27.5:**
**The 'lobe city and string of beads' model for urban development**

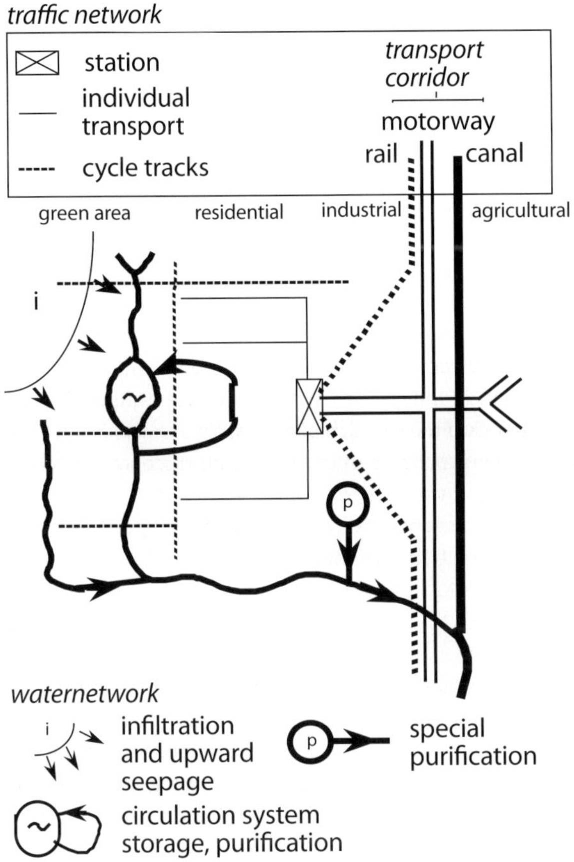


**Illustration 27.6: The general guiding model of the Two Network Strategy**

ants, built in the 1960s (Van Eijk, 2002). Here, I will discuss the use of the Two Network Strategy as a tool for making a regional plan for the northern part of the Randstad: the rim-like urbanised area around Amsterdam, The Hague, Rotterdam and Utrecht in the western part of the Netherlands.

## The Warande case

*Warande* is an old Dutch word for a delightful place situated on the edge of a park or hunting area. We chose this name for a plan that had to include a spatial development structure for the north wing of the Randstad, between the cities of Amsterdam, Haarlem and Leiden, including Schiphol Airport. The central map shows Leiden in the South, the North Sea shoreline in the West and the Amsterdam agglomeration and harbours in the North. About 2.5 million people live in this area, including the bigger cities. The Warande Plan focuses on the interstitial area between the cities. The sharp political debate between proponents of rapidly developing the airport and their opponents, who are defending the environmental qualities of the region, demonstrates the powerlessness of an object-oriented ecological strategy when confronted with powerful economic interests. Alternatives need to be found for this deadlock debate (Van Eeten, 1999:113-143). However, the airport is not the only controversial issue in the area. Chaotic urbanisation and the construction of more and more infrastructure is threatening the so-called Green Heart of the Randstad, a relatively open green countryside, designated by official policy to remain open. Despite this, the cities expand and the villages grow in the Green Heart and further fragmentation of the existing green areas is very likely.

The Warande Plan is founded on the traffic network, that being the carrier of the economic processes. A second carrier is the surface water and groundwater network, because this mainly supports the green countryside. These two networks carry the economic and ecological processes, and simultaneously function as a backbone to create conditions for richly contrasted urban landscape; an attractive Warande between the Randstad and the Green Heart. Illustration 27.7 shows the main elements of the plan (Tjallingii, Snijders & De Boer, 1999).

The Warande Plan concentrates traffic into two corridors, named after the A4 and A9 motorways (see upper-left map). The main railways, motorways and regional roads are concentrated within these corridors and the best sites for the future development of Schiphol Airport are also here, just south of the junction between the A4 and the A9. The corridors offer two different scenarios with respect to expanding Schiphol:

1. If the growth of the airport is limited to its present location, then it will be important to concentrate the runways and the noisy fly zones into the two corridors;
2. If further expansion is agreed, then the corridors could be used as links to a possible new satellite airport.

The transport corridors carry the main business areas. The A9 links the Amsterdam harbour area with the south-east business districts of the city. The A4 carries the rapidly expanding business area near Hoofddorp and an intensively farmed area to the south. This will enable new area development. The concentration of the main roads and railways within the two corridors provides an alternative for present plans that will cause further fragmentation of the whole landscape. The concentrated corridors will create better conditions for reducing noise and pollution. The Warande Plan cuts down on the number of barriers for wildlife and for local traffic, cyclists and pedestrians. Although the new barriers will be bigger and more difficult to cross, because they will be concentrated into a small area,

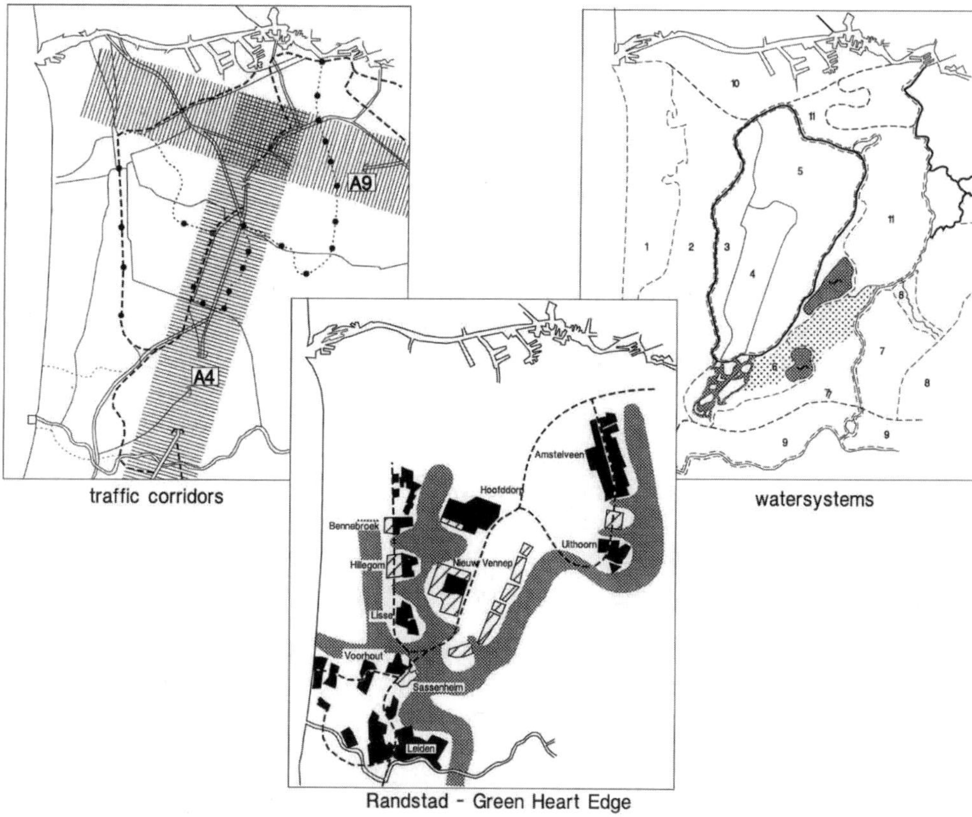


**Illustration 27.7: Traffic and water networks as carrying structures for the northern Randstad: The Warande Plan (Tjallingii, Snijders & De Boer, 1999)**

the planned investments for bridging them will be more easily realised. In some places, building green bridges for pedestrians, cyclists and wildlife has been proposed, to facilitate movement from one green area to another. In the upper-left map, where the A4 is indicated, the high-speed train tunnel that has already been planned will be positioned differently to allow a valuable and vulnerable green network to cross the transport corridor. Another tunnel is being planned near the A9 sign on the map. Here, both a residential area and a green zone have to cross the transport corridor.

With a few exceptions, residential areas are located separately from the infrastructure corridors. The central map shows the existing settlements in black, and possible expansions as hatched zones. The residential areas will be further developed, like a necklace of beads, along local railway and light-rail routes (the dotted lines on the upper-left map). In this way, public-transport will have a fair chance of competing with individual motorised traffic. This will encourage the use of more environmentally friendly means of transport, the forms that use less energy, make less pollution and cause less congestion.

The aim of the water plan (upper-right map) is to keep and use rainwater for as long as possible, thus preventing floods downstream. At the same time, seasonal storage in the polders can reduce the intake of polluted or eutrophicated river water in summer. Storing rainwater in the ground and in surface waters requires the careful analysis of local hydrological conditions. The plan proposes different subsystems. These are numbered 1 to 11 on the map. Each of these subsystems has its own sources of clean water and its own sources of pollution. The local hydrology offers opportunities for making differentiated land-use plans, using the guiding models as helpful tools. The cities of Haarlem and Amstelveen, for example, have used the circulation model to plan seasonal storage ponds along the urban fringe that are linked to a system of canals in the built-up area. The Amstelveen water scheme is briefly discussed in Tjallingii (2000:110-111). The Warande Plan sets the conditions for these local plans, but also points to the regional opportunities. In the coastal dune area (labelled '1'in the upper-right map), drinking-water production limits the prospects for wildlife, because pumping lowers the groundwater tables and the artificial infiltration of river water, causes high phosphate contents. The Warande Plan proposes inundating a number of small polders in the dotted zone of Subsystem 8. Here new lakes for drinking-water production can replace the existing sub-optimal agriculture. This will contribute to the visual quality of the border between the Randstad and the Green Heart. It will also create interesting new options for wildlife and recreation in the dunes. Thus the water processes will create contrast and identity in the urban-rural landscape that will benefit not only people, but also plants and wildlife. Moreover, in a strategy geared towards making urban and rural land-use more ecological, managing the water system is a way of monitoring and learning from experiences, so that further improvements can be made.

The central map demonstrates how the Warande Plan structures the chaotic urbanisation in the area. The border between the Randstad and the Green Heart becomes more beautiful, a true *warande*, accentuated by new, functional lakes in the water systems. The length of the border is extended by developing green lobes between the villages and suburbs, creating a string of bead-like settlements. The connection between the green lobe and the Green Heart is enabled by a tunnel and other constructions that allow the transport corridor to cross it. Improving the quality of the green areas creates more attractive locations for living and working. The carrying structures of water and traffic become a framework for sustainable urban development which offers opportunities for new residential areas to be built, business areas to be expanded and even allows the airport limited growth. But while the framework is durable, the infill remains flexible within the limits of a set of rules that emerge from the spatial and environmental principles of the Two Network Strategy.

# Discussion

## The goals and means of the Two Network Strategy

Taking the first two questions raised in the introduction together, I will first discuss the goals and means of the Two Network Strategy, as illustrated by the Warande case. What is the goal of a strategic tool? The Two Network Strategy is described as a tool for making plans. In the Warande case, these are strategic plans, i.e. plans used in drawing up a framework for operational policy decisions (Mastop, 1997:807). Strategic planning is 'decision centred' (Faludi, 1987) and can be assessed by looking at how well it influences decision-making. This is different from operational planning or project planning. Typically, an operational plan draws a straight line to a fixed point and makes a commitment to reach that point within a certain period, and for a fixed budget. The Warande Plan is an example of how a strategic plan can frame further operational decision-making about building roads and houses and realising green areas with certain specifications. In urban planning, there is a general uncertainty about future developments, and tension due to conflicting interests. Under such conditions, one of the first aims in urban planning is to try and reach consensus about the general aims and the basic spatial structure. The goal of the Two Network Strategy is to include ecological conditions within this basic structural plan.

As can be seen in the making of the Warande Plan, the Two Network Strategy guides the structuring of the plan in three ways. The development of the water and traffic systems is directed towards resource conservation, energy saving and pollution prevention (the 'flow' aspects). The ecological potentialities of the local landscape are used to enhance the contrast between dynamic and quiet zones, to increase the variety of habitats and to strengthen the identity of the new urban landscape (the 'area' aspects). The plan generates a framework for discussion among the stakeholders, which may lead to a structural plan that support further negotiations concerning operational decisions (the 'actor' aspects). The guiding models can be viewed as the means for bridging the gap between the general aims of sustainable development and the more concrete regional goals. To reach more specific targets for biodiversity, for housing, or for any other issue, operational plans are required which have more specific goals and means.

## Is the Two Network Strategy an alternative?

To what extent is the 'process-oriented' approach an alternative for defensive, 'object-oriented' approaches? In the light of the difference between strategic and operational plans, clearly both carrying structures and operational targets need to be met. The Two Network Strategy, as a tool in a 'process-oriented' approach, has, of course, to be combined with concrete operational 'object-oriented' targets, either defensive or offensive. Thus, a strategic approach in general, and the Two Network Strategy in particular, is not an alternative to concrete operational plans; both are needed. What matters, however, is the sequence. If, in the Warande case, one starts with a landscape evaluation map, the planning problem will be defined as saving nature at one location and sacrificing it at another. This may sharpen controversies rather than create co-operation in finding solutions. Moreover, designating spaces alone does not create a good plan. Fences alone do not protect nature. Planning practice has proved this approach to be unsuccessful. A critical discussion of landscape evaluation (Tjallingii, 1996:36-39) demonstrates the weaknesses of this defensive object-oriented approach. Yet, authors like De Groot, 1992 and Costanza *et al.* 1997 still strongly advocate this way of landscape planning. If, on the other hand, one starts with a process-oriented strategic plan, then the basic conditions for maintaining or developing landscape values are addressed. These basic conditions structure the plan that could be fine-tuned in later stages of planning. Starting with the Two Network Strategy is thus an alternative that creates options for synergy between land-use functions, by looking at the carrying structures. Together, the guiding models provide a toolkit of conceptual instruments for the making of plans. The object-oriented approach tends to focus on the testing of plans. If there are no good proposals, however, even the best appraisal criteria cannot produce a good plan.

## Two or more networks?

Why water and traffic networks? There are other relevant networks, of course: urban and regional planners talk about 'urban networks' or 'network cities', and landscape ecologists draw maps of 'green networks'. In the Netherlands, it is official government policy that a 'National Ecological Network', a system of ecological corridors throughout the country, will be created (Ministerie van Landbouw, Natuur en Visserij, 1990). The Warande Plan demonstrates that, in making a plan, all the relevant aspects of landscape and land use, nature and culture are important, and will indeed play a role. The crucial point is where to start. Starting with traffic and water means first viewing the problem from a distance, away from the direct interests of the 'green' and 'red' areas. The proposals for water and traffic processes create a spatial and functional framework for the basic systems that serve all interests. This framework makes it possible to weigh up interests further. The dominant criterion is whether the proposals for 'green' or 'red' development fit into the *durable* framework. In design workshops with stakeholders and professionals, a fruitful approach has proved to be to start with water and traffic networks. Inviting the actors to generate ideas and make sketches of long-term carrying structures for spatial development helps to create a climate for co-operation, without neglecting the short-term interests (Van Eijk, Tjallingii & Van den Top, 2001).

One of the dilemmas of sustainable development is related to pattern and process. If the objective is a durable pattern, as in nature and landscape conservation, then a drastic change in economic and social processes is required; a change from production driven processes to those geared towards conservation. However, where sustained production-driven processes are the objectives, as in agriculture, manufacturing and services, then a drastic change of urban and rural landscape patterns is inevitable. The Warande case shows how, by using water and traffic as a basis for planning, this can lead not to museum landscapes, or to those of concrete and asphalt, but to a mutual adjustment of processes and patterns. Making change more ecological may lead to a modern functional, yet sustainable, landscape in which new cultural identities can be created from the potentialities of that landscape.

## City and nature, antagonism or partnership?

The antagonism of city and nature is real if the city is defined as 'buildings and asphalt' and if nature is interpreted as wildlife. I argue for wider definitions. The city is the complex system of buildings and open spaces, of 'red' and 'green', a system that cannot escape the workings of climate, soils, water and living organisms, whether we like it or not. These are the workings of 'nature' in a broad definition. The Two Network Strategy and the related 'guiding models' are tools for ecologists, urban planners and others who are working together to make the urban system more ecological. In the long-term, that will be the most sustainable basis, both for the city and for the nature defined as wildlife.

In regional planning, the so-called antagonism between the city and nature usually equates with the contrast between town and countryside. The more urbanites prefer to live in suburban or rural environments, the more this contrast will disappear. The Two Network Strategy is a useful tool for those planners and politicians who are exploring ways of retaining and developing contrast, not only between red and green, but more general, between dynamic urban life and quiet places. The urban quality of life depends on the opportunity to live on the border of these two worlds.

## References


Aalbers, C. & J. Jonkhof, 2003, *Planning on principle*, Aeneas, Boxtel, The Netherlands

Breheny, M. & R. Rookwood, 1993, Planning the Sustainable City Region; In: Blowers, A. (ed.), *Planning for a Sustainable Environment*, Earthscan, London, pp. 150-189

Cook, E.A., 2000, *Ecological Networks in Urban Landscapes*, PhD Thesis, Wageningen University, Wageningen (Neth.)

Costanza, R., R. d'Arge, R. de Groot, S. Farber, M. Grasso, B. Hannon, K. Limburg, S. Naeem, R. O'Neill, J. Parvelo, R. Raskin, P. Button & M. van den Belt, 1997, The value of the world's ecosystem services and natural capital, *Nature*, 387, 15th May, pp. 253-260

De Groot, R.S., 1992, *Functions of nature, evaluation of nature in environmental planning, management and decision making*, Wolters-Noordhoff, Groningen (Neth.)

Faludi, A.,1987, *A Decision-Centred View of Environmental Planning*, Pergamon Press, Oxford

Hall, P., 1996, *Cities of Tomorrow*, Blackwell Publishers, Oxford

Hough, M., 1984, *City Form and Natural Process*, Routledge, London/New York

Hough, M., 1995, *Cities and Natural Process*, Routledge, London/New York

Lynch, K., 1981, *Good City Form*, MIT Press, Cambridge Mass./London

Mastop, H., 1997, Performance in Dutch spatial planning: an introduction, E*nvironment and Planning B: Planning and Design*, 24, pp. 807-813

McHarg, I., 1969, *Design with Nature*, Doubleday & Co., New York

Ministerie van Landbouw, Natuur en Visserij, 1990, *Natuurbeleidsplan*, Den Haag

Ministerie van Volkshuisvesting, Ruimtelijke Ordening en Milieubeheer, 1997, *Discussienota Nederland 2030*, Den Haag

Ministerie van Volkshuisvesting, Ruimtelijke Ordening en Milieubeheer, 2001, *Ruimte maken, ruimte delen: Vijfde Nota over de Ruimtelijke Ordening 2000/2020*, Den Haag

Spirn, A. W., 1984, *The Granite Garden*, Basic Books, New York

Spirn, A.W., 1998, *The Language of Landscape*, Yale University Press, New Haven/London

Tjallingii, S.P., 1995, *Ecopolis, strategies for ecologically sound urban development*, Backhuys Publishers, Leiden (Neth.)

Tjallingii, S.P., 1996, *Ecological Conditions*, PhD Thesis, Delft University of Technology

Tjallingii, S.P., 2000, Ecology on the edge: landscape and ecology between town and country, *Landscape and Urban Planning*, 48, pp. 103-119

Tjallingii, S.P., H. Snijders & T.W. de Boer, 1999, De Randstad als Warande, *S&RO*, 80/2, pp. 9-15

Van Eeten, M., 1999, *Dialogues of the Deaf: defining new agendas for environmental deadlocks*, PhD Thesis, Delft University of Technology, Eburon Publishers, Delft

Van Eijk, P., 1997, Integraal Waterbeheer; In: *Handboek RO en Milieu*, Samson HD Tjeenk Willink, Alphen a/d Rijn, pp. 161-179

Van Eijk, P., 2002, Integraal en duurzaam waterbeheer; In: *Handboek RO en Milieu 2001-2002*, Kluwer, Alphen a/d Rijn, pp. 257-275

Van Eijk, P., S.P. Tjallingii & M. van den Top, 2001, Workshops for Sustainable Urban Development, *Nordic Journal of Architectural Research*, 14, no. 4, pp. 45-60

Van Schendelen, M., 1997, *Natuur en Ruimtelijke Ordening: een symbiotische relatie*, NAi publishers, Rotterdam

Zonneveld, W. & M. Dubbeling, 1996, *Visie Ecopolis, De Strategie van de Twee Netwerken*, Ministerie van Volkshuisvesting, Ruimtelijke Ordening en Mileiubeheer, Den Haag

# 28 Building a Multinodal Metropolis: A Short Guide

Marc Jacobs

## Introduction

Scientists and professionals in urban and regional planning and urban and regional design still disagree. Some support the proposition that the Dutch metropolis or urban area Randstad, also known as Randstad-Holland or Deltametropolis, (Ill. 28.1) does not exist (De Boer, 1996). Others seem to be strong advocates of the concept 'Deltametropolis – the new logo for the Dutch metropolis' (www.deltametropool.nl). I strongly believe that the disagreement and ongoing discussion should be interpreted as a miscommunication.

Those who oppose the idea of the Randstad argue that this urban area is not a metropolis. Their argument is based on a definition that is strongly related to the classical or even nostalgic 19th century idea of a metropolis (Blumenfeld, 1967; Van der Cammen, 1988; Gordon & Richardson, 1996); an opinion in which the international influence of a city, and its content (e.g. the level of the urban amenities) have to be taken into account, and compared with other supposed metropolises.

The supporters of the Deltametropolis have a different approach. They admit that the current international position of the Randstad is rather weak, and that the different cities that comprise the Randstad do not offer much in themselves in comparison with cities like Berlin, Milan, or even Paris and London. Although the Randstad is a minor player on the metropolitan battlefield, the supporters of the Deltametropolis argue that the western part of the Netherlands has an enormous potential for becoming a strong network-metropolis (also known as a multinodal or polycentric metropolis).

The real discussion about the Randstad is how to define this opportunity. How real is it? Which possibilities can be designed and defined, and how do we reach the goals? More specific within the context of urban and regional design: what specific spatial interventions are required to be able to create this multinodal metropolis? I strongly believe that an answer to this question cannot be obtained from the ongoing – almost political – discussions. It should be sought in the scientific basis of planning and urban and regional design. To put it shortly: in order to step forward, we need more science in planning and design.

**Illustration 28.1:**
**The Randstad (Neth.) (Klaasen & Jacobs, 1999)**

In this chapter, I will discuss how multinodal urban systems develop, how urban areas become transformed within this process, and what spatial conditions, or prerequisites are required for this to take place. Based on scientific knowledge, the types of spatial intervention required to build a multinodal urban area are identified. In the concluding section, several trends and developments that may affect urban transformations are discussed. I will also position the scientific approach to urban planning and design within the context of Dutch spatial planning, and more specifically within the field of Dutch urban and regional design.

## The centre paradigm

The lack of up-to-date knowledge on the development of multinodal (multicentred, polycentric or polynuclear) metropolises hampers the discussion about the Randstad or Deltametropolis, and even more standard urban areas (Klaasen, 2004). This seems strange as there is an enormous amount of scientific knowledge about the development of metropolises. On the other hand, it is understandable, as most scientific knowledge on this subject is based on the single-centreparadigm. This paradigm states that every urban system has one centre that functions on the highest level possible, given the size of that system and the position of the area in which it is located. For many years, this paradigm has been the foundation of numerous spatial, spatio-economic and economic-geographical theories (Jacobs, 2000; Berry & Kim, 1993; Richardson, 1988; Wadell & Shukla, 1993). In research into the development of multinodal urban areas, this paradigm has been set aside (Jacobs, 2000). A centre, simply speaking, should not be seen as a somewhat nostalgic and mythical higher good, a bonus for good city planning, but as a concentration of urban and regional activities; nothing more and nothing less. Any urban area in which we can find two or more concentrations of urban and regional activities is considered to be multinodal. Theoretically, it can even be argued that an urban area does not have to have a concentration of urban and regional activities at all, as in the case of complete sprawl.

It is obvious that the concentration of activities will be related to the area itself. Urban and regional activities are related to an urban area or urban region. This brings us back to the discussion about the Deltametropolis. We need to question whether, on the scale of this prospective metropolis, there are any activities related to its size. We also have to question whether there are any relations specifically in the area of the Deltametropolis that result in flows of people, goods and information. In other words: can the Deltametropolis be considered an urban system? And if not, can the Deltametropolis be considered a potential urban system? This final question might be the most important, because it will be the starting point for urban and regional planning and design.

## Three points of departure for developing a multinodal system

Based on research into the development of multinodal urban areas, we can conclude that a multinodal urban system develops in several steps (Jacobs, 2000), and that the centres of which it is comprised, regardless of whether they have specialised or mixed economies, have to have certain spatial conditions. If these conditions were absent, then a centre, or rather a concentration of urban and regional activities, would not come into being. Their presence, on the other hand, does not automatically bring about a concentration of urban and regional activities. The relation between the spatial conditions and the occurrence of a concentration of urban and regional activities is conditional, not causal. In the following paragraphs, the steps in the development of a multinodal urban system and the influence of the spatial conditions are explained.

There are several points of departure for the development of a multinodal urban system. The most basic situation is that where a single (nodal) urban system transforms itself into a multinodal system. This situation is shown in Illustration 28.2. Another point of departure is when a multinodal system becomes a model for the development of a completely new town or urban area. A more common point of departure is when two or more cities grow (functionally) together into one new multinodal urban system. This development is shown in Illustration 28.3.

Our first point of departure is where a single urban system changes from nodal to multinodal. We can recognise this process when new centres start to grow up, in addition to the existing centre. There are many examples of this type of development. Although written from a normative point of view, some of them are well described by Joel Garreau in his book *Edge City* (Garreau, 1991). In fact, almost all of the cases in his book are examples of the development from nodal into multinodal. Next to the real-life cases, there is also a theoretical case within this point of departure, that of the single nodal system that loses its concentration of urban and regional activities. The final result in such as case would be the complete sprawl of those activities. We can hardly use the term 'multinodal' to describe such a situation, because the term nodal supposes at least one concentration, node or centre. Nevertheless, in both the real-life situation and the theoretical case, the forces behind the developments are the same.

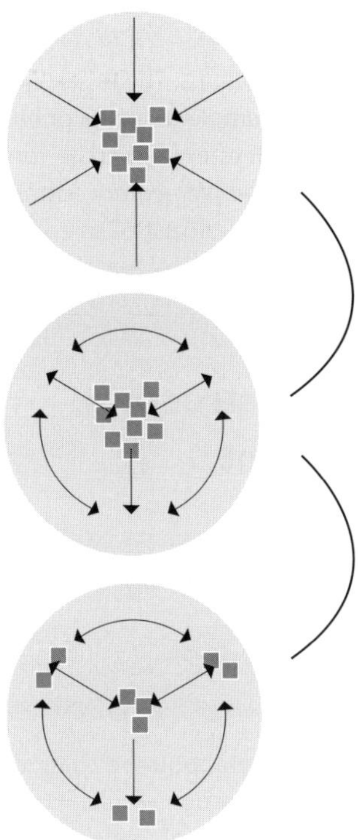

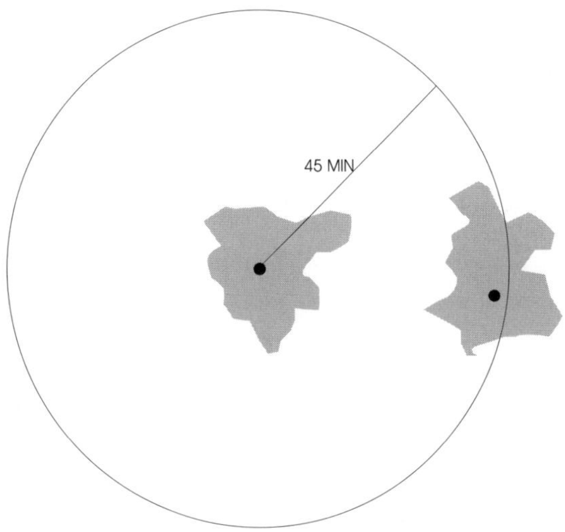


**Illustration 28.3: Two cities within each other's influence, growing into a multinodal system (Jacobs, 2000)**

**Illustration 28.2: The transformation from a single urban system to a multinodal system (*ibid.*)**

In our second point of departure, the multinodal system is used as the model for constructing a new town. The Dutch town of Almere is often named as an example of this kind of development. It is, however, a confusion of tongues; a confusion based on the clear distinction between morphology and functional-spatial structure. Almere consists of several large urban areas, each consisting of approximately 10,000 dwelling units. These areas are spatially divided by green zones. So, morphologically speaking, Almere is indeed made up of several areas. Formally, we should call this 'multizonal'. The term multinodal is reserved for functional-spatial structures. A multinodal urban system means that the system is based upon a multinodal structure, in other words, that there are multiple concentrations of urban and regional activities. When we observe the situation in Almere as it was in 1990, its centre was the only concentration of urban and regional activities. So we conclude that, in 1990, Almere was nodal. Recent development – some due to interference with the Amsterdam-Utrecht urban complex- has led to the development of commercial areas and business districts outside the city centre. From this observation one could conclude that Almere may gradually develop into a multinodal urban system.

Our third point of departure is the development of a multinodal urban system as a result of the growing together, functionally, of two or more previously separate urban systems. During the previous century, it was mainly this type of process that led to the emergence of numerous multinodal urban systems in the Western world. The effects of this type of process are also clearly recognisable in some parts of the Deltametropolis area. This process basically consists of the transformation of the urban structure by the interaction, interference and system development (conurbation formation) of two or more urban systems to form a new urban system. This new urban system has a different spatial and functional structure from the ones out of which it grew, and the nodes (centres) of the previously separate urban systems, whether nodal or multinodal, are now part of the larger system. The process of the growing together of two cities is shown in Illustration 28.4.

The process of conurbation formation has many variants, depending on how it begins. The urban systems may be roughly of the same size with respect to number of inhabitants, employment or public amenities, or they may be very different. The nodes that make up the urban structure may also be very different. In that case, the resulting urban structure may consist of so-called *complementary centres*. Conversely, where the nodes are rather similar, then the resulting structure may consist of so-called *competing centres*.

## Steps in the development of a multinodal urban structure: interaction, interference and system development

As described above, the development of a multinodal urban system by the growing together of several cities would seem to explain the Dutch planning problem of the Deltametropolis. We have seen that this type of process consists of three major steps.

### Interaction

First of all interaction (see also Ill. 28.4). Separate urban systems interact with neighbouring cities, as well as with other parts of a country and even with the rest of the world. Interaction results in the exchange of goods, people, information and money, and it is local, as well as regional, national and international. Indicative of a further internationalisation of the economic system is increasing interaction on a supraregional and international level. The traditional concept of the urban and regional-economic system, as shown by Christaller (Christaller, 1950) is clearly outdated. Interaction is a normal phenomenon which hardly has any effect on the spatial structure of an urban system. Without interaction, an urban system would be autarkic – a closed system. But since the existence of autarkic urban systems has not been proven, we should consider urban systems to be open, always having relations with other elements or systems outside their own.

The spatial structure of an urban system consists of the functional elements of that system, located at different places within the system, combined with the relations among the elements of that system. The elements can be business districts, residential areas, commercial areas, the major hospitals, universities or other educational institutes, theatres and so on. The essence of an urban system is that it has a daily pattern, and a, secondary, weekly pattern (Klaasen, 2004). Therefore, the elements that are part of

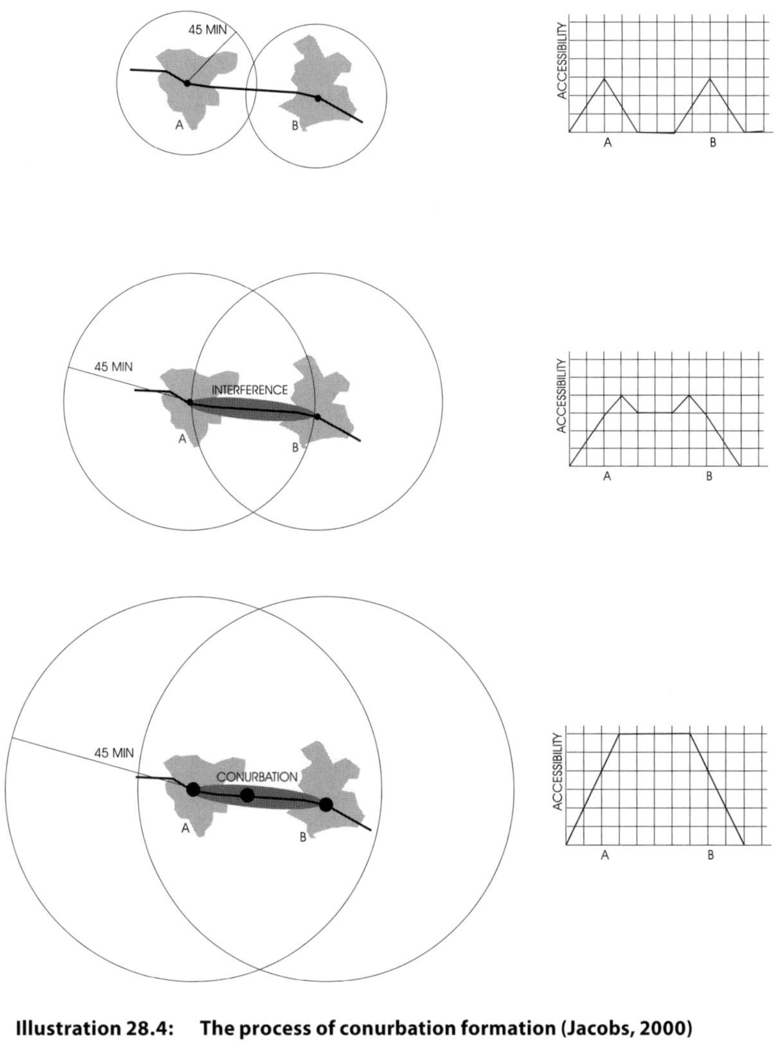


**Illustration 28.4: The process of conurbation formation (Jacobs, 2000)**

the spatial structure, generate a daily or weekly flow of people, goods and information, and may even be considered unique, existing because it is supported by the presence of the urban area.

It is important at this point to clarify the meaning of infrastructure. Infrastructure is basically a synonym for roads, paths, railway lines, data lines and so on, and should not be confused with the relations between elements in a spatial structure. The spatial structure is a means for observing, modelling and understanding urban reality; it is a theory-based abstract connotation of urban reality. The relations within an urban system – the relations on which the spatial structure is based – can be observed by its resulting flows of people, goods, information, and so on. Infrastructure can be, and often is, the physical area in which the flows can take place. Nowadays, an enormous exchange of information takes place between simple notebook computers with hardly any physical infrastructure. What it does indicate, though, is that there are strong relations between the spatial elements (e.g. hot-spot locations) and therefore that there is a strong spatial structure.

When an urban system interacts with other, external, urban systems or elements, this interaction has no implication for the urban system's own spatial structure. Interaction has no effect on the elements comprising that system, or on their position. What defines their position is the potential relations among the elements within the system. What can affect an element however, are the external relations of an urban system, which most of the time consist of the relations between the element and elements or systems outside the urban system. For example, firms located in a business district can be affected by developments in the steel or oil markets far away from these firms' own urban system. The developments affect the firm as part of the business district, though do not necessarily influence its locational behaviour within the spatial structure of the urban system. This is because the relative position of the location, its 'location value' (Klaasen & Jacobs, 1999), does not change due to these external economic factors.

## Interference

When interaction increases strongly, and the resulting flows equal or even exceed the flows within an urban system, then the spatial structure starts to be affected. This phenomenon is called *interference* (see Box 28.1). When external relations become the dominant flows, not only is the position of elements within an urban system affected, but also the spatial structure itself.

Many examples of interference can be found in both the Randstad-Holland, and in many other Dutch urban areas. Numerous retailcentres for furniture and home decorating, as well as business areas, benefit from their position within an urban system, as well as their position in relation to other adjacent urban systems. When we take into account the usual range of activity in an automobile-based society, interference can be expected if two urban systems are located roughly 30 km or less apart, as the crow flies, based on the distance between their (historic) urban cores.

Interference does not offer more in terms of employment, quality and level of amenities, or, for example, the size and quality of retailing within an urban system. If we consider the prosperity of a nation, and more specifically a region, then we can argue that the urban and regional activities will be of a certain size or quality. For example, a Dutch city of approximately 40,000 inhabitants will be the proud owner of a small theatre (500-1000 seats) and one or two museums (not including local museums). A Dutch city of 200,000 inhabitants will have several theatres, and several museums. Suppose now that

In the pre-automobile era, the cities Minneapolis and Saint Paul in the State of Minnesota, USA, were separate urban systems, interacting with each other through trade and a (fairly limited) flow of people between the two cities. The increase in automobile ownership and use increased the individual range of action. As a result, the flows of people between the two urban systems increased, and slowly led to the development of the Midway area between the two cities. This Midway area specialised in retail that benefited from the adjacency of two consumer markets. The position of these retailing companies is clearly related to the urban system of Minneapolis as well as to that of Saint Paul.

**Box 28.1: Example of *interference* (Jacobs, 2000; Adams & VanDrasek, 1993)**

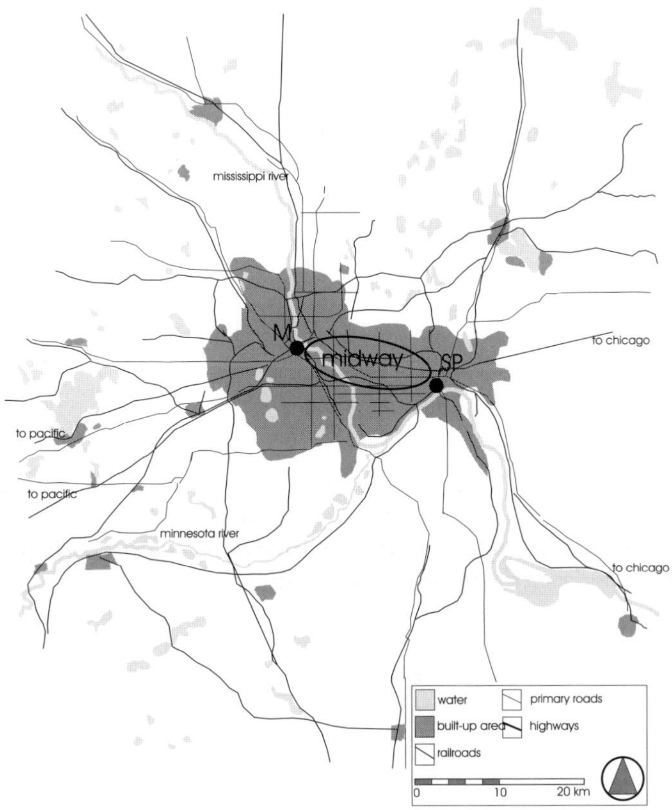


**Illustration 28.5: Minneapolis-Saint Paul: Midway (Jacobs, 2000)**

there is interference among five urban systems, each of them with a population of 40,000. This will not increase the number of theatres or museums, or any other element of the urban systems. It will merely affect the relative position of those elements, and thereby the spatial structures. Where there is interference, the total is no more than the sum of the parts. This conclusion is strongly related to the Niek de Boer's 'basic rule of the big city', which relates urban size to amenities (De Boer, 1996; Klaasen, 2004). According to this rule, an urban system with 100,000 inhabitants will have more amenities, and a higher quality or uniqueness, than the sum of 10 urban systems of 10,000 inhabitants each.

### System development

When the flows between two (possibly interfering) urban systems increase, they can affect the properties of the urban elements. In that case, not only the position, but also the size or quality of the urban elements may be affected. In most cases, new (higher level) urban elements will start to appear. This situation can occur when both urban systems are completely within each other's range of action. The term we coin for this process of system development is *conurbation* (see Box 28.2). Conurbation is defined as the (functional) merging of two or more urban systems into a new system.

In the Netherlands, the first indication of a conurbation can be found in the south wing of the Randstad, an area known as the South Wing. One indication of this is the location of the Swedish furniture retailer Ikea at the motorway exit to Delft. Because of the size of this Ikea (still the only so-called InterIkea in the Netherlands), and also given the period in which it was established (during the 1980s), it is plausible that it has survived at this location because of the size of its catchment area, which is the whole of the South Wing. The universities and also the many offices in this South Wing are activities on a scale larger than those in the separate urban systems of Delft, Leiden, The Hague, Rotterdam, Dordrecht or Gouda. As a result, traffic flows in the South Wing have increased in the past decade from 100,000 cars/day in 1993 to 164,000 cars/day in 2001 (Ministry of Transport, Public Works and Water Management, 1993, 2001).

As already discussed, the development of a new urban system, larger than that of the 'donor' systems, and also with a higher level of activities, proceeds by interaction, interference and conurbation formation. Two remarks should to be made, however. First of all, it can be argued that it is possible for a conurbation to form without interference in a preceding period. If accessibility increases dramatically, this condition alone is sufficient for a conurbation to occur. Building bridges and tunnels are examples of spatial interventions that can promote conurbation formation without a preceding interference stage (Marchetti, 1988). The elimination of political barriers, like the Berlin Wall, can also lead to conurbation formation without a preceding interference stage. Secondly, we should realise that a conurbation does not necessarily lead to a multinodal urban structure. The new urban system may also be single nodal.

## Prerequisites

By analysing how urban areas develop and their changing structure, and by differentiating the stages into interaction, interference and conurbation, we can gain insight into this process. However, insight alone does not give us the tools needed for urban and regional design and planning. To create these tools, knowledge of the underlying processes of transformation, and the specific spatial conditions

Automobile ownership and use strongly increased after the 1940s in the Minneapolis-Saint Paul area. This made locations within a 45-minute driving distance reachable on a daily basis from both urban systems. It was not long before new theatres, sports-stadiums, and several office complexes were located there. There were also two Major League stadiums at these locations, the numbers of which, throughout the US, are limited by licence. Only metropolitan areas above a certain size can be granted such a licence. Minneapolis or Saint Paul would both have been too small for such a licence, but combined, as a new urban system, this provided them with an opportunity to gain the licence.

**Box 28.2: Example of *conurbation* (Jacobs, 2000)**

is required. Based upon the insight into these processes and spatial conditions, designs can then be tested to validate their supposed effects. Insight also enables us to design scientifically valid prototypes for urban areas.

In the research into four large multinodal urban areas (Jacobs, 2000), increased accessibility clearly seems to be the force behind the development process and the transformation of the spatial structure. This research analysed the change in the spatial structure of the metropolitan areas of the South Wing of the Randstad-Holland (NL), Frankfurt am Main-Wiesbaden-Mainz (D), Minneapolis-Saint Paul, and Dallas-Fort Worth (US). In all cases, the change in the spatial structure, and also the conurbation formation, seemed to be interrelated with increases automobile use and building new infrastructure. In most cases, empirical evidence was found, proving that interference had started at the time when the automobile was becoming more common as a mode of transport. Conurbation formation begins when the road network is nearing completion (i.e. the radial and tangential net), when the automobile has become the most dominant mode of transport in an urban region. So it becomes clear that change in the spatial structure, and processes such as conurbation formation, lead to a change in how accessibility is differentiated. Changes of this nature are made possible by the shift from tram to automobile, by changing the road network, or by changing the morphological layout of the urban area, for example, by building suburban areas. However, it is not clear whether the major force behind suburbanism has been that suburban living was made possible by the automobile and the roads, or whether it was the deep-rooted desire for suburban living – a latent wish ever since Ebenezer Howard's book *Garden Cities* was published.

One of the most important conclusions of this research is that the concentration of urban and regional activities – in centres within a nodal or multinodal structure – are located only in areas that can be accessed from 80% of the entire urban region, within 45 minutes. This 45 minutes is not indicative, but has been a constant value for over a century (Levinson & Kumar, 1994). Knowledge about the location of urban and regional activities in relation to the required accessibility of this location, teaches us a lot about how to shape the spatial structure, and in effect, how to design an urban region. By conducting design-research, the 45-minute/80%rule, gives us the possibility of analysing the effect of proposed changes in the spatial pattern or infrastructure on the spatial structure of an urban region.

Illustrations 28.6 to 28.9 show the result of design-research carried out on the South Wing of the Randstad-Holland (NL) (Jacobs, 2000). In this design-research, the effect on the spatial structure of changes in the spatial pattern and infrastructure – both possible within the context of Dutch spatial planning – was analysed.

The design-research, as shown in the above mentioned illustrations, shows that a significant change in the spatial structure of a large urban area (in this case, of roughly 3 million inhabitants), requires significant changes in the urban pattern, or in the infrastructure. In areas of this size, a single new highway or an urban extension of 100 hectares would not have much effect on the spatial structure.

The so-called 45-minute/80%rule gives us another possibility for design-research. Because the rule is strongly based on (relative) accessibility, we can investigate the effect on the spatial structure of changes in mobility and splits in types of transportation used (modal split). For example, we can formulate the environmentally plausible hypothesis that the automobile will lose ground, and that new forms of public transport will become dominant within the modal split. This will also mean that most of the urban area will be serviced by this mode of transportation. Within this hypothetical experiment, we can also assume that the average speed of this mode of transportation will reach 100 km/h. Given these points of departure, it is likely that there will be interference between the urban systems and, though related to the distance between urban systems, also conurbation formation. If we project this hypothetical case onto the area of the Deltametropolis (Randstad-Holland), and given the correct network configuration, several locations will become accessible from 80% of the Deltametropolis area within 45 minutes. Having said this, we can conclude that it would be possible to create an urban system, in this case multinodal, the size of the Deltametropolis. Put simply, it would require a new form of transportation that meets the necessary specifications, and 40 years of development. It would not be just a matter of increasing the speed of this mode of transportation, but also its position in the modal split and therefore the size of its catchment area for which it provides a service.

In the current Deltametropolis there are no locations accessible within 45 minutes from at least 80% of the urbanised area. The initiative to build a public transportation system in this area (the so-called RandstadRail), offers better opportunities for travel from station to station. The service area of this system, and also its average speed, does not seriously compete with the automobile, and the effect on the modal split in the Deltametropolis will also be limited. This means that, given the distances, the modal split and the current road network, conurbation formation on the scale of the Deltametropolis is very unlikely to occur.

## Is accessibility losing ground?

Research on multinodal urban development has shown that accessibility is still the major factor for situating urban and regional activities. There are, however, trends and developments conducive to a more subtle approach to accessibility. First of all, the role of 'the office'. Several research projects (Hughes, 1991; Hessels, 1992) have shown that accessibility is the major factor in the locational behaviour of office firms. However, within the research field, there is growing doubt if this conclusion still stands nowadays (Occelli, 1998). Lacks of differentiation in accessibility, and the foot-loose character of firms, seem to counteract accessibility as an important factor. There is also the argument that the office, as we know it, will hardly exist in 20 years from now. Traditional organisational concepts based on a headquarters, regional headquarters and back-offices do not seem to exist anymore among multinationals. Large companies and organisations operate more and more as networks of smaller units.

**Illustration 28.6:**
**South Wing: Dispersed urban development (sprawl) (Jacobs, 2000)**

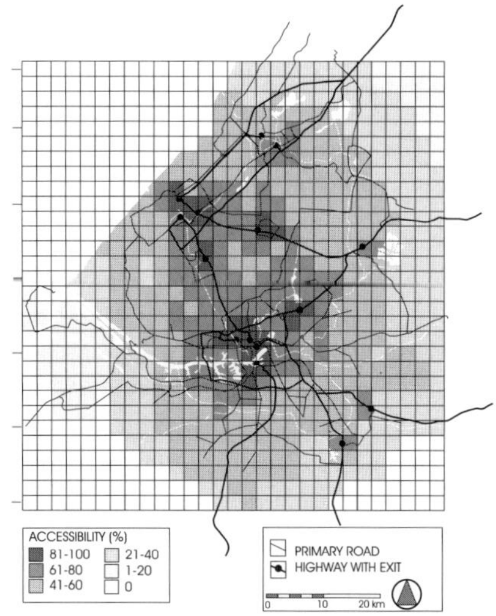


**Illustration 28.7:**
**South Wing: Analysis of the sprawl model (*ibid.*)**

**Illustration 28.8:**
**South Wing: Concentrated urban development (*ibid.*)**

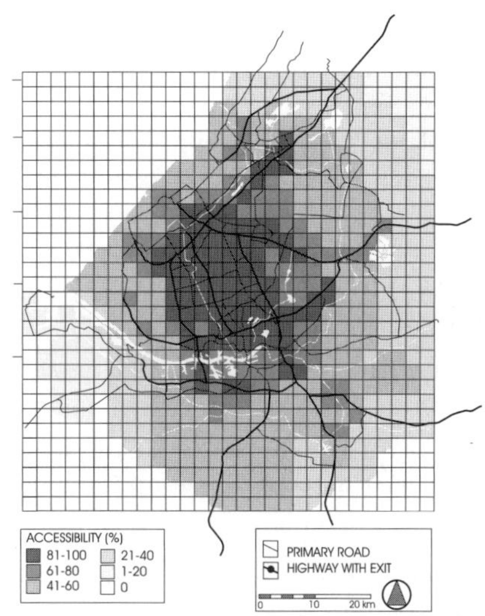


**Illustration 28.9:**
**South Wing: Analysis of the concentration model (*ibid.*)**

This results in a very loose relation between the city and the firms located there. This also means that the traditional competition between locations within an urban area has nowadays turned into competition between urban regions.

The second aspect that could affect the role of accessibility and the development of concentrations of urban and regional functions, is the role of public space. This space has traditionally been the area outside the buildings - basically the street, square and pavement. Whyte (1988) clearly shows the difference between the US and North-Western Europe in the role of public space. In the US, the distinction between the office and the public space has already faded, and also that between the office and the semi-public domain - hotel lobbies, libraries, airports, plazas, and so on. Current trends indicate (Drewe, 1998; Mitchell, 2000; Mendonca, 2004) that traditional activities in an urban area, such as living, shopping, working and so on, are losing their direct and unique relation with function-specific buildings. We have also learned to shop on the Internet, we work at home, while travelling, and in the park, in addition to the office. Clearly, this is not possible for every type of work. The shift to a service-oriented and information-based economy makes these shifts more and more possible. As a result, the 'standard' accessibility of a location is losing its significance and is being replaced by aspects such as the quality of public space. If this trend continues, the mono-functional office location that we see along our main roads looks as though it is heading for a dead-end. However, that other factors besides accessibility will become increasingly important is a conclusion that does not influence the possibility of using these factors as a design-tool in, for example, design-research (Klaasen & Jacobs, 1999).

Accessibility is related to the size of the area from which a certain location can be reached, given a certain amount of time. Clearly, accessibility is influenced by congestion. Locations which were very accessible 10 years ago, have become inaccessible, seemingly captured in the urban gridlock. So far we have learned about how unplanned urban development, based on the opportunities offered by the automobile, in the end lead to congestion and decreasing accessibility, not to mention parking problems. Every location becomes equally inaccessible, and gridlock is our doom scenario. From an objective point of view, we can argue that gridlock is just a symptom of an urban system that slowly outgrows the automobile era; one that will see the evident need for more space-efficient transportation systems. Cities like New York (Manhattan), London, Tokyo and also Shanghai-Pudang, would not be able to exist without their subway systems. Gridlock symptoms are also evident in the Deltametropolis area, even though it has a very limited capacity for main roads compared with other Western areas. Nevertheless, even after building new roads, the limits of the Deltametropolis as an automobile-based urban area will soon be reached, and the Randstad Rail initiative is insufficient, in its present form, to provide enough transportation in this area. Without new and tested initiatives, the Deltametropolis will remain a would-be metropolis, imprisoned by the limitations of the automobile.

## Towards a short guide to introduce a sense of science

The increasing scientific knowledge about urban development and its underlying processes such as transformation or conurbation, and even knowledge about the relation between spatial interventions and their effects, does not necessarily mean that this knowledge will be applied by practitioners of town and urban planning and design. From a scientific point of view it should be unacceptable that major urban extensions of the past decade in the Netherlands, such as those at Leidsche Rijn (Utrecht), Ypenburg (The Hague), IJburg and Zuid-As (Amsterdam), and Kop van Zuid (Rotterdam) are insufficiently based on the scientific field of urban and regional planning and design. Most of these plans are related to Dutch national planning policy, but although this makes it easier to gather funding for them, the plans are often questionable from a scientific point of view. We can conclude that most major plans in the Netherlands that arise from this national planning policy, are thrown into the force-field of local opportunism and created using a common method of 'design and discuss'. In this method, it is often practitioners who conduct the discussions, with hardly any input from the scientific field of urban and regional planning and design.

It is the task of local governments to realise national policy by planning the urban transformations and extensions. These governments tend to be opportunistic, and the urban plans and designs more often than not become the subject of local politics. It can be argued that the lack of scientific argumentation for good designs makes it even easier to place a design within the political arena. Clearly, space is scarce, so I believe it should be the democratically chosen representatives of society who are given the task of distributing space. Let us not forget that an urban or regional design is, above all, a plan of how the urban system is build up, and to a far lesser degree a plan for distributing space.

Klaasen (2004) clearly defined the scientific possibilities for urban and regional design. Practitioners, however, face two fundamental issues. First of all, spatial planning may be the last real issue on the political agenda of local governments in the Netherlands, and probably in the Europe of the European Community. Many fields of policy are now regulated by international, national and even regional acts and policies. Spatial planning is much harder to regulate; one might argue just because of the lack of applied scientific knowledge in this field.

Secondly, because of the societal position of urban and regional planning and design, practitioners are not forced to be scientific. If we go to a hospital for surgery, we expect the surgeon to be trained, but also that the method used has been scientifically tested, and prescribed medicine too. If we buy a car, we expect the product to perform within the specifications, and also that this performance (especially with regard to safety) has been scientifically tested. However, if we extend our town by say 10% of the built-up area, then we do not have such expectations. I would argue that the lack of public information concerning the scientific possibilities, combined with the political aspects, will continue to remain a disadvantage for planning and design as an applied science. Nevertheless, I would challenge the universities and scientific institutes to position themselves more critically towards their field of practice.

## A short guide

I began this chapter with a discussion about the Randstad-Holland, a discussion split by proponents anxious to gain from envisaged potential, and the scepticism of pragmatists. There may be truth in both parties' argumentations, but there is still the question of how to build a multinodal metropolis. Scientific research has revealed what basic spatial interventions are required to start a process of interference between urban systems, and also what is needed to bring about conurbation formation. For those who aspire to building a multinodal metropolis, a suggested guide is given briefly below:

*Step 1*: First we need to build a model of the real situation, because otherwise it would be too complex to grasp. This model should contain the current nodes and infrastructure, and the built-up areas. We also need information about the current traffic situation (how the modes of travel are split, travel times, congestion spots and levels).

Based on this information, we then need to measure relative accessibility based on a 500 m$^2$ grid (or smaller). Next we need to analyse the measured relative accessibility and test the correlation with the current spatial structure. This will also reveal whether or not there is on-going interference or conurbation formation.

*Step 2*: First we define the extreme parameters of the design, e.g., complete sprawl or maximum concentration; infrastructure in a grid network or concentrated along one or two lines. Based upon these parameters, consistent designs can be made, i.e., designs that do not contravene with, for instance, the 45- minute/80% rule. Next we analyse the design within the actual situation (i.e., within the model); then assess which design will have the most effect, and what kind of effect that will be.

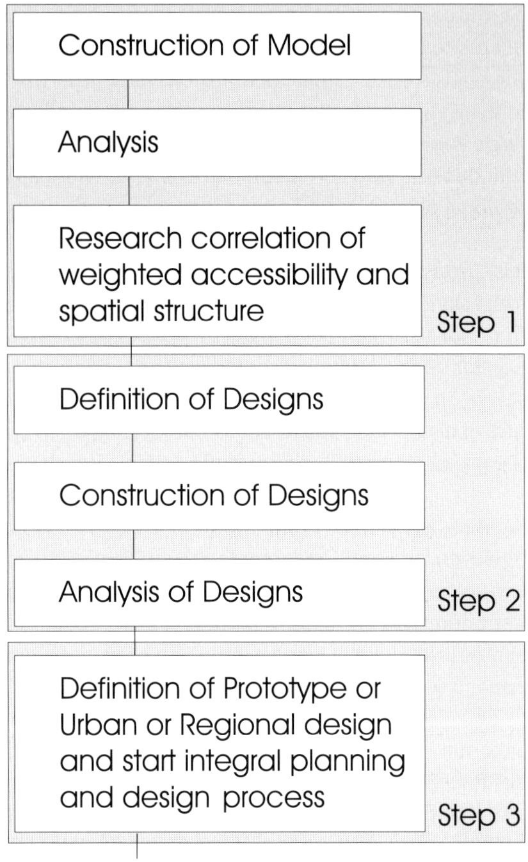


**Illustration 28.10: A short guide**

*Step 3*: With scientifically-based insight into the spatial interventions and their probable effects, an urban or regional design can then be made. The scientific justification needs to be formulated and presented along with this design. The following step would then be to start an integrated planning process in which the design is related to many sectoral goals, demands and limitations. Preferably, this will result in an urban or regional plan that can be used as a basis for many detailed urban plans and designs, and/or building plans.

## Conclusion

It is far too complex a problem to give a simple solution in this short chapter for how the Deltametropolis should be developed within its Dutch context, and all the steps and points of attention required. Here all we can do is to provide some very rudimentary elements that should evolve scientifically within laboratory conditions. Urban and regional planning and design is very complex, and also its field of operations. The urban area is probably the most complex of human constructions. In fact, we have made it so complex, we can hardly understand it! Professionals in the field of urban and regional planning and design should consider it an obligation to take great care of the town and city, and to view their profession very critically and preferably scientifically. The short guide above shows how simple it can be to tackle a design problem scientifically. Any urban or regional plan can be given a scientific base by using existing scientific knowledge founded upon scientifically sound design research.

Because planning the Deltametropolis is an extremely complex task, large amounts of time and money will be involved. Therefore it is all the more important that the Deltametropolis is developed on a sound scientific basis. Without it, its planning will fall back into the political arena, to be tossed back and forth, like a ball in a game, prior to every election period.

The Deltametropolis presents a very interesting challenge for planners. It is a prospective multinodal metropolis; with the potential to become a serious contender in the international arena, but before it can truly develop in that direction accessibility has to be drastically improved and there has to be a major change in traffic and transportation in this part of the Netherlands. It will take at least 40 years to build the Deltametropolis and the spatial investments will cost many billions of euros. This is reason enough to use a little scientific evidence to help further its case.

## References


Adams, J.S. & B.J. VanDrasek, 1993, *Minneapolis – Saint Paul, People Place and Public Life*, University of Minnesota Press, Minneapolis

Berry, B.J.L. & H.M. Kim, 1993, Challenges to the Monocentric Model, *Geographical Analysis*, Vol. 15(1), pp. 1-4

Blumenfeld, H., 1967, *The Modern Metropolis*, MIT Press, Cambridge Mass.

Christaller, W., 1950, Das Grundgerust der Raumlichen Ordnung in Europa, *Frankfurter Geografische Hefte*, 24, pp. 1-96

De Boer, N.A., 1996, *De Randstad bestaat niet*, NAi, Rotterdam

Drewe, P., 1998, In search for new spatial planning concepts inspired by information technology, congress paper, *Cities and metropolis: breaking or bridging*, French Ministry of Housing, Transportation and Public Works, La Rochelle

Garreau, J., 1991, *Edge City, Life on the New Frontier*, Anchor Books, Doubleday, New York

Gordon, P. & H.W. Richardson 1996, Beyond polycentricity: The dispersed metropolis, Los Angeles 1970-1990, *Journal of the American Planning Association*, Vol. 62 (3), pp. 289-295

Hessels, M., 1992, *Locational dynamics of business services*, Rijksuniversiteit Utrecht, Utrecht

Hughes, M.A., 1991, Employment Decentralization and Accessibility, *Journal of the American Planning Association*, Vol. 57 (3), pp. 288-298

Jacobs, M., 2000, *Multinodal Urban Structures: A Comparative Analysis and Strategies for Design*, Delft University Press, Delft

Klaasen, I.T., 2004, *Knowledge-based Design: Developing Urban & Regional Design into a Science*, Design/Science/Planning Series, Delft University Press, Delft

Klaasen, I.T. & M. Jacobs, 1999, Relative Location Value Based on Accessibility: Application of a Useful Concept in Designing Urban Regions, *Landscape and Urban Planning*, 45, pp. 21-35

Levinson, D.M. & A. Kumar, 1994, The Rational Locator: Why Travel Times have Remained Stable, *Journal of the American Planning Association*, Vol. 60 (3), pp. 319-332

Marchetti, C., 1988, *Building bridges and tunnels, the effect on the evolution of traffic*, International Institute of Applied Systems, Luxemburg

Mendonca, M., 2004 *Telecommunications and the city*, progress report PhD research Spacelab, Faculty of Architecture, Delft

Ministerie van Verkeer en Waterstaat, 1993, *Verkeersgegevens Jaarrapport 1993*, Rotterdam

Ministerie van Verkeer en Waterstaat, 2001, *Verkeersgegevens Jaarrapport 2001*, Rotterdam

Mitchell, W., 2000, *E-topia, urban life Jim, but not as we know it*, MIT Press, Cambridge Mass.

Occelli, S., 1998. Revisiting the Concept of Accessibility: Some Comments and Research Questions; In: Janelle, D. & D. Hodge (eds), *Measuring and Representing Accessibility in the Information Age*, Varenius, Washington

Richardson, H.W., 1988, Monocentric vs Polycentric Models: The Future of Urban Economics in Regional Science, *The Annals of Regional Science*, 22 (2), pp. 1-12

Van der Cammen, H. (ed.), 1988, *Four metropolises in Western Europe: development and planning of London, Paris, Randstad Holland and the Ruhr Region*, Van Gorkum, Assen/Maastricht

Wadell, P. & V. Shukla, 1993, Manufacturing location in a polycentric urban area, a study in the composition and attractiveness of employment subcentres, *Urban Geography*, 14 (3), pp. 277-296

Whyte, W.H., 1988, *City, Rediscovering the City Center*, Doubleday, New York

**website**

www.deltametropool.nl, november 2004

# 29 Olympic Thoughts in Urbanism: The New Charter of Athens

« I wanted to plan a city, but it had plans with me,
creating a dream from which I will one day awaken. »
Konrád, *The City Builder*

Rypke Sierksma

How should one explore the future of spatial planning? The present cities, especially the metropolitan areas, are characterised by complex spatial, social and economic conditions, which are less and less easy to grasp using theoretical constructs. The gamut of related and unrelated scientific disciplines that study urbanisation also contributes to this complexity. One is reminded of the proverbial Babylonian confusion of tongues. So many actors are involved, so many differences in culture, interest and discipline, not forgetting the occasional hypes. They blur the debate, yet spur it on. To focus all these efforts, what is the best way to deal with this?

The customary way of bringing various bits of knowledge together is to write and publish books and articles, organise meetings and conferences and, more recently, use Internet for telecommunication and to disseminate information. Not long ago, a new, modern-day phenomenon appeared: the charter. A trend that began slowly, the publication of such new charters has gained amazing popularity. Recent examples are *The Aalborg Charter on 'European Cities and Towns'* (1994), *Carta di Megaride 94,* (1994), *Charter on 'Women in the city'* (1994) and a proposal for a *Charter on 'Mankind's Responsibilities'* (2001). One of the aims behind these charters was to collect knowledge, however, first of all they make an appeal to urbanism professionals, in order to organise them into common action.

Why a charter? Charters have always appealed to potential subscribers, either individuals or organisations, who are happy to pledge their loyalty to the principles and theses of a particular charter. In medieval times, loyalty was the dominant principle of social binding; it was the cohesive force that held together the various stages in the feudal mode of production. Charters arose at that time as a means of organising medieval hierarchies and strengthening loyalties. Take the *Magna Charta* of 1215, for example, which laid down the privileges and rights of the barons, the church and freemen. What is surprising is that the phenomenon of the charter has suddenly resurfaced in the 20th and 21st century – in a world, that is steeped in a quite different mode of production: capitalism, primarily characterised by contractual obligation rather than loyalty. Where the *Magna Charta* reinforced existing loyal, yet democratic relationships between king or prince and 'his' aristocrats, recent charters, by contrast, are characterised by rather sublime, Olympic abstraction and generalisation.

In the second half of the 1990s, the European Council of Town Planners (ECTP) took the initiative to prepare a document containing a set of guiding principles for the next century. The ECTP wanted to make a statement, and they did this by replacing the original 1933 *Charter of Athens* with a *New Charter of Athens*. There are now two new versions – one written in 1998, a second one in 2003. By rewriting the Charter every five years, ECTP hopes to keep a lively discussion going about the future of cities and the discipline of urbanism.

The decision to replace the 1933 Charter by a *New Charter of Athens* seems to have been geared primarily towards strengthening the internal coherence of the profession of urbanism. Urban planners are invited to pledge themselves to new principles and certain types of proposed strategic action. Interestingly enough, CIAM, and especially the clique within CIAM comprising Corbusier, Giedion and Sert, both had the same objective. One should not forget that the sole renewal in the first Athens Charter was that traffic problems gained dominance in town planning theory. Already, decades earlier, people such as Unwin had researched and practiced 'guidelines' for urbanism. The new documents – both CIAM's *Statements* (1933) and Corbusier's popular version of *La Charte d'Athène* of 1943 – aimed at unifying their 'brothers-in-art'.

However, quite often it is forgotten that Corbusier tampered with the results of the CIAM congress – instead of merely popularising, actually falsifying them. Instead of 'statements', Corbusier deliberately called them dogmas (*points de doctrine*). It is precisely this type of 'theoretical fraud' that leads me to view these new charters as expressions of professional politics. Corbusier had failed to persuade the majority of people within CIAM to use his special vision on urbanism as the core of the official document. By producing his own version – a very successful little book, supported by a truly modern advertising campaign, and illustrated with lots of pictures that focused on *his* perspective only – Corbusier tried to regain an ideological grip on colleagues.

One thing is clear. Many have accused CIAM of positions taken by Corbusier in his *Charte*, that did *not* appear in CIAM's *Statements* of 1933. For example, whereas the 1933 document explicitly centred on linking environmentally, and in both directions, 'separate areas' for housing, work and recreation, Corbusier wrote in his 'doctrines' about the 'key functions' of 'sectors', each with its own 'territory' and 'autonomy'. When fierce criticism of 'Modern Urbanism' was voiced years later, Corbusier's terminology and his intention were unjustly ascribed to the original CIAM document (Van der Woud, 1983:71-76, also the chapter on Town Planning).

More generally, one can conclude that the function of both modern *and* post-modern charters is mainly ideological; to contain and guide the work of professionals. With what success, one may ask? Obviously, CIAM participants could not come to an agreement in 1933, and although Corbusier's ideological *coup d'état* attracted much attention at the time, he did not get many subscribers either. He can also hardly be said to have 'caused' the post-war reconstruction of towns and housing by his *Charte*, as this can be far better explained by a series of social and economic factors at work at that time.

It is also clear that, like Corbusier, the authors of *The New Charter of 1998* try to gain an ideological grip on their professional colleagues. In the 70s, similar attempts had been made to influence architects, by referring to them as 'generalists', suggesting that the architect should function as the chief engineer in building projects. These new charters, it seems, are now trying out something similar for the profession of urbanism. After decades, during which the profession of urbanism has struggled to maintain its position in university Faculties of Architecture, this is not surprising.

The intention behind the statement that, since 1933, "the role of the professional town planner as a well-informed coordinator and mediator is crucial" (European Council of Town Planners, 1998:395), is to link self-evidently, the town planner's educational certificate with their coordinating power. The New Charter claims that, for some decades now, spatial planners "have played a *prominent* role in the development of techniques of popular participation, that are considered to be an integral part of the process of spatial planning", but also mentions that the town manager has to handle the problems involved "in *partnership* with other disciplines" [italics by the contributor]. Without further comment, the Charter adds the *non sequitur*, that "the execution of the urban agenda must of necessity result in more and better educated planners" (*ibid.*:397, 400-401, 403).

Note that the '98 Charter first writes about 'town planners', subsequently uses the term 'spatial planners', and, finally, simply talks of 'planners', *tout court*. By so doing, the appeal of the document grows step by step. Nonetheless, it finally concludes that "the primary role of *town* planning" is to "provide a spatial framework for the future management and development of the city". ECTP's statement that "town planning, being a grown-up profession, will play an unmistakable as well as indispensable role in deciding what the new town agenda will be" is then added as a sort of incantation (*ibid.*:403).

What all this amounts to is that the New Charter of 1998 is claiming for the *town* planner a *decisive* role in a technocratic programme. According to this text, it is engineers who should determine spatial policy, *not* politicians! This equates quite closely with the 'generalist' claim that was made for architects. Happily, the 2003 version of the Charter, in its passage on 'the commitments of planners' is somewhat less pretentious, claiming that "the planner's role evolves following the development of society and of planning laws and policies" (European Council of Town Planners, 2003:22).

Crucial in my argumentation is the need to identify the true object of town planning in the New Charters. "It includes" no less than "housing, employment and welfare facilities. The core aims of spatial planning are given as furthering social sustainability, recognising trends leading towards diversity and pluralism, and providing an adequate response to the needs of diverse segments of the population" (European Council of Town Planners, 1998:397). The 2003 version, in which the notion of the *Connected City* features dominantly, again broadly defines the 'limits' of the town planner's task. It expects "the city of the 21st century" to "restore ties of social solidarity" (European Council of Town Planners, 2003:17). It also claims that "the connected city" requires "a new approach to governance involving all stakeholders and tackling social problems, such as unemployment, poverty, exclusion, criminality and violence", as well as "securing better access to educational, health and other social facilities" (*ibid.*:5).

By suggesting a whole new approach to urbanism, a rupture is introduced between the contents of the old 1933 Charters and the New Charters, implying that, in '33, town planners did *not* yet think in terms of connections – whereas in fact they did (though our concept of 'connections' has indeed been updated in 'systems theory'). At the same time, the ideologies of both New Charters obscure the fact that they actually ascribe *more* power to the town planner than Corbusier and CIAM had done. This was 'achieved' by the ECTP fictitiously enlarging the town planner's objective in intervening. It is benevolent to suppose that the town planner has a grip on both social segregation and disintegration, but this statement of faith remains unsubstantiated by argument. This is typical of much of town planning theory (for example Drewe & Hulsbergen, 1998:41). Half conscious of this, could this be the reason why the authors of the 2003 version called their text "a *vision* of a network cities", explicitly refusing the term 'utopia' or any other 'outlandish projection of technological innovations' (European Council of Town Planners, 2003:1,3).

The notion of 'network' is straightforwardly linked to the 2003 Charter's 'theoretical' framework of 'The Connected City'. However, the text states rather unclearly: "A network of cities ...will become connected in a multitude of meaningful and functional networks". In the Connected City, there is "opportunity for all to live and work in proximity, connected to well-maintained elements of cultural and natural heritage" (*ibid.*:1, 12). Although the authors are obviously convinced that the Connected City will produce "a breakdown of the isolation between parts of the city *and* to achieve retention and continuity of character, in the face of the impersonal trends of homogenisation" (*ibid.*:13), they do not provide the reader with an argument as to how and when. Like other sentences in the Charter, such phrases border on pleonasm and tautology, and seem to indicate an incapacity to understand what is involved *theoretically* in 'urban *connections*'.

What is at stake? Always and everywhere, *each* town is 'connected'. Between the various elements of any urban system there were and are certain connections or relations. These may change in the course of history; they may grow more, or less, complex – they are there, nevertheless. Existing terms, such as 'urban fabric', 'network', 'connectivity', 'nodality' etc., describe such connections. Whatever theoretical or ideological names we invent for these relationships, the concept of *system* must have primacy within our analysis, not some vague notion of 'connectedness'. However, for town planners, who are involved theoretically and practically, this is not always clear.

Considering the desired *coup d'état* by what Konrád in his novel aptly called the 'City Builders' – hidden in the subtle changes in the terminology of the new Charters, as analysed above – the incantation, if not abuse, by city planners of 'systems theory' in the last decade can be better understood. Systems theory in this case has primarily not a theoretical, but the ideological function of coordinating an ongoing debate within the confines of a shared terminology. However, there is an abyss between what systems *theory* actually performs, and the *practical* demands and expectations of spatial planners, who use its terms.

First of all, it is good to remind the reader that the notion of 'urban system' in recent analysis does not refer to the system of *one* specific town, but to urban units that are 'connected'. Self-appointed 'systems theoreticians' amongst urban planners define the objects on which they work very broadly – so broadly, in fact, that the definition tends to encompass the whole of society, the "total environment" (European Council of Town Planners, 1998:400). Perhaps, this is the result of an unconscious Hegelianism – 'everything connecting to anything else'. For instance, the theoretical introduction of time budgets in urban analysis makes an immensely complex object of such city systems. It is good to quote Prigogine and Stengers here, as they are knowledgeable in this respect: "The evolution of complex systems shows how difficult it is to steer or control a development that is determined by many factors in mutual interaction" (1985:221, 214-221). In his great summary *opus*, the famous systems analyst Luhmann (1999:1053) mentions the city once, and this only in the context of his analysis of political systems.

Secondly, systems theory qualifies systems *in general* – something, which should warn those who expect it to be of immediate practical help in spatial planning. There is no doubt that a certain intervention has *very* many effects; the problem, however, is to ascertain which effect influences what. How, in short, can one ensure in urban analysis that a certain intervention resembles either the wing beat of a butterfly at the mouth of the river Amazon, causing a storm over Tokyo – or rather the proverbial stone, which drops into the bath tub, not merely spilling a lot of water over the edge, but taking the baby with it... According to systems theory, systems tend to 'reduce their *own* complexity'. As systems analysis can only explain something *post hoc*, 'urban interventions' are not very much more than the connection of only a little bit of systems knowledge with a lot of causal *imponderabilia*. It is not by chance that much of the research from a systems analysis perspective is focused on traffic planning, which, after all, is *the* urban element that can be manipulated most 'easily'.

Thirdly, systems analysis teaches the urban practitioner that, for his intervention to have the desired effect, a whole *series* of other interventions is necessary, all of them carried out in a given order and of a type determined by the chosen intervention. But then, no society could ever afford such a planner's approach! Consequently, system restrictions such as these are crucial for the spatial planner, who, in the words of the New Charters, is supposed to become the overall, central manager of a world, which according to the very same theory of The Connected City no longer *has* a centre. Only, perhaps, an intense growth of political authoritarianism might give such grand planning a positive impulse. Is it a coincidence that Koolhaas (2004), in order to realise his mega XL 'urban' designs, prefers authoritarian countries such as China and Singapore to Western democracies.

Fourthly, it remains unclear what a city that is supposedly undergoing a 'successful' series of urban interventions, will actually look like. The urban aesthetics that fed the undertakings of people such as Sitte, Wagner and Berlage is gone – they still considered a part of a city as an entity, an architectonic object to be fashioned and formed. The New Charters do indeed emphasise the conservation of urban monuments. However, they overlook the fact that such an approach must, of necessity, make museums of the older cities, attracting hordes of tourists, who finally cause them to erode and become travesties of themselves. From the point of view of economic production, Karl Marx once called this the process moral erosion – *moralischer Verschleiss*...

There are also some everyday linguistic problems to be tackled. In urbanism – urban design and planning - we may distinguish two basic attitudes – the 'physical-spatial pattern orientation' and the 'societal-process orientation' (Klaasen, 2004). Some of the terms of the pattern-oriented and process-oriented design have the same definitions, although each has its own terminology. Because of this, the terms describe different things and the definition means something different in each terminology. Each approach also uses and develops its own kind of spatial models (*ibid.*:86). Consequently, there is an ongoing confusion about the substance of the discipline, its terminology and the societal significance of urban and regional design. Some examples are given in Illustration 29.1.

All in all, the New Charters that have been studied have a high logo-magical content and offer sublime prospects and great promises to potential subscribers – a little bit like trying to sell shares in something unknown. They pay little attention to the real, systemic 'contradictions' involved in the objects on which planners have to work, and ignore the theoretical difficulties in understanding them. Instead, with a little help perhaps from systems theoretical terminology they set their reader's mind at ease. The town planner is left in good faith to muddle through, hoping that not too much will go wrong.

| **Term** | **Physical-spatial pattern orientation** | **Societal-process orientation** |
|---|---|---|
| **Residential milieu** | Dwelling type and housing density | Type of (potential) spatial activity pattern |
| **Density** | Dwelling per hectare | Users per hectare (people use amenities) |
| **Urban region** | Continuous distribution of urban elements; associated with term 'inner city' | Coherent whole of cyclic urban processes; associated with term 'urban regional centre' |
| **'green' norm** | Area of 'green' space per dwelling | Accessibility, use value and informative value of green space |
| **Zoning** | On basis of functions | Primarily on basis of accessibility |

**Illustration 29.1:** **Some examples of terms with different meaning depending on the chosen design approach (Klaasen, 2004:87)**

Had the recent Charters contributed to a critical discussion amongst town planners about new tasks and key issues, they might have fulfilled a function – even *with* their defects. Perhaps the next re-written version may still inspire such debate. However, if once more it will float in a heaven of Olympian visions, aiming simply at attracting new 'subscribers' to their professed faith, the writers may well fail once more.

## References


Drewe, P. & E.D. Hulsbergen, 1998, Een veranderende maatschappelijke context, nieuwe stedebouwkundige opgaven; In: Bekkering, H. *et al.*, *Stedelijke transformaties*, Delft University Press, Delft, pp. 37-54

European Council of Town Planners, 1998, The New Charter of Athens 1998; translated as 'Nieuw handvest van Athene 1998', *Planologisch nieuws*, 1999, 19, 4, pp. 395-408

European Council of Town Planners, 2003, *The New Charter of Athens 2003, The European Council of Town Planners' Vision for Cities in the 21st Century*, Lisbon, http://www.ceu-ectp.org/e/athens

Klaasen, I.T., 2004 , *Knowledge-based Design: Developing Urban & Regional Design into a Science*, Design/Science/Planning Series, Delft University Press, Delft

Koolhaas, R., 2004, Naar de utopie, *De Volkskrant*, Kunstkatern, 25-03-2004, pp. 6-8

Luhmann, N., 1999, *Die Gesellschaft der Gesellschaft*, Suhrkamp, Frankfurt

Prigogine, I. & I. Stengers, 1985, *Orde uit chaos*, Bakker, Amsterdam

Van der Woud, A., 1983, *CIAM, Housing Town Planning*, Delft University Press, Delft

# Epilogue

> « The quality of the places where we live and work, and our immediate social networks, still matter deeply to us, even in a world of increasingly sophisticated communications, mobility and globalised systems. Planning is at the heart of shaping that quality. »
> Town and Country Planning Association, 1999, p. 43

The most ancient human settlements – even though they might have practical drawbacks according to our standards – show clear signs of spatial planning (Hawkes, 1994; Scientific American Staff Ed. *et al.*, 2005). This suggests that spatial planning is not a recent invention, but a basic feature of human nature. Increasing spatial complexity and lack of space might explain the growing interest in making the discipline more independent.

Nowadays, in the year 2005, spatial planning is a much used term, judging by the frequency of its occurrence on the world-wide web and in a large number of textbooks, courses and policy documents. As a subject, spatial planning is included not only in architecture, but in various other university courses, within disciplines that seem to compete with one another (even within the Delft University of Technology), for instance in planning, and (social) geography.

Increasingly, spatial developments are gaining European dimensions, and with the expansion of the European Union, this tendency is expected to become even more pronounced. Study programmes are also becoming attuned to Europe; nowadays, students commute between universities in different countries to get their degree. This diversity can be considered as an enrichment of the area of study. At the same time, it stands to reason that the sometimes very different operational definitions can cause confusion, because differences in tradition, culture and legislation have a great influence on the problems and procedures of spatial planning.

The New Charter of Athens (European Council of Town Planners, 2003), that has been discussed from various perspectives in different chapters of this book, is an effort to formulate not only a collective framework for the various planners and designers, but also to act as an instrument for enforcing discipline within the field. Research that has been carried out in preparation for the new Chair of Spatial Planning and Strategy in the Faculty of Architecture at Delft, for the information-gathering round and consultations among related disciplines, has shown that there are large differences of opinion about what the content and limits of that Chair should be (Brandes, 2005). The scientists and other professionals who were interviewed, expressed a wide range of opinions: from a wide to a very narrow demarcation of the content and area of application; from a national to an international orientation; from a purely technical interpretation of the design to one in which societal developments are viewed as a context for spatial planning; from being restricted to a regional scale to an emphasis on the importance of all spatial scales. There are also people who would like the research to be positioned outside the Chair, in contrast to the current opinion so far that a guarantee for up-to-date, in-depth education is the Chair's own research programme. From the chapters written by members of the Chair of Spatial

Planning, it will be clear that they view personal research as an integral part of their academic work, as do the other contributors to this book.

Language and terminology appear to create a lack of clarity, both within and outside the universities and especially between different European language areas. None of the survey respondents expressed any doubts about the societal relevance of spatial planning, quite the opposite, in fact, in view of the generally felt increasing complexity of society's spatial and functional-spatial demands. The variety of opinions about the content and future of the Chair of Spatial Planning and Strategy probably stems from the differences of opinion regarding the actual content of societal complexity, although some of these opinions may have been influenced by self-interest, perceivable here and there in the background. After all, when a Chair becomes vacant, this opens up possibilities for influencing the scientific field of force.

The developments of the last three decades have shown that spatial planning is a dynamic, practical field of science: theoretical, methodological and applicable. Dynamism will also be a feature of the future, if only because part of the working field of the discipline of spatial planning will be to formulate and design answers to societal assignments that will become more complex rather than simpler.

One of the aims of this book, Shifting Sense, is to position the discipline of Spatial Planning within the Faculty of Architecture at Delft and to set cornerstones for formulating future outlooks. We realise that the number of contributors, within and even more so outside the Faculty, could easily have been larger. However, in our opinion, the important themes for the future have been sufficiently well described for their implications to be understood. With hindsight, it can be seen that although themes such as cross-border planning, new technologies, urban regeneration, new forms of partnership, time–space reflections, and so on, may suddenly be a focus of interest in spatial debates, these developments have been going on for some time already. It is the task of the Chair of Spatial Planning and Strategy to identify developments, to help define urban planning and design assignments, and to train new generations adequately.

## References

Brandes, E., 2005, *Consultatieronde Ruimtelijke Planning en Strategie*, Department of Urbanism, Technische Universiteit Delft, Delft

European Council of Town Planners, 2003, *The New Charter of Athens 2003, The European Council of Town Planners' Vision for Cities in the 21st Century*, Lisbon, http://www.ceu-ectp.org/e/athens

Hawkes, J., 1994, *Atlas of Ancient Archeology*, Michael O'Mara Books Ltd., London

Scientific American, Staff Ed. *et al.*, 2005, Mysteries of the Ancient Ones, *Scientific American*, Special Edition, Vol. 15, No. 1

Town and Country Planning Association, 1999, *Reinventing Planning*, TCPA, London

# References Index

## A



## B



## C

## D



## E



## F

## K



## L



## M

## V



## W



## Z

# About the Contributors

**Prof. Dr. G.R.G. Allaert**, Department Civil Engineering, Mobility and Spatial Planning, University Gent, Belgium
georges.allaert@ugent.be

**Ir. H.M. Blank**, Managing Director, BVR Rotterdam, urban development, landscape and infrastructure consultants
hilde@bvr.nl

**Prof. Dr. Ir. L. Boelens**, Department of Human Geography and Urban and Regional Planning, Faculty of Geosciences, University of Utrecht; Director,Urban Unlimited, urban and spatial planning consultants, Rotterdam
l.boelens@geog.uu.nl

**Dr. Ir. J. Bredenoord**, urban and regional planner; private consultant for municipalities, Amersfoort; worked in Nicaragua and Peru
bredenoord@planet.nl

**Prof. Dr. J. den Draak**, Associate Professor of Spatial Planning, Faculty of Architecture, Delft University of Technology; part time Chair of Sociology of Spatial Planning, Vrije Universiteit, Brussels, Belgium - both until 1997
draa1@freeler.nl

**Prof. Dr. P. Drewe**, Chair of Spatial Planning, Faculty of Architecture, Delft University of Technology
p.drewe@bk.tudelft.nl
pauldrewe@wanadoo.nl

**Prof. Dr. G. Dupuy**, Directeur, Centre de Recherche sur les Réseaux, l'Industrie et l'Aménagement, Université Paris 1 Panthéon Sorbonne, Paris, France
gdupuy@univ-paris1.fr

**Dr. Arch. A.M. Fernández-Maldonado**, Assistant Professor in Spatial Planning, Faculty of Architecture, Delft University of Technology
a.m.fernandez-maldonado@bk.tudelft.nl

**Dr. Ir. J.C. Goedman**, Directorate General for Spatial Planning, Netherlands Ministry of Housing, Spatial Planning and Environment (VROM), The Hague
jan.goedman@minvrom.nl

**Ir. R. Houterman**, Assistant Researcher in Spatial Planning until July 2005, Faculty of Architecture, Delft University of Technology; currently working at Kristal NV, project developers, Amsterdam
robinhouterman@hotmail.com

**Dr. E.D. Hulsbergen**, Associate Professor of Spatial Planning, Faculty of Architecture, Delft University of Technology
e.d.hulsbergen@bk.tudelft.nl

**Dr. Ir. M. Jacobs**, town and regional planning and design consultant; Director, Town and Regional Planning Consultants, The Hague
marc@trpc.nl

**Dr. I.T. Klaasen**, Associate Professor of Spatial Planning, Faculty of Architecture, Delft University of Technology
i.t.klaasen@bk.tudelft.nl

**Ir. I. Kriens**, Senior Advisor in Spatial Planning, Faculty of Architecture, Delft University of Technology; Urban Design Consultant, Delft
i.kriens@bk.tudelft.nl

**Dr. F. Magrinyà**, Associate Professor, Department of Infrastructures, Transportation and Territory, Technical University of Catalonia, Barcelona
francesc.magrinya@upc.edu

**Drs. M.J. Moens-Gigengack**, Assistant Professor in Technical Ecology and Methodology, and Assistant Professor in Landscape Architecture, Faculty of Architecture, Delft University of Technology
m.j.moens-gigengack@bk.tudelft.nl

**Prof. E. Philibert Petit**, Professor of Architecture and Urbanism, Instituto Tecnologico y de Estudios Superiores de Monterrey, Santiago de Queretaro, Mexico
ephilibe@itesm.mx

**Prof. Dr. P. Roberts**, Chair of Regional Planning, Department of Civic Design, University of Liverpool; Chairman of the Board of the Academy for Sustainable Communities, UK
sueedwds@liv.ac.uk

**Arch. R.C. Rocco**, PhD Researcher in Urban Renewal and Management, Faculty of Architecture, Delft University of Technology
r.rocco@bk.tudelft.nl

**Dr. Ir. R.M. Rooij**, Assistant Professor in Spatial Planning, Faculty of Architecture, Delft University of Technology
r.m.rooij@bk.tudelft.nl

**Drs. H.J. Rosenboom**, Assistant Professor in Spatial Planning, Faculty of Architecture, Delft University of Technology
h.j.rosenboom@bk.tudelft.nl

**Prof. Dr. N.A.Salingaros**, Department of Applied Mathematics, University of Texas at San Antonio, USA
salingar@sphere.math.utsa.edu

**Ir. J. van Schaick**, PhD Researcher in Spatial Planning, Faculty of Architecture, Delft University of Technology
j.vanschaick@bk.tudelft.nl

**Prof. Ir. J.M. Schrijnen**, Chair of Regional Design, Faculty of Architecture, Delft University of Technology; Director of Space and Mobility for the Province of Zuid-Holland
wagner@pzh.nl

**Dr. R.J. Sierksma**, Assistant Professor in Typology of Buildings, Faculty of Architecture, Delft University of Technology
r.j.sierksma@bk.tudelft.nl

**Dr. Ir. P.L.M Stouten**, Assistant Professor in Urban Renewal and Management, Faculty of Architecture, Delft University of Technology
p.l.m.stouten@bk.tudelft.nl

**Drs. M.H.H.K. Tacken**, Guest Researcher in Spatial Planning, Faculty of Architecture, Delft University of Technology
m.tacken@hccnet.nl

**Dr. S.P. Tjallingii**, Associate Professor of Environmental Design, Faculty of Architecture, Delft University of Technology
s.p.tjallingii@bk.tudelft.nl

**Ir. J. Vogelij**, Director of Spatial Development with Royal Haskoning, Amsterdam; President of the European Council of Town Planners
vogelij@wanadoo.nl
jan.vogelij@royalhaskoning.com

Previously published in the series Design/**Science**/Planning

Klaasen, I.T., 2004, *Knowledge-based Design Developing Urban & Regional Design into a Science*

Fernández-Maldonado, A.M., 2004, *ICT-related Transformations in Latin American Metropolises*

Restrepo, J., 2004, *Information Processing in Design*

Salingaros, N.A., 2005, *Principles of Urban Structure*